Policy that works
for forests and people
Real Prospects for Governance and Livelihoods

James Mayers and Stephen Bass

**International
Institute for
Environment and
Development**

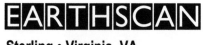

Sterling • Virginia, VA

First published by Earthscan in the UK and USA in 2004

Copyright © International Institute for Environment and Development, 2004

ISBN 1–84407–096–4

Design by Eileen Higgins
Original layout by Bridget Tisdall
Illustrations © Christine Bass
Cover design by Andy Smith
Cover photo © James Mayers
Printed and bound in the UK by CPI Bath

For a full list of publications please contact:

Earthscan
8–12 Camden High Street
London, NW1 0JH, UK
Tel: +44 (0)20 7387 8558 (main)
Fax: +44 (0)20 7387 8998
Email: earthinfo@earthscan.co.uk
Web: **www.earthscan.co.uk**

22883 Quicksilver Drive, Sterling, VA 20166–2012, USA

Earthscan publishes in association with WWF-UK and the International Institute for Environment and Development

A catalogue record for this book is available from the British Library

Library of Congress Cataloging-in-Publication Data

Mayers, James.
Policy that works for forests and people : real prospects for governance and livelihoods / by James Mayers and Stephen Bass.
    p. cm.
Includes bibliographical references and index.
ISBN 1–84407–096–4 (pbk.)
1. Forest policy. 2. Forest management. 3. Sustainable forestry. I. Bass, Stephen, 1958- II. Title.
    SD561.M37 2004
    333.75–dc22

                                                          2003021891

This book is printed on elemental-chlorine-free paper

Contents

List of Figures, Boxes and Tables

Country profile tables

Foreword

The first edition of *Policy That Works for Forests and People* was published in 1999, at the end of a five-year project coordinated by the International Institute for Environment and Development (IIED). The authors James Mayers and Stephen Bass drew on the work of teams from six developing countries, as well as a wide range of briefer studies on particular policy innovations and/or long-festering, but instructive, policy inertia.

The book was soon established as a key text for students and professionals. Others beyond the forest sector also began to realise that the ingredients of 'living' policy processes, highlighted in the book, were highly relevant for them too. After selling out its initial print run, the book is now being re-issued by Earthscan, and I am delighted that there will be an opportunity for many more people to read a book which is refreshingly outside the usual confines of dry academic policy analysis.

At the beginning of the Policy That Works project in 1995, it was clear that, despite many internationally agreed policies and the genuine efforts of many governments, success in developing countries was uneven or thin on the ground. There was some progress in both documenting truly destructive forestry practices and beginning to put a halt to them. There were many plans but little evidence of how to improve the lot of the millions of people who depend on forest products and services or whose livelihoods are otherwise affected by forests, such as forest dwellers, indigenous people and farmers.

The Policy That Works project set out to find those success stories, cases where most stakeholders were getting a fair deal, and where equitable and sustainable benefits were being delivered. National research teams documented what appears to have worked, and under what conditions. What factors were conducive to effective change on the ground? Where did this lead to supportive change in policies? What processes led to such policy decisions being taken?

The project recognised that it was vital to involve researchers from different disciplines in the national teams, along with advisory groups from different areas and levels of forestry policy and practice, including government, the private sector and civil society. These combinations enabled the teams – in

Pakistan, Papua New Guinea, India, Ghana, Zimbabwe and Costa Rica – to look at forestry from new and innovative angles, to challenge assumptions about established practice and to inspire each other to think differently.

The teams discovered some consistent *characteristics of good policy* which seem to apply in most contexts. There is no 'one size fits all' panacea. The key was to recognise that what was needed was not countless prescriptive lists of restrictions or desiderata but straightforward policy and flexible institutional arrangements that people believed in and that motivated them to act.

These characteristics of good policy were all about how to work effectively with all the many different forest stakeholders whose aims and needs are often wildly different and competitive. By bringing them all together and establishing processes in which the views of more than just the 'usual suspects' are valued and respected, these diverse groups of people can agree on a common vision for their shared resource. All the processes that work share the determination to confront potential conflicts and deal with them openly and fairly. The result of these processes is realistic policy based on strong partnerships and linkages, mutual understanding and a real desire to act at every level, from local to international.

No country is starting from scratch on policy. *Policy That Works for Forests and People* acknowledges this reality, and aims to help those involved to pause where they are now, and to think clearly about how to get to a position where policy is working for everyone. It is worth highlighting here the four critical steps identified by Mayers and Bass to make that transition happen: (i) *recognise multiple valid perspectives and the political nature of the game;* (ii) *get people to the negotiating table;* (iii) *make space to disagree and experiment;* and (iv) *learn from experience, get organised and fire up policy communities.*

This is an agenda that engages with imbalances in power – it is highly political. *Policy That Works for Forests and People* is such a valuable contribution to the politics of forestry because it starts from what we already know, studies and identifies lessons from examples of best practice, and in practical ways guides readers to start transforming their own policy and institutional environments.

Significant progress in the development of international forest policy has been made since the 1992 Earth Summit in Rio de Janeiro. The United Nations Forum on Forests (UNFF) was established in 2000, and the results of intergovernmental deliberations held under the auspices of UNFF and its predecessor *ad hoc* bodies, the Intergovernmental Panel on Forests (IPF) and the Intergovernmental Forum on Forests (IFF), along with the 'Forest Principles' negotiated at Rio, now constitute a comprehensive international and national agenda for action on forests. Three overarching principles have emerged and now govern the forest agenda for action. First, that forests provide multiple benefits and have multiple constituencies; consequently for both policy development and the implementation of forest policy, the interests of various special interest groups must be reconciled through an open, transparent and participatory decision-making process. Second, that forest policy is cross-sectoral in nature and that cross-sectoral policy harmonisation is a critical element at the national level. Furthermore, at the international level, fostering synergies among various intergovernmental agreements and conventions that impact on forests is equally essential. Third, that the environmental benefits and services provided by forests have expanded the scope of forest policy, which now also includes transboundary, regional and global considerations. It is now widely recognised that the scope of forest issues is both national as well as global, and that national forest policy development and implementation can no longer be pursued in isolation from regional and global contexts.

Forests are no longer considered as nature's factory that provides only wood. To receive a range of economic, social, environmental and cultural benefits sustainably, forests need to be managed as ecosystems. National forestry programmes (NFPs), as agreed by the IPF, are formulated through an open and participatory process and integrate the range of cross-sectoral considerations as well as geographic dimensions. NFPs are considered as fundamental to sustainable forest management. The forest community now faces the challenge to provide a range of these benefits from forests for human well-being at all geographic scales. The contribution of forests needs to be examined not only for providing wood and non-wood forest products, but also other benefits such as wildlife and water as well as sustainable livelihoods, rural development and poverty reduction. *Policy That Works for Forests and People* makes a significant contribution towards the future challenges associated with forests.

J.S.Maini

J.S. Maini, O.C., Ph.D.
Former Coordinator and Head of the Secretariat to the UN Forum on Forests

Key messages

Forests and people on the world stage

We are used to being told that forests are good for us all. Certainly, the range of benefits that can be derived from forests and trees are legion. But there are costs too, and no-one thrives on forest goods and services alone. Forests must also be transformed, in some places, to make way for farming and settlement to meet other needs. In theory, policy should be able to ensure some kind of balance so that forests are conserved, developed – and cleared – in the most suitable places.

But policies that affect forests are a reflection of the dramas being played out on dozens of stages at the same time. It is difficult, and perhaps meaningless, to attempt to understand what is happening to forests and the people who depend upon them without seeing the bigger picture of political and economic realities – from pressures for local control, to globalisation of markets, capital flows and technology, to rising inequality.

In some places, forests and people are doing well. But others are experiencing continuing decline in quantity and quality of natural forests, where conventions for using forests, based on trust and a sense of fairness, are eroding. The results are cronyism, gangster methods and the predatory business practices of timber kings; poorly-resourced, inflexible forestry institutions; one-sided forest revenue shares; and loss of 'location' through forest evictions or nomadism in forest employment. For those who can afford it, insurance and armed guards in protected enclaves are available. Many of those who cannot seek ways of opting out of a global economy which is overwhelming them; losing commitment to legal and non-violent norms of behaviour, and increasing demands for local autonomy.

If policy is going to work for forests and people – to produce forests that people want and are prepared to pay for – it needs to engage with these political and market realities. Finding out how this can be done is the challenge addressed in this report. We aim to discover what it takes for policy to provide a working, trusted, guiding framework – a process for tackling forest problems and delivering equitable and sustainable benefits. Our work is based substantially on consultative, multi-disciplinary country studies led by local professional teams in six developing countries: Costa Rica, Ghana, Zimbabwe, Pakistan, India and Papua New Guinea. We also draw on studies of: Sweden, Scotland, Australia, Portugal, and China; international forest policy processes; and the interactions of the private sector in policy processes.

The policy play – a deceptively simple plot

Policy is what organisations do. Policy has content – in the form of policy statements and policy instruments – and it has process – policy-making, implementing and reviewing. We need to understand the complicated area between policy pronouncements and practice, and to explain the difference between what people say they will do and what people actually do. And policy is not only the business of government – but of civil and private organisations too. 'Real world' policy (in contrast to formal policy documents) is the net result of a tangled heap of formal and practical decisions by those with varying powers to act on them.

Forest policy is no longer the main influence on forests and forest stakeholders. Bigger effects are often produced by policies that influence demands for forest goods and services, and those that determine the spread of farming and settlement. So we need to bear in mind the prices of farm, energy or mining products; the cost of capital (interest rates); and the cost of foreign exchange – all these shape the effects of the above policies. Many of these policies are, in turn, influenced by international processes and market movements.

Thus we must also watch the international forestry stage – on which some very grand sets have been erected over the last few years. Is this effort genuinely forging useful consensus, or is it doomed to failure because of the unconquerable diversity of forest values amongst the players, and the irrelevance of the plot to local circumstances? And, given the increasing influence of the (international) private sector in forest policy, how can this introduce the knowledge, capital and technology for good forest management – and close the doors to continued forest asset-stripping?

Recurring themes – conflicting intentions, murky practices and muddling through

There are many policy players and a lot of enthusiastic spectators. But, as we shall see, there are also many key people who are not allowed to come to the show, whilst others don't bother or can't afford it. Since policy positions, statements, practices, and even outcomes, are based fundamentally on value judgements, there are no absolute, 'true stories' in policy. Instead, we have found it useful to identify what appears to have worked for most stakeholders under known *conditions* – what contextual factors are conducive to effective policies; and, given a context, what *processes* lead to policy decisions that are agreed to be sound; and (although secondarily in this study), what policy *contents* and *instruments* have proven useful.

Changing power... over time. Power is manifest by participation in real decisions or, in other words, the degree of influence on policy. Where policy is inert it is usually because weighty institutions are 'sitting on it'. But such institutions can and do change, given time. Indeed, policy is often more susceptible to change than has been assumed. In Costa Rica, government's main forest policy tools – financial incentives for reforestation – used to benefit only larger landowners, and were generally insensitive to other people's motivations for forest management and conservation. The main losers were the smallholders, who collectively own about two-thirds of the country's land. However, the shortcomings of the incentives system generated considerable debate, and stimulated the formation of smallholder forestry organisations at local level. These eventually federated at regional and national levels and were able to exert enough influence over the policy process to swing the incentives programme significantly in the smallholders' favour.

Pushing formal policy reform. A range of technocratic approaches have been used around the world to bring about comprehensive policy change. The impact of some approaches has been a mixed blessing. Some have lasted only as long as donors prop them up, and many have benefited only a few. However, some approaches have kicked off considerable stakeholder engagement which has, in turn, generated novel institutions with real motivation for sustainable forest management. In response to a widespread perception of crisis in the forestry sector of Papua New Guinea, a national programme involving wholesale policy and institutional change, and a range of donor-funded projects, began in the late 1980s. But the programme over-estimated the power of the state to regulate customary land – which covers most of the country – and the instruments deployed were not flexible enough. A new forest revenue system could not cope with the wide differences in forest type and the range of deals between companies and local people. However, the process of debate brought many stakeholders to the table and resulted in an increased recognition that state roles, along with the roles of others, need to be negotiated.

Reinventing state roles. The imperatives of financial belt-tightening, and the demands for more social and environmental benefits from forestry are putting pressure on government in many countries. In the past, government has often sought, to varying degrees, to be forestry player, manager, owner, referee and coach. Recent pressures tend to focus government – often reluctantly – on the last two of these roles, whilst private sector and civil society actors take over the other roles. But this is often a painful process, and its results cannot be guaranteed. In India, federal and state-level forest agencies have different decision-making powers and are often fighting with other sector agencies for institutional turf. As a result, policies often become paralysed in practice. However, over the last decade, national and state-level policy resolutions have supported each other in formalising many joint forest management agreements between forest departments and local people. In some locations this has translated into little more than a new strategy for government to reassert control over forest land. But in others, an interface between local people and government staff has developed, which may yet lead to a flexible match of government roles to the ecological and social environments in which they operate.

Linking the people who change things. Many initiatives to change policy and institutions are premised on 'rational' arguments about objectives and roles which 'make sense'. But old institutional ways are found to persist because these initiatives fail to get to grips with people's real motivations. Even those fired up to change things often founder because of institutional cultures that reproduce inertia. Yet innovative managers and other 'new foresters' of various kinds do sometimes 'break through' from government and NGO backgrounds. They tend to be characterised by their ability to: see the big picture, take on tactical battles, use a mix of 'insider' and 'outsider' traits in their institutions, make alliances, and use these alliances to tackle bigger issues. In Zimbabwe, the Forestry Commission's traditional approach to forestry extension, based on woodlots of exotic species, was criticised by NGOs. These criticisms were listened to because certain of the Commission's senior managers had good connections with the NGOs. Experiments with natural forest management followed, with the support of astute donors, and these built on government–NGO links. This resulted in the emergence of broader alliances, led by the Commission, and a policy approach providing for a wider range of forest extension efforts.

Looking beyond the forest reserves. Traditionally, forestry has focused on a reserved forest estate, often under government control and management. As a result, forestry institutions were missing the real action – on farms and mixed farm-forest landscapes – where a wide range of forest goods and services are being used, nurtured or abused. There is ample evidence that farmers will grow trees and take responsibility for private forests and woodlands, but government's enabling role is key. This often means paying more attention to smallholder forestry. In Pakistan, government forestry departments traditionally focused their efforts on the remaining natural forest area, and on attempting to control a 'timber mafia' that has controlled the market and kept timber prices high. Meanwhile farmers were all but ignored despite having demonstrated – given improved information and a little support for organisation – that they are adept tree-growers. A shift in policy emphasis has begun, and price liberalisation is now being examined with a view to providing incentives for woodfuel production by many small farmers rather than timber production by a favoured few.

Improving learning about policy. One of the key elements of a policy process that 'stays alive' is its ability to link directly to experiments with new ways of making things work on the ground. Local projects allowing stakeholders enough slack to investigate alliances and roles can be vital learning grounds – but they only really become useful on a significant scale if they seize the attention of at least some of the current power-brokers or 'policy-holders'. In Ghana a forestry departmental unit was set up with a specific mandate to develop understanding of local capabilities for forest management, and to undertake experiments which modified foresters' roles in relation to those of other local stakeholders. The innovations in the experiments undertaken and the communication skills of the unit staff were very effective in attracting the interest and support of senior ministerial and departmental staff. These policy-makers were keen to associate themselves with the experiments and this association catalysed considerable learning amongst other 'high-level' staff. The results are now being seen in a broader process of institutional and policy change in favour of local forest management capabilities.

Dealing with tensions in devolution. Decentralisation is the proclaimed way forward for forestry in many countries. However, this often involves confused or conflicting objectives, sometimes from the same stakeholders: saving money for the central authority, or empowering the people? transferring land and incentives to promote large forest industries or encouraging farm foresters? These tensions may take the lid off a Pandora's box. Whilst much may be said for the centre strengthening its effectiveness through deconcentration, to do so at the expense of the periphery's forest management capabilities is a step backwards. There are worries that just this may be happening in some decentralisation programmes. Experience in West, Central and Southern Africa, India and China suggests, again, that experimentation is generally the best way forward – trying through experience to come up with spreadable models.

Building policy communities. Those engaged with a policy process on a regular basis constitute a policy community. Such a community needs to be able to channel the ideas of all those who are important to the prospects for sustainable forest management – the stakeholders – onto the policy stage, and disseminate the outputs. Mechanisms are needed which can recognise who has power (to help or hurt the cause of good forestry) and capability (actual or potential), and which can engage with them. If the process is too broad-ranging it will be unworkable; too narrow and the ideas will be the wrong ones. In Sweden, where a strong public interest in forests prevails, government has put high priority on access to good information in the policy process. The forest authority's major role is disseminating guidance and information about policy and how to implement it, while another body was set up specifically to act as a brokering agency between forest owners, users and researchers. Membership of this body covers most of Sweden's forests. By channelling its members' needs to researchers and, in turn, making research information useful, a high degree of engagement of forest owners and users in influencing and implementing policy has been achieved.

International forestry shows – hot tickets and dull side-events

National policy processes are an opaque mix of decisions, both overt and covert, often with murky pasts and uncertain intents. In contrast, international processes tend to be relatively easy to understand: they have involved more or less clear, time-bound, written policies with well-documented participation and decisions – although the interests of powerful groups similarly prevail. Some international policy initiatives appear promising, although all of them need to evolve further:

- Some of the *multilateral environmental agreements* which focus on specific global forest services, and include (under-utilised) implementation provisions – but which need informing about good forestry and need to be better recognised in key trade fora

- The *criteria and indicators* processes, which encompass the main elements of sustainable forest management, and allow for local interpretation – but which need application to the key areas of trade, investment and multilateral environmental agreements

- The process of developing and implementing *certification*, which can provide real incentives for good forestry – but which needs to continue to improve its 'fit' with local policy, livelihood and land-use realities, so as to solve real forest problems and not merely service the needs of particular markets

- Country-led *national forest programmes*, which could be major vehicles for reconciling pressures of globalisation and localisation – but which need to be built on local knowledge and institutions as well as the internationally-agreed elements such as the Intergovernmental Panel on Forests Proposals for Action

- Focused *regional* agreements, which offer the right political and operational level for integration of local and international needs – but which need to ensure they are strongly purpose-led, not to become vehicles for other agendas

It appears that we are reaching the limits of what can be achieved by intergovernmental effort in the forest sector alone. By the same token, the really big extra-sectoral problems – world trade rules, debt, foreign investment, technology access, etc. – can only really be dealt with intergovernmentally. They are too big for the forest sector alone to handle effectively.

Policy instruments – argument is healthy

Both forest practice and the balance of power between stakeholders have often changed significantly through implementation of, and/or reaction to, policy tools such as log export bans, certification and national plans. It is often argument over particular policy instruments that brings people together in the first place.

Policy instruments are even more context-specific than policy processes. However, it is possible to make some conclusions about those policy instruments which serve not only as implementation tools, but also as means to feed back to the policy process itself. Two such instruments are:

- *Mechanisms for increasing local negotiating capacity, through legal, financial and information means*: 'Public interest' objectives for forests need to be balanced against conflicting private interests through location-specific negotiation. Similarly, only through negotiation can potentially good forest managers at local level – currently marginalised from the policy process – hope to achieve the capacity to protect their interests in the long term. In such contexts, experience in Papua New Guinea suggests that state agencies should take the lead to: scrutinise the plans of developers; publish model contract provisions; legislate for court review of manifestly unfair contracts; and create finance arrangements, where local groups can borrow against future income to pay for professional advice.

- *Property rights changes*: Such changes are difficult, but not impossible with practice. Local security of resource tenure, by itself, is not sufficient to ensure long-term sustainable forest management. When customary tenure is not backed up by sufficient local institutional strength – both to be able to deal with outsiders, and to maintain the local side of the bargain in any deals made, the long-term management of any piece of forest land cannot be guaranteed. But it can be done! New legislation, in places as diverse as Ghana, China and Scotland, is tipping the balance in favour of more control of trees and forests by local farmers and communities. Here too, improved formal tenure is only part of the story. The considerable technical problems of integrating timber and forest trees with agriculture also needs to be addressed – hence the close linkage of tenure change with research and experiment, and with information, extension and support systems.

Each of the above policy instruments are, effectively, 'power tools'. They both implement policy and increase its information base and reliability, by providing feedback. In so doing they are instruments of change, helping to unblock situations of entrenched excessive power and stifled creativity.

Characteristics of good policy

In the last decade, policies for forestry and land use have become more numerous and complicated. They limit stakeholders – rather than free them to practise good forestry. They do not seem to 'fit' well, even with the rather limited number of over-structured and under-resourced institutions charged with implementing them. We need to turn this around – we need straightforward, motivating policies that people believe in and organise themselves to implement. This will enable the emergence of a greater diversity of more flexible, still learning and better integrated institutions.

'Policy inflation, capacity collapse' syndromes are paralysing the world of forests. They need replacing by simple, agreed policies with vision, and with strong capacities to interpret and implement them. This requires engagement with the varied actors demanding specific forest goods and services, and with those in a position to produce them – not just engagement amongst authorities and élites. Good policy will:

- Highlight and reinforce forest interest groups' objectives
- Provide shared vision, but avoid over-complexity

- Clarify how to integrate or choose between different objectives
- Help determine how costs and benefits should be shared between groups, levels (local to global) and generations
- Provide signals to all those involved on how they will be held accountable
- Define how to deal with change and risk, when information is incomplete and resources are limited
- Increase the capacity to practise effective policy
- Produce forests that people want, and are prepared to manage and pay for

In short, effective real world policy connects local action to plans and programmes through integrating institutions and top-bottom linkages. These linkages comprise information flows, debate and partnerships. As the linkages strengthen, so also does the mutual understanding amongst stakeholders.

Seven desirable processes to achieve good policy, and four key steps to put them in place

Wherever we look, there are recurring themes in the processes of policy-making and implementing: the way some people are involved while others are not; the common requirement for institutions which integrate people in varied ways; the way institutional capacity and practice tend to defy policy aspirations; the special power of some policy instruments which are not mere implementors of policy, but actually help to improve the policy process itself; and the ways in which these things change over time. Some of the processes which help to achieve good policy include:

1. *A forum and participation process*: to understand multiple perspectives and needs, to negotiate and cut 'deals' between the needs of wider society and local actors, and to initiate partnerships.

2. *National definition of, and goals for, sustainable forest management*: focusing on the forest goods and services needed by stakeholders, and on broader sustainable development objectives.

3. *Agreement on ways to set priorities* in terms of e.g. equity, efficiency and sustainability, as well as timeliness, practicality, public 'visibility' and multiplier effect. This will require methodologies such as forest valuation and organised debate. Without agreed approaches to setting priorities, an overly-comprehensive 'wish-list' policy may arise but be ineffective.

4. *Engagement with extra-sectoral influences on forests and people*: using strategic planning approaches, impact assessment and valuation, but also emphasising the active use of information and advocacy to influence broader political and market processes.

5. *Better monitoring and strategic information on forest assets, demand and use*: as the 'hidden wiring' which allows a continuously-improving policy process.

6. *Devolution of decision-making power to where potential contributions for sustainability is greatest*: decisions are best made and implemented at the level where the trade-offs are well-understood and there is capacity to act and monitor.

7. *Democracy of knowledge and access to resource-conserving technology*: openness to information from all sources, and communication of both information used in policy-making and information on policy impacts, are vital processes for empowering effective forest stewardship.

This list of desirable processes for some will be Utopian. The more important challenge to address is likely to be: how do we get there, from where we are now? We outline four critical steps to make the transition to the kinds of policy process described above.

Step one: Recognise multiple valid perspectives and the political nature of the game. Policies are based on assumptions. The challenge is to promote recognition of different conceptions of what the problems and priorities are. People's priorities for forests should be judged not on whether they are 'true' or 'rational', but on the level and degree of social commitment which underlies them – who 'subscribes' to them, and what impacts that has.

Step two: Get people to the negotiating table. Each group of actors needs to present their priorities in ways which they can 'sell' to others. Current inequities, forest asset-stripping or stakeholder stalemate may persist because of poor knowledge amongst stakeholders of each others' perspectives, powers and tactics, and the potential for change in these.

Processes which help identify and build shared vision or consensus on key goals can be effective. Cross-institutional forestry working groups in Ghana and Zimbabwe, the Sarhad Provincial Conservation Strategy in Pakistan and the Joint Forest Management institutional support network in India, have all made notable progress on this. However, multi-stakeholder processes in forestry which assume that societal consensus is possible have often grossly under-estimated the time and resources (of goodwill and money) needed to generate or refine such a shared vision, and especially to get the necessary power transfers to make the vision a reality.

Step three: Make space to disagree and experiment. Where policy involves people with completely different levels of power and resources, with a history of disagreement, consensus can be illusory, disabling or merely a sham. In some contexts, 'consensus' ends up as synonymous with 'conventional wisdom' – remaining stuck with its patchwork of anomalous or untested assumptions. Emphasis on consensus can lead to cynicism and disengagement from policy as people feel unable to change things, and may thus impede creativity and innovation. Where people are at odds with each other (but not actually at war) on the methods or content of forestry or policy, it can result in greater richness of debate and of needed checks and balances. It can allow the interplay of groups with differing objectives to flag errors and provide corrections.

Non-consensus-based approaches are often needed, which can accept dissenting views. Such approaches may temporarily manage conflicts, but they seldom permanently resolve them. Collaborative management approaches in forestry are in some cases – such as in Ghana, Zimbabwe and parts of India – being treated as *collaborative learning* processes. The learning element is critical: policy experiments cannot be whims, but require deliberate monitoring by stakeholders with different views, and an open process to consider adaptation and review.

Step four: Learn from experience, get organised and fire up policy communities. It has been said that, since human understanding of nature is imperfect, human interactions with nature should be experimental. Forestry actions and policies should thus be treated as experiments from which we must learn. Good policy helps 'learners' from different groups to come together, to pose questions, solve problems and evaluate information for themselves. It allows local experimentation and initiative to thrive and aggregate at national and international levels. Experiments with different forestry pilot projects and trials of policy tools are vital for stakeholders to explore each others' claims, make mistakes, learn, and make changes for themselves.

This can help to move the policy process out of the exclusive hands of foresters and consultants, spread information, and allow mutual recognition amongst stakeholders of power, claims and potential. Improved understanding leads to improved potential to change policy for the better. Some people will need to be empowered to make positive contributions, whilst others may need to be restrained from wreaking havoc, and clear tactics are needed for this. In some cases this will mean working directly with the current 'policy-makers' to improve policy where opportunities arise. Well focused, often highly detailed, analysis may be needed to get the mix of policy instruments and options right. In other situations, effective policy work requires pointing to new information, challenging deeply-held assumptions and contributing to a new vision of what policy should be aiming for. It is becoming increasingly apparent in many forestry contexts that this requires collaborating on analysis and organisation with those who are currently marginalised from the policy process, so that they can 'muscle in' on policy in the future. We discuss some of the tactics for analysing and influencing policy in Annex 1 of this report.

Summing up – linking the corridors of power to local reality

To sum up, the four 'steps' describe a learning, adaptive process brought about by a regular forcing open of the policy debate by stakeholders and their ideas, and a continuous sharpening of priority problems and proven solutions. A premium is placed not on one-shot 'planners' dreams' but on step-wise approaches that notch up shared experience – making visible progress and building momentum for broader change.

To improve policy, we need to unite decision-making with its consequences, such that policies, plans and strategies are not separated from practice, but are linked to it. This means that they benefit or suffer from it; that they learn from it; and that they improve it. Both policy processes and instruments are needed to make such links. Good policy becomes defined, and refined, through experience of those who have the potential to deliver good forest management and work for equitable livelihoods – often the very people who are marginalised by current policy processes. The challenge for all those who can get their teeth into policy for forests is to find the right 'power tools' for the right people. They will then make their own policy space.

There is a common perception amongst foresters that the fate of forests is determined by forces beyond their control. In the face of these extra-sectoral influences, foresters are inclined to declaim a 'lack of political will', retreat into their shells and encourage the illusion of stability: if the determining forces are beyond control, it is appropriate to ignore them. Yet foresters do often have considerable powers, and these confer responsibilities. Foresters can make progress which engages and tackles some extra-sectoral influences. *Policy That Works* showed that much progress has, in fact, been made by policy processes learning from local solutions to forest problems, both indigenous and project-driven. It has also been made by local user groups and farmers coming together to tackle local forest problems, and by 'policy-makers' giving them the chance to experiment. This has widened the ownership of policy and formed larger policy communities.

The type of work now needed is collaboration on analysis and institutional change *with* those who are currently marginalised from the policy process, so that they can present their views and experience, and make their claims, more effectively. In a sense, this means turning the conventional approach on its head, i.e. we need more policy process challenges for the powerful, and policy *content* analysis for the marginalised. It also implies that work needs to be better targeted such that policy-makers can learn, and be subject to checks, balances and incentives from below, e.g. due process/ diligence.

Almost every aspect of forestry is a political activity. All those who want forest goods and services need to find ways to act on this reality, rather than shy away from it. 'Policy that works' is not a dream about 'saving' forests, or 'halting deforestation', or 'afforesting the earth', all of which would match the desires of only a few. Neither is it about introducing comprehensive and logical master plans for all forests and people, and then expecting everyone to comply quietly and implement 'the plan'. This approach does not recognise historical and political contexts and the ways in which real change is made in practice. Rather, we should aim for a unity of theory and practice – constructive engagement with each other in processes of debate, analysis, negotiation, and the application of carefully-designed instruments of policy – from taxation to certification to extension. Forestry can and should be an activity which changes the political environment for the better.

Acknowledgements

This report draws on the efforts of a great many people involved in the *Policy That Works* project over the period 1995 to 1999. Country teams – of professionals of different disciplines in six countries – worked with great wit, sweat and imagination to form the core of this work. These teams comprised: Javed Ahmed and Fawad Mahmood in Pakistan; Colin Filer and Nikhil Sekhran in Papua New Guinea; Arvind Khare, Seema Bathla, S. Palit, Madhu Sarin, M. Satyanarayana, NC Saxena and Farhad Vania in India; Nii Ashie Kotey, Johnny Francois, JGK Owusu, Kojo Amanor, Raphael Yeboah and Lawrence Antwi in Ghana; Calvin Nhira, Sibongile Baker, Peter Gondo, JJ Mangono and Crispen Marunda in Zimbabwe; Vicente Watson, Sonia Cervantes, Cesar Castro, Leonardo Mora, Magda Solis, Ina T. Porras and Beatriz Cornejo in Costa Rica.

A wide range of people provided information and insight as members of advisory groups for each of the country teams. The individuals are listed in the country study reports and thematic reports in the *Policy That Works* series – (see details in the series box on the reverse of the title page in this report). We would also like to thank a very large number of people – necessarily in an impersonal manner, for which we aplogise – who provided vital information and responded to our pestering queries about key policy developments (or stasis) around the world. Also falling into the category of those who have shaped the project's findings through their diverse inputs are large numbers of students, resource people and policy practitioners taking part in several teaching and training courses with which the authors have been involved. These include events at: the Oxford Forestry Institute; University College London; International Agicultural Centre, Wageningen; and the Centre for International Forestry Research, Indonesia.

We much appreciate vital inputs provided at key stages by the following people who prepared thematic studies, made linkages with other key initiatives, provided key ideas, or participated in project-wide workshops in the UK and Zimbabwe: Bruce Aylward, Pippa Bird, Brian Brunton, Jane Clarke, Hannah Cortner, Simon Counsell, John Dargavel, Penny Davies, Annie Donnelly, Ko van Doorn, Olivier Dubois, Jason Ford, Irene Guijt, Mary Hobley, Caroline and Bill Howard, John Hudson, David Kaimowitz, Peter Kanowski, Jan Joost Kessler, Bill Mankin, Pedro Moura-Costa, Elinor

Ostrom, Manuel Paveri, Duncan Poore, Simon Pryor, Bjorn Roberts, Ian Scoones, Anthony Smith, John Spears, Ian Symons, Marc Stuart, William Sunderlin, John Thompson, Koy Thomson and Freerk Wiersum.

We also thank the following reviewers of a draft of this overview document: Javed Ahmed, Neil Byron, Carol Colfer, Richard Dewdney, David Edmunds, Colin Filer, John Hudson, Peter Kanowski, Arvind Khare, Nii Ashie Kotey, Fawad Mahmood, Calvin Nhira, Cathrien de Pater, Nick Robins, and Vicente Watson.

Elaine Morrison at IIED provided key inputs at many stages of the project and helped to keep the whole show on the road throughout.

In bringing this overview report together the authors found it very useful to have the chance to debate, refine and practise some of the ideas developed through the project with colleagues involved in ongoing work in the forestry institutions of Malawi, South Africa, Grenada and Himachal Pradesh, India.

The project was made possible through the financial support of the Department for International Development (DFID) of the United Kingdom and the Ministry of Foreign Affairs, Netherlands Development Assistance

The opinions reflected in this report are those of the authors and not necessarily those of IIED, DFID or NEDA.

Acronyms and abbreviations

AKRSP	Aga Khan Rural Support Programme
APKINDO	Indonesian plywood industries association
AusAID	Australian Agency for International Development
BCSD	Business Council for Sustainable Development
C&I	Criteria and Indicators
CAP	Common Agricultural Policy (of the EU)
CAR	corrective action request
CBD	Convention on Biological Diversity
CBO	community-based organisation
CEO	chief executive officer
CF	Commission Foncière (Land Tenure Commission, Niger)
CFB	County Forestry Board (Sweden)
CFMU	Collaborative Forest Management Unit (Ghana)
CIFOR	Centre for International Forestry Research
CITES	Convention on International Trade in Endangered Species
CSD	Commission on Sustainable Development
DFID	Department for International Development (United Kingdom)
DGIS	Ministry of Foreign Affairs (the Netherlands)
EC	European Commission
EFI	European Forest Institute
EMAS	Environmental Management and Auditing Scheme
EMS	Environmental Management Systems
ENDA	Environment and Development Activities (Zimbabwe)
EU	European Union
FA	Forestry Adminstration (Sweden)
FAO	Food and Agriculture Organisation of the United Nations
FC	Forestry Commission
FCA	Forestry and Conservation Act (Sri Lanka)
FCCC	Framework Convention on Climate Change
FD	Forestry Department
FSC	Forest Stewardship Council
FSMP	Forest Sector Master Plan
GATT	General Agreement on Tariffs and Trade

GB	Great Britain
GDP	gross domestic product
IC	institutional change
IDA	International Development Agency
IDS	Institute for Development Studies
IFF	Intergovernmental Forum on Forests
IFOAM	International Federation of Organic Agricultural Movements
IIED	International Institute for Environment and Development
ILO	International Labour Organisation
IMF	International Monetary Fund
IPF	Intergovernmental Panel on Forests
ISO	International Organisation for Standardisation
ITTO	International Tropical Timber Organisation
IUCN	The World Conservation Union
JFM	Joint Forest Management
JUNAFORCA	National Smallholder Forestry Assembly (Costa Rica)
KIDP	Kalan Integrated Development Project (Pakistan)
MBI	market-based instrument
MEA	multilateral environmental agreement
MLF	Ministry of Lands and Forestry (Ghana)
MTK	Maa- ja metsataloustuottajain Keskusliitto (Central Union of Agricultural Producers and Forest Owners, Finland)
NBoF	National Board of Forestry (Sweden)
NCS	National Conservation Strategy
NEAP	National Environmental Action Plan
NFCAP	National Forestry and Conservation Action Programme (PNG)
NFI	national forest inventory
NFP	National Forestry Programme [also National Forest Plan]
NGO	non-governmental organisation
NSDS	National Sustainable Development Strategy
NSSD	National Strategies for Sustainable Development
NTFP	non-timber forest product
ODA	Overseas Development Administration (United Kingdom)
OECD	Organisation for Economic Cooperation and Development
P&C	principles and criteria
PAC	Political Action Committee (USA)
PNG	Papua New Guinea
PTW	Policy that works for forests and people
RDC	rural district council (Zimbabwe)
RSA	Republic of South Africa
SAP	structural adjustment programme
SCS	Scientific Certification Systems (USA)

SD	sustainable development
SEK	Swedish Krone
SFM	sustainable forest management
SGS	Société Générale du Surveillance
SINAC	Sistema Nacional de Areas de Conservación: National System of Conservation Areas
SLU	Swedish University of Agricultural Sciences
SRDFP	Scottish Rural Development Forestry Programme
TBT	Tariff Barriers to Trade
TDC	Tribunaux Départementaux de Conciliation (District Conciliation Tribunals, Burkina Faso)
TFAP	Tropical Forest Action Programme
TUC	timber utilisation contract (Ghana)
UK	United Kingdom
UNCED	United Nations Conference on Environment and Development
UNDP	United Nations Development Programme
UNEP	United Nations Environment Programme
UNFF	United Nations Forum on Forests
WBCSD	World Business Council for Sustainable Development
WCED	World Commission on Environment and Development
WCMC	World Conservation Monitoring Centre
WCFSD	World Commission on Forests and Sustainable Development
WHC	World Heritage Convention
WRI	World Resources Institute
WTO	World Trade Organisation
WWF	World Wide Fund for Nature
ZERO	A Regional Environment Organisation (formerly Zimbabwe Environmental Research Organisation)

1 Introduction

1.1 What this report is about

What do we intend by titling a book *Policy That Works for Forests and People?*

People seek a startling variety of goods and services from forests: from timber to wild foods, from spiritual inspiration to medicine for the body, from landscape beauty to burial grounds. Local people may number the benefits as crucial - poor people in rural areas may depend almost entirely on forests for their livelihoods. Other benefits are more significant at national level - supplies of timber for sustaining a forest industry, and water supplies from major watersheds. Yet others are a more recent concern - the conservation of biodiversity processes, now known to be important to the world as a whole, and to the global climate regulation that forests provide.

In spite of forests' beneficence, no-one thrives on forest goods and services alone. The 'natural capital' of forests must also be transformed, in some places, to make way for farming and settlement to meet other needs.
How are these varied interests integrated, or traded off in practice? Markets, behavioural norms, regulations, and incentives determine the mix between forest and non-forest, and how forest goods and services are mixed in any one forest. All of these are, in turn, influenced by policy. Some interest groups are more powerful than others, both within a certain locality and between the levels from local to global. If unchecked by policy, the private interests of powerful groups may undermine the overall public interest, or the needs of disadvantaged groups.

What kind of policy can do this? Clearly, we are talking not only about forest policies, but also those other policies which influence demands for forest goods and services, and those that determine the spread of farming and settlement. Construction policies and export policies influence timber demands. Conservation and tourism policies influence biodiversity demand. And forests are cleared because farmers receive 'signals' from food prices and agricultural input policies. All such policies need informing about the real potentials of forests, the social and developmental needs of people, and the principles of good forest management.

'Policy that works for forests and people' results, therefore, in forests being conserved, developed – and cleared – in the most suitable places. It produces a desirable distribution of the goods and services that people need, taking special account of the needs of the poor and of future generations. It supports local groups who share the costs and benefits of forest management equitably. In short, it will produce forests that people want and are prepared to pay for.

In this sense, policy that works is not a dream about 'saving' forests, or 'halting deforestation', or 'afforesting the earth', all of which would match the desires of only a few. Neither is it about introducing a comprehensive and logical master plan for all forests and what they should be producing, which would not recognise historical and political contexts and the ways in which real change is made in practice. Rather, it concerns the constructive engagement of stakeholders in processes of debate, analysis, negotiation, and the application of carefully-designed instruments of policy – from taxation to certification. It is about real people and de facto power, the significance of which is discussed in Section 2.

This report is particularly concerned with developing countries, where dependence on forests for rural people's livelihoods is high, where forests are required for industrial development, and where forest clearance may still be needed. It assesses stakeholder engagement in policy, the outcomes of this, and the impacts on human and ecosystem well-being. It attempts to assess the policy causes of forest problems, which include:

- continuing decline in quantity and quality of natural forests;
- logging companies and land speculators seeking out national authorities with weak abilities to protect public forest benefits;
- over-concentrated control and inequitable access to forests;
- ill-informed public and consumers; and
- poorly-resourced, inflexible forestry institutions.

The report is based substantially on consultative, multi-disciplinary country studies led by local professional teams in six developing countries: Costa Rica, Ghana, Zimbabwe, Pakistan, India and Papua New Guinea. The approach of these country studies is discussed in Section 3, and their findings are elaborated in Section 4.

Many of the policy processes and instruments used in developing countries also have counterparts – or alternatives – from industrialised countries. Furthermore, developing countries are substantially influenced by international policy processes, and need to make their own influence better felt in these fora. There has been enormous effort put into international forestry initiatives over the last few years to identify common goals, much of it geared to pinning down the components of 'sustainable forest management'. Is this effort genuinely forging useful consensus, or is it doomed to failure because of the unconquerable diversity of values and local circumstances? And, especially for countries with a comparative advantage in timber production, there are increasing influences from the (international) private sector in forest policy. How can this introduce the knowledge, capital and technology for good forest management – and close the doors to continued forest asset-stripping? Hence the report also draws on IIED-led and commissioned studies on:

- Sweden, Scotland, Australia, Portugal, China
- international forest policy processes
- the interactions of the private sector in policy processes

These lessons are integrated with the discussion of policy themes in Section 4. A special section on the international processes is offered in Section 5.

Section 6 explores the overall findings. It draws almost as much from the cases of policy that has not worked – or almost worked – as from the clear examples of success. As such, many opportunities are noted amongst the problems. The findings are, naturally, context-specific, but there are common themes which are addressed. In particular, wherever we look, there are recurring themes in the processes of policy making and implementing: the way some people are involved while others are not; the common requirement for institutions which integrate people in varied ways; the way institutional capacity and practice tend to defy policy aspirations; the ways in which these things change over time; and the special power of some policy instruments which are not mere implementers of policy, but actually help to improve the policy process itself.

In the process of working with country teams, we faced many methodological challenges. Many of the policy analysis approaches cited in text books proved overly academic, and were not useful for engaging ordinary stakeholders in identifying 'what worked'. Thus we have taken the opportunity to bring together some of the frameworks and methodologies which we have found to be useful. These are offered in Annex 1, as a rough 'tool kit', in the hope that we can encourage others to conduct practical, purposeful research and debate.

1.2 Why focus attention on policy?

Policy matters a lot. Around the world, people arguing on behalf of either forests or particular groups would agree with this statement. But that's where the agreement is likely to end. When it comes to what policy actually is, what type of policy matters, and what needs to be done, it all depends on who you are and where you stand. It is not surprising that people are not good at agreeing about policy.

Forests are in trouble in some places, but trees are being nurtured in others. Much of what is happening to forests and trees is unique to particular conditions of place and time – you cannot generalise. Neither can you generalise about what should be done – the history of policy and planning in various sectors of development is all too full of the disastrous results of attempts to force common solutions on very different local realities. Forest policies have encouraged massive Eucalyptus afforestation to improve timber supplies – fine where other forest goods and services are in good supply, but not so fine where diverse rural livelihoods require varied landscapes. Log export bans can support local forest industry – but they can also bolster inefficient local processing, meaning that forests have been stripped for few lasting gains.

Although contexts differ, the relationship between forests and people, and the way policies are interwoven with it, show some strikingly common themes between many countries. We believe there is a need to investigate these common themes. Thus, our focus is on:

1. *contextual factors* which commonly present the right conditions for good policy, and those which do not
2. *relationships between actors, and the integrating institutions and processes* which commonly help effective policies to emerge and be implemented
3. *certain policies and policy instruments* that appear to be generally successful in making the transition to sustainable development

We concentrate more on the first two themes, as these tend to have more elements shared across countries than specific policies or instruments, and are more fundamental to meeting the challenges faced in many countries today. Whilst the contexts and arguments generated by bad policies can sometimes speed the development of processes which produce good policies, it is more common that bad policy entrenches positions and stifles development of such productive processes. Good policy rarely emerges without good process. Thus, identifying appropriate contexts and mechanisms for policy change is generally the bigger challenge.[1]

Whilst we believe that improvement in policy process, and thus policies, is fundamental to improving the position of forests and people, there is resistance to this approach. There appear to be three related criticisms:

- Policy is abstract, but reality is local – it is more important to understand and support local forest livelihoods, local capacities to manage forests, and investment in forest business

- Forestry and policy are both constrained by power and politics, and so policy on issues that really matter is too difficult to change

- Participatory strategies, programmes and agreements can be developed in ways that avoid political baggage, and can thus make more immediate progress

Thus, compared to 'real life', the world of policy seems at times irrelevant and vacuous. There are those who argue that you don't improve policy by analysing it.

Box 1.1 What people often say about policy

- Policy is all hot air and no action
- Policy is just a smokescreen to cover up what 'they' are really doing, or not doing
- Policy is just a list of wishes, there is no money or capacity to implement it
- Policy is always overridden by powerful interests and big international forces
- Policy doesn't matter, what matters is what people actually do to forests
- Policy can never truly empower people at local level because it comes from government
- Policy cannot be fine-tuned enough for different conditions
- Policy is a political business, and politicians don't listen, so researchers knowing 'what works' won't improve anything

---

1 Other work by IIED concentrates on specific policies and instruments (e.g. Mayers *et al*, 1995; Bass and Hearne, 1997; Landell-Mills and Ford, 1999; and the six *Policy that works for forests and people* (PTW) country case studies).

These criticisms are valid to the extent that they point to some realities about policy, but each has flaws. Attention to local practices and livelihoods is vital, yet local actions will have limited impact and life-span unless wider institutional, political and legal constraints are tackled. Power and politics do change over time – even the Berlin Wall and President Suharto fell eventually – and it is people chipping away at problems who make this happen. National forest programmes and plans often last only as long as donors prop them up, and remain anodyne wish-lists precisely because they are not politically engaged.

What we take from these worries about policy is that the *separation* of policy and practice is helpful to neither. This is all too evident in examples where people, who might otherwise make practical steps towards sustainable development, are weighed down carrying 'excess policy baggage'. These are the people who are expected to put all the latest policies, decided by ministers, markets and intergovernmental conferences far away, into practice – often with ever-shrinking human and financial resources. In these situations, new policies are nothing but a burden, obstructing anything practical. They are the equivalent of financial inflation – lots of paper produced without real capacities to back it up.

Internationally, more attention has been given to making policy for forests over the last ten years than ever before. Both legally binding obligations and market mechanisms have proliferated. These have come, not only from national forest authorities, but also from environmental authorities, industry groups and NGOs. These policies are piling up around people who might be in a position to manage forests. Do they help? Or are forest managers getting buried under a suffocating heap of confusing and contradictory obligations? We need to tease apart the policies in this heap: to recognise who are the winners and losers under different policies; to recognise what catalyses forest removal and – in contrast – what sustains flows of forest goods and services; to reveal policy red tape in its true colours; and to understand the ingredients of truly enabling policy practice.

Whether forests and trees are planted, nurtured or removed depends on:

- the decisions people make
- the power they have to act
- the resources available to them
- the knowledge, skills and information people have

All of these factors in turn are influenced by beliefs and 'paradigms' about what makes for progress. Post-World War Two notions of development have been associated with forest removal, indicating that forests have little role in society-building; these notions tend also to be associated with formally-educated urban people. Sometimes these beliefs are enshrined in, and protected by, the policy process itself.

Thus policy of various kinds shapes both the available options and the balance of power between the different people who make decisions. Current policies in many countries send signals that favour only a few interests in forest use or clearance. These policies do not require those interests to cover the associated environmental and social costs, and it is ordinary people who bear the brunt of most of these costs. Furthermore, environmental and economic change is occurring which is beyond the control of particular groups of people, yet policy processes are not well set up for collective decisions about how to deal with rapid change.

Some decisions made in the 'corridors of power' – be they in government, companies, or civil organisations – are clearly matters of life and death, such as the transmigration policies which sent people to live in Indonesian and Amazonian forests, with little help on how to survive there. Policies can foster liberating land and resource redistribution – or they can maintain unjust ownership of forest land, such as the Amazonian land accumulation by ranchers in the 1970s and 1980s. Policies can wrap people up in red tape, but if they are got right, they might release creativity and innovation – witness current trends in both India and Vietnam, which appear to be beginning this transition.

Governments, civil society groups and far-sighted private sector interests in an increasing number of countries are looking for answers to tackling the policy causes of forest problems. If policy and the processes for reforming it are 'right', they can pave the way for solving problems which are created by the excessive power of some people, and the lack of influence of others. Yet policy pronouncements alone will not do the trick: capability to change is vital, but is often weak.

The report argues that some people who think they are working to get policies right are wasting their time, or doing more harm than good – perhaps by setting irrelevant targets for percentages of national land area to be afforested, or by aiming to halt deforestation at all costs. To be frank, policy work is often done by the wrong people in the wrong place. Those who are in a position to make policy really work – by ensuring it reflects local conditions while also being able to integrate local, national and possibly global needs on an equitable basis – are often not involved at all. This report aims to demystify policy for those profoundly affected by it, but currently not engaged with it, to show why engaging with policy is important, and to show how it can be done. Policy does matter, and if it is currently doing the wrong thing it can be changed for the better.

2 Forests, people and power – the scene, the players and the drama

2.1 Forests – why people get so fired up about them

Forest issues are big issues in many countries. Large areas of land are usually involved, large amounts of money may be made, and many livelihoods may be connected to forests. Forests have often formed the power base for many governments and social groups. Finally, forest issues can be highly contentious, because specific groups of people understand them and value them very differently.

The values which people place on forests vary greatly, depending upon the culture and social group in question, and the roles which forests play in their livelihoods and quality of life (Table 2.1).

Table 2.1 A spectrum of social values associated with forests

Social Values	Forest Management Provides	Typical Trends and Conditions
Livelihood basics Staple food	Carbohydrates and protein for forest-dwelling communities Fuelwood for cooking	• Dwindling in most areas due to penetration of rural markets by new products, taste changes, reduced supply, high labour costs, loss of traditional knowledge • Still key for remote developing country communities
Supplementary food	Variety/ palatability to diet through meat, fish, fruit Seasonal buffers/ famine foods	• Important in many developing countries experiencing economic/ climate uncertainty and food insecurity, especially for marginalised groups including women
Health	Water supplies Climate moderation Medicine Vitamins and minerals	• Water and climate roles increasingly critical in most countries; global markets for these services developing • Medicinal value important in many forest communities, associated with culture
Shelter	Poles, thatch	• Important in practically all forest communities, strongly associated with culture
Economic security Main income	Forest products for sale	• Where incomes are rising generally, most people move

	Forest services e.g. tourism for sale	away from forestry; a few (richer) local groups specialise in forestry • Where incomes are declining, and where many are landless, still high dependence on forests • Community systems/ rules for management of common pool resources tend to be in decline • Globalisation, taste changes, and market price fluctuations constrain investment in SFM
Supplementary income	Forest products for sale	• Very common for rural communities, especially where pressure on farm land makes farm income inadequate - here, access to natural forests/ fallow is key • Farmers with adequate farm land but labour constraints often plant trees • Government/ corporate control of land is reducing access by poorer groups
Savings/ social security	Timber stocks Land value	• Traditionally key for periodic expenditure e.g. dowry, feasts • Forest stocks more important for their timing, not the size of income
Risk reduction	Biodiversity Multiple products Soil conservation Water conservation	• Increasing realisation of the value of diversification, but as yet inadequately developed markets • Risk reduction should be seen in context of whole-livelihood/ farm system
Cultural and social identity Cultural, historical, spiritual and symbolic associations	Forest landscapes Forests as sacred groves Individual species and their products	• Many developing country cultures are forest-based, suffering tensions when faced with Western cultures based on forest removal and standardisation of products • Symbolic association with forests may affect decisions: prestige, unequal land distribution or rebellion
Social identity and status	Forest as source of power from ownership/ cultivation/ clearance Ability to pass forest on to future generations	• Outside involvement in local forests can draw on, and exacerbate, local power inequities
Quality of life Education/ science	Biodiversity conservation Means of access to forest	• Where livelihood and economic security are obtained largely by non-forest means, forests are valued largely for their 'quality of life' attributes - as in most developed countries
Recreation	Biodiversity conservation/ control Forest-based facilities	• Where livelihood/ economic security are closely dependent on forests, there can be big clashes with outsiders' 'quality of life' demands
Aesthetic values	Landscape design and management Biodiversity/ conservation	• Seen as a key area for government intervention, but markets are also developing • Local/ government partnerships are key

Source: Bass (1999 forthcoming)

These values are not static, but they change over time. Key change factors appear to be:

- Shortages of, and opportunities to develop, non-forest capital (physical, financial, human/ individual, and social/ community capital) by liquidation of forest capital where this is more abundant

- In a similar vein, the absolute and relative scarcity of particular forest resources will affect the types of forests which are valued; where timber is in short supply, plantation or intensively-managed forests may be more highly valued

- Scientific discovery, education and technology change will affect what can be done with forests – or with the alternatives that can be gained by forest removal

- Access rights, and resources such as labour that are available to exploit forests

- An individual's allegiance to certain groups and their shared values; this can be altered by campaigning and by the political influence of certain groups on other groups

- Tastes and associated communications and media links – the prevalence of individual consumerism has changed values away from those of the community

- All of the above are influenced by changes in political culture, national development conditions, and an individual's income

Policy ought to reflect the many values of stakeholders; furthermore, it needs to make decisions that integrate them – or to make choices between them where integration proves impossible. It is obviously a considerable challenge to understand, and then to make decisions about the relative weights of the values of groups, institutions and corporations from local to global levels. Three trends are helping here:

- The development of more inclusive systems of forest policy and management decision-making, such as Round Tables, national or local forest fora, and stakeholder liaison groups: consultation amongst different groups about their values seems to be on the increase (even if as yet this is more about controlling the behaviour of rowdy groups than about significant participation in decision-making and management)

- The trend for certain markets to recognise certain social and environmental values of forest management for which producers were previously unrewarded, e.g. the market for certified forest products and the emerging market for carbon storage;

- Work by economists to categorise values and to develop methodologies to assess them on a single (financial) scale. These methodologies remain contentious (it is not so much the financial magnitude which counts as who bears the cost and who gains the benefits), and indeed they have so far made little difference to major policy decisions, especially those regarding private forests. Nonetheless, the economic categorisation is useful information science, if not yet decision-making science (Bass, 1999 forthcoming):

  o *direct use values:* where the value is derived directly from the forest, either in a consumptive manner (timber, nuts, fodder, game, fish, etc) or a non-consumptive manner (tourism, recreation, etc)
  o *indirect use values:* where the value derives from environmental services such as watershed and social protection, carbon sequestration or biodiversity protection, rather than from the forest directly
  o *passive use values:* where value is accorded by the mere fact that the forest exists (existence value), or for the future possibilities that the forest represents (option value), or for the ability to pass the forest on to future generations (bequest values)
  (Gregersen *et al*, 1997)

Perceptions of forest problems are clearly linked with people's different forest values. There are several global reviews which cover forest problems, although some are restricted to certain values. The most useful of these reviews include:

- FAO's State of the Forests Report (the latest is 1999), with a broad range of facts and figures about changing forest areas and uses (FAO, 1999a)

- The review by the World Resources Institute of relatively untouched 'frontier' forests and the forces acting on them (WRI, 1998)

- WCMC's analysis of threatened tree species (Oldfield et al, 1998)

- The report of the World Commission on Forests and Sustainable Development (WCFSD), based on regional civil society hearings, is perhaps the closest to date in illustrating the range of concerns about threatened forest values (WCFSD, 1999)

We have drawn on this global material to summarise the forest problems facing the world today (Box 2.1).

Box 2.1 Problems facing forests and people

The main physical forest problems are:

Declining quantity and quality of forests. This is because wood, fuel and food are being harvested at rates faster than forest regeneration; because remaining growing stock is often poorly managed; because fire control may be inadequate; and because many forests are being cleared to make way for other land uses. Developing countries lost 16.3 million hectares of diverse natural forests from 1980 to 1990, and gained 4.1 million hectares of simpler plantations (FAO, 1997). This trend continues (Palo and Uusivuori, 1999).

Environmental degradation of forest areas. Forest exploitation and clearance can create interlinked problems, notably soil erosion, watershed destabilisation and micro-climatic change. These threaten the soil and water base for agriculture. About half of the precipitation in the Amazon basin arises from forest evapotranspiration, and already there are areas which are drying out - which can be correlated with deforestation. Industrial air pollution, particularly common in some temperate forests, also reduces forest health.

Loss of biodiversity. These problems are contributing to a rapid reduction in ecosystem, species and genetic diversity. The present rate of species extinction is estimated to be between 1,000 and 10,000 times the historical (pre-10,000 years before the present) rate (Wilson, 1988). One thousand species are recorded as having become extinct since the year 1600 (Groombridge, 1992). These include 77 tree species, obviously an under-estimate. More than 8,700 tree species, equivalent to an estimated 9 per cent of the world's trees, have recently been assessed as globally threatened (Oldfield *et al.*, 1998). This lowers the world's biological potential for improving material, food and medicine production and increases vulnerability to environmental, economic or social change. With tropical forests being perhaps the major repository of biodiversity, forest abuse in tropical regions is causing particular concern.

Climate change. Forests play a major role in carbon storage - they and their soils store about 2-3 times as much carbon as is in the atmosphere. And they act as an important brake on global warming by absorbing 15-25 per cent of annual anthropogenic carbon emissions. Thus, the loss of forest ecosystems results in both a source of emissions and a reduction in the natural sink. Small reductions in the sink provided by forests are likely to cause large increases in the radiative forcing effect of each unit of CO_2 emitted by fossil fuels. Recent predictions by the UK's Met. Office and Institute of Terrestrial Ecology indicate that forest sinks themselves are vulnerable to the effects of climate change, such that forest die-back on a vast scale could occur from the 2040s onwards, leading to significant emissions during the second half of the next century. Thus it is probable that the cumulative effect of global forest loss and environmental degradation will be a net contribution to regional and global climate change.

This could bring many problematic side-effects: sea level rise, risks to food supply, declining productivity of soils and the quality and quantity of fresh water, inviability of current land uses and protected areas, and human health problems.

The main costs for people are:

Loss of cultural assets and knowledge. The culture and knowledge of many peoples, which are not always documented, and which have evolved through long periods of nurturing the forest, are diminishing as forest area, access, and traditional rights are reduced.

Loss of livelihood. Forest loss and degradation are affecting the livelihoods of forest-dependent peoples - particularly poorer groups who may not have significant agricultural land, and who depend on forests for 'social security'. For example, common complaints about forest conversion amongst poor groups in West Africa include the loss of access to bush meat (which is a very important source of protein) or its commercialisation (Falconer, 1990).

Rising inequality. Increasing concentration of control and access to forest wealth in fewer hands is removing development options for the majority of people in many countries. Those who lose their forest livelihoods and become marginalised may be forced to create - and themselves suffer from - social and economic problems elsewhere, such as in cities.

Loss of forest asset base for national development. Asset-stripping of forests for short-term gains wipes out any potential for forest-based strategies for sustainable development. The problem at present is that such strategies - which might be based on sustainable timber production, tourism, water production, or selling environmental services - are not well-known. Here, it is the task of policy at least to generate conditions to explore such options.

Policy successes might therefore be defined in terms of finding ways of (a) realising the key social values of any forest tract, and producing these alongside as many other values as possible (Table 2.1) and (b) avoiding the problems and costs as above. There are several examples, some of which we go on to analyse: farmer tree growing in Africa; forest regeneration in Central America; private tree growing in USA and Sweden; joint forest management by communities and authorities in India; improving corporate environmental practice; and multiple use of forest conservation areas in Belize.

It is not difficult, however, to find isolated successes. But they can turn out to be dangerous. The 'cult of the success story' is a kind of policy hysteria where 'models' are widely replicated, without understanding the institutional, cultural or policy conditions which allowed the original example to be a success in the local environment. Hence a third ingredient of policy success might be (c) understanding and responding to local institutional, cultural or policy conditions.

Generally speaking, there are prevalent policy conditions which constrain the 'replication' of what appear to be successes. We have described elsewhere some of the typical features of the policy problem for forests and people (Bass *et al*, 1997; Mayers and Bass, 1998). These are summarised in Box 2.2.

Finally, since policy is about facing the future in circumstances of uncertainty and necessarily limited information, an ingredient of success is (d) building in resilience and adaptability.

Box 2.2 Typical features of the policy problem for forests and people

1. Perceptions of forest 'crisis' and resource scarcity lead to normative government statements of policy and the imposition of cumulative layers of formal control

2. Policy mechanisms are too complex and incoherent to function well, i.e. to seek, analyse and respond to information on multiple values, problems and opportunities; to be responsive; to be equitable; and to set a long-term vision

3. 'Sectoralism' and polarisation of views increase with the uncertainty and complexity of forest issues

4. Pro-timber emphasis has concentrated wealth in some actors and retarded the development of democratic institutions connected to forest use

5. Those who have captured the benefits of industrial forestry have not paid the full costs - financial, social and environmental

6. There is a gulf in communications and perceptions between local forest actors and national 'policy-makers'

7. 'Informal' policy, i.e. decisions and actions reflecting power structures and domination by certain actors and interests, is often more significant than formal policy

8. Much centralised, formal forest policy is, in practice, disintegrating as locally-controlled, multiple-use forestry rises in importance

9. Policies from outside the forest sector have more influence than 'forest sector' policy. Yet most efforts to deal with forest problems do not acknowledge this reality. They remain 'forestry solutions'. Thus they are limited in their effectiveness.

2.2 People – the policy players and the spectators

A typical 'cast of characters'

In subsequent sections, we will be examining a number of policy processes and their outcomes. In all of them, there tends to be a bigger cast of policy actors than first meets the eye.

It is obvious that the national forest agencies are frequently the principal policy actors – but we cannot repeat past mistakes by assuming they are the central actors who control the plot. In the current era of globalisation, another government agency, responsible for trade and macroeconomic policy, for example, may prove the more influential. In a climate of decentralisation, local government agencies may be taking up forestry roles. 'Pro-active' companies concerned about securing the land and resources for long-term plantation development may be less the 'targets' of policy than active setters of policy – they want policy conditions to be conducive to good business, and to be stable over the life of their investment. Private sector associations may set their own policies of self-regulation in order to secure a long-term future. National NGOs, academics and researchers may prove to be critics and sources of ideas – all faculties which help to turn the 'wheels' of policy. International donors, consultants and NGOs may wade in with new paradigms and conditionalities, backed up by the weight of their purse or by international agreements. Finally, some community institutions may make surprise appearances – carrying compelling information from the field, or bringing 'higher level' policy-makers face-to-face with stark realities.

Whilst, on the surface, these actors tend to be clearly-defined institutions with obvious mandates, the institutions reflect multiple identities – depending upon who their representative is, or the phase in their development. Ultimately, it is the real people within institutions – as charismatic leaders or as 'yes-men', as cooperative or difficult people – who are the actors who really lead the plot. Nonetheless, there are certain characteristics which institutions tend to reflect:

Forest authorities

Forest departments and other lead governmental forest agencies commonly grapple with a number of roles, which sometimes contradict or compromise each other. These include:
- income-generating roles e.g. earning timber revenue for the state
- political roles e.g. controlling territory or certain groups of people
- developmental roles e.g. supporting rural development, or the development of sectors (industry, agriculture or energy)

- social roles e.g. local (community) development
- environmental roles e.g. biodiversity or water conservation
- investment-attracting roles, which tend to result in big companies being favoured over communities (Bass *et al*, 1998)

In many developing countries with significant forest assets, financial and developmental roles have been uppermost, and forest institutions have evolved around them. Typical key features of problems facing forest authorities are:

- Roles from the past can remain fossilised. The procedures that had been developed in order to exercise these roles can become ends in themselves, irrespective of current needs. Challenges to these roles tend to reinforce a 'fortress forestry' mentality.

- Pressures from both international and local actors are, however, starting to put new (or renewed) emphasis on how environmental and social roles can be developed and paid for.

- Dealing with multiple needs: there is a perennial institutional problem of how to be accountable for trade-offs – within a single organisation or different groups?

- Weak political status – forest authorities tend to have low ability to influence other government bodies which determine land allocation, land use, and its profitability.

- There is often a technical bias to problems and their solution, resulting from the lack of political or policy engagement in the past. This may render the authority blind to underlying power and rights issues, or at least uncertain of how to deal with them.

Central government agencies
Authorities in charge of financial, macroeconomic and major development planning issues, and those in charge of the most commercially important land uses, e.g. agriculture ministries, can have a very strong influence on forests. These are the authorities which negotiate structural adjustment conditions – which may end up down-sizing the forest authority. They are the bodies which determine financial flows into the country, and who gets the biggest deals. They determine the prices which farmers receive, and hence the incentive to clear forests to plant crops. They plan the major infrastructure investments – which might either support sustainable forestry

processing, or drive new roads across forests that enable settlers to make of the forests what they will. And they define the financial roles and expectations of the forest authority – and consequently how much it is obliged to earn from forest production, or can spend on conservation.

Indeed, forests themselves are frequently placed under authorities (ministries) which implicitly or otherwise stress only some forest values – agriculture, rural development, land-use planning, tourism, wildlife – or a few combinations of these. This emplacement can change over time, which is often at odds with the long time-frame needed for forest policy.

In recent years, there have been attempts to integrate forest-related social and environmental concerns into the work of these powerful bodies. Environmental impact assessment, national conservation strategies and environmental action plans have all clarified the links between forests and other sectors. But it is rare that they have changed the powers available to these bodies. Hence 'extra-sectoral' issues are invariably seen as threats to forests – whereas they might, with better integration of institutions and incentives – become new sources for SFM.

Local government agencies

In countries as diverse as Bolivia, Mali and Ghana, authority for using forests, and sometimes their allocation, is being transferred to local government. This follows a global decentralisation trend. The notion is that forests can then be better integrated with the needs of local livelihoods and industry, and be handed to the most effective local stewards. But this notion was rarely cooked up by the local authorities themselves. There are many transitional pains – most local authorities do not have the forestry skills. Whilst this can be relatively easily rectified, incentives for forest abuse – the interaction with local politics and patronage which bedevils local government – is a more difficult issue. Decentralised forest management may help with real local needs, capabilities and trade-offs, but is yet to be fully transparent or accountable. This is less an argument against decentralisation per se than it is against seeing decentralisation as a quick fix. The tensions in decentralisation and devolution are explored in detail in Section 4.7.

'Pro-active' companies

Private sector actors are introduced more fully in Section 5. Suffice it to say here that they cover a range of characters, from the asset-strippers to the real investors, from the politically powerful multinationals to the almost innumerate community enterprises. They are all motivated to some degree by profit, and especially by the need for a secure operating environment.

Some companies get involved in policy specifically to maintain the *status quo*, or to change policy in ways to ensure their continued ability to produce fibre for profit (sometimes at the cost of forest management for social and environmental services). Other companies have introduced codes of practice in response to environmental and social concerns, whilst more ambitious enterprises have made public pledges and introduced third party certification in efforts to improve transparency. Some companies are strongly influenced (whether they like it or not initially) by environmental NGOs and discriminating purchasing policies, where products from well-managed forests are sought by buyers. Market and regulatory pressure have been key in encouraging movements towards the more ambitious strategies. However, pressures to improve the private sector's social performance have proven less effective (perhaps also revealing something of the class background of many environmental organisations).

Companies tend not to act *en masse*, however. Private sector associations tend to be weak. Yet there are also examples of well-organised associations (USA, Sweden and Costa Rica). If these engage transparently with other actors, the outcomes can be positive.

National NGOs
NGOs range from rich, internationally-connected bodies to small outfits with a half-life of a year or so; from those who represent a big membership to those which promote the views of just a minority; and from campaigners to facilitators. Their ease of operating and, to some extent, their policy influence are strongly conditioned by government and its ability to foster civil society.

In many countries, policy change can be traced back to environmental and forest people's NGOs. They are rarely present in the group which 'writes' policy, with the exception of new 'soft policy' instruments such as certification. But they have placed new issues firmly on the agenda through their varied tactics. As such, they have had very strong catalytic effects, – usually (but not always – because of the pursuit of rather singular interests) towards positive social and environmental ends. NGOs' abilities to lead to change depend very much on their strategic skills, which range from media relations, to political lobbying, to litigation, to influencing public opinion, to working with technocrats and people in the field. Sometimes their goals may be compromised by their needs for financial or political survival. (Figure 2.1)

Consumers
Consumers' demands make up a significant indirect factor in determining how forests are managed, but one which forest policy has so far not

Figure 2.1 External pressures on a forestry organisation

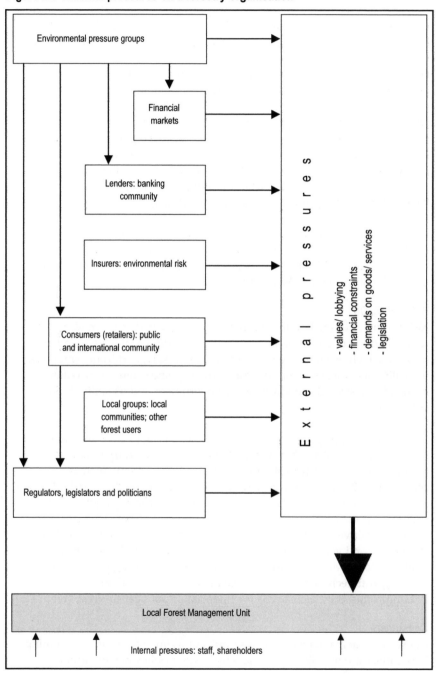

integrated sufficiently. In the past, demand-side policies have simply sent signals to producers about the quantity of forest products required to satisfy market needs. Now, retailers and others are starting to take a more active role in influencing the environmental quality of the forest products they sell by developing `brands' of certified or eco-labelled forest. Efforts to tackle the expansion in waste, for example by promoting paper recycling, are also having repercussions for forest management. Looking ahead, it is possible that rapid innovation in information and communications' technologies could start to damp down demand for paper and other wood products. As a result, current trends in demand cannot be treated as sure signs of future movements. To be fully effective, those involved in forest policy must become much more aware of the forces which shape the patterns of consumption for forest goods and services.

Academics and researchers

Academics are often held in particular positions of trust by the other groups, and thus can be key for clarifying positions (e.g. commissions of inquiry) or for developing options for change. However, scientists have their own pet theories and preferred forest values; and thus (overtly or otherwise) weave their own myths and stories. Simple narratives, in particular, tend to substitute quite quickly in policy-makers' minds for what would otherwise be complex webs of problems, and hence are important in setting agendas. Sometimes this is helpful, but other times – and this is especially evident in many of the stories of 'deforestation causing recent floods' (Thailand, China) – the simple story is not necessarily the correct one.

This approach – of being the knowledgeable authority whose story or 'spin' is held in the highest esteem – is also a tactic of NGOs. Both academics and NGOs may also be heavily influenced by the science and agendas of outside or international groups and donors. Again, this can be helpful; but it can also end up substituting for real local inquiry.

Development assistance 'donors'

Since the Cold War thawed out, the key underlying goal of international aid has shifted from geopolitical control to market competition, and to maintaining economic growth and reducing poverty in countries of the South and East. Politicians have a harder job to convince voters that aid does not 'disappear down a rat hole' (Senator Jesse Helms). Stricter conditions on performance and reporting are introduced. Under the label of 'good governance', the role of supposedly inefficient and corrupt governments is being reduced and donors are allocating their funds on contractual terms, fitting in with privatisation policies which are now the fashion in the North.

Pressure is being put on countries of the South and East to introduce multi-party politics, slim-down their bureaucracies ('downsizing and decruiting'), be more transparent and accountable, create greater space for civic action and reduce military expenditure. The economic prescriptions which go along with this are commonly known as structural adjustment. All of these activities can and do have policy interventions attached: policy studies, participation in policy decisions; and conditions regarding the scope of policy change. The activities themselves may generate new and powerful policy narratives which change the debate.

Some donors can act as useful links between local realities (based on their field project experience), national policy processes (through their – considerable – enthusiasm for comprehensive strategy approaches such as national forest plans), and global processes (through supporting developing country professionals to take part in activities such as conferences of the parties to environmental conventions).

In spite of their policy concerns, donors tend to employ technical and managerial approaches, in contrast to facilitating and developing relationships between groups – the stuff of policy – which is assumed to have uncertain outcomes. It also tends to gravitate towards project-based work dominated by professionals and specialists, usually consultants. Several studies have shown that the aid system has become over-tilted towards a self-serving professional class isolated and detached from both domestic and 'target' constituencies (Fowler, 1998, reports on some of these).

Consultants
The influence of consultant 'experts', paid for by governments or donor agencies, in shaping policy can be high. Increasingly, donor agencies are using independent consultants in policy-oriented work, rather than those from higher-cost public bodies. Contracts are often short, and terms of reference tight. Incentives structures – money and future work – in such an environment are likely to favour consultants who give the answers the donor would most like to hear, not questioning underlying assumptions, nor fighting against donors' favoured means, such as action plans.

Furthermore, even longer-term physical proximity of consultants to local reality 'in the field' does not necessarily equate with better results. "The enthusiastic field officer who prides him- or herself on being 'in touch with the grassroots', heroically ranging over the countryside in four-wheel drives and joining in back-slapping drinking sessions with village heads, is a familiar figure in the annals of rural development worldwide" (Lohmann, 1998). Yet, the information, understanding and relationships which this figure, like anyone else, develops, are shaped by experience from elsewhere and by self-interest. In these situations, it is quite likely that unsubstantiated assumptions will be recycled and 'off the shelf' policies or intervention packages will be promoted, such as standard formats which are inadequately rooted in local reality.

International NGOs

These bodies may display some of the characteristics of donors or local NGOs. In part, it depends whether they are advocacy- or service-oriented, and whether their intervention is backed up by political power, financial power, and/ or knowledge. The bigger NGOs – and notably the World Wide Fund for Nature (WWF) – may have more political power than some government bodies. They invariably want emphasis on social and environmental services of forests, and seek binding policy commitments, targets and monitoring. Global 'themes', campaigns and 'targets' tend to dominate, these themes often reflecting Northern stakeholders' perceptions and demands. International NGOs can 'shake up' local policy processes – quickly legitimising local NGOs, community-based organisations (CBOs) and certain academics. The earlier national conservation strategies, e.g. in Nepal, Pakistan and Zambia, had the effect of suddenly opening doors to groups who had not had access to policy debates before. However, international NGOs can also silence some local actors, or filter who is a legitimate player in a policy process in ways which are largely unaccountable.

Community-based organisations

CBOs' most outstanding characteristic is their emphasis on field operations, dealing with local conditions and exposed to multiple local needs. They have therefore had to develop pragmatic ways of negotiating the policy 'minefield'. This can provide good evidence about whether policy conditions help or hinder sustainable forest management. Where trust is built up with policy-makers, this can have the effect of providing arenas for experimentation at the interface of policy areas . However, they are rarely central to policy processes. If they are, it is often as much because of the charisma and/ or social standing of their leaders as it is because of their evident successes (e.g. IUCN and the Aga Khan Rural Support Programme in Pakistan). There is often a presumption amongst policy actors that CBOs effectively represent their constituencies. This may not always be the case, and indeed it can be damaging if a CBO is brought in to discuss policy when a real community representative is required.

The spectators, the policy 'target groups', and those who don't come to the show

There are several 'mainstream' forest actors who are not normally present in policy processes, yet they represent a substantial majority of people involved in forestry. These include non-progressive forest companies, communities, and farmers.

Logging industry and non-progressive forest producers

As yet, in most contexts, there is no clear profit motive for companies to engage in sustainable forestry on any type of land, public or private. Markets currently, and increasingly, reward short-term behaviour where policies and laws permit this, and companies stripping away forest assets is the norm in many countries, to the fury of other stakeholders. Most of the worst conflicts in forestry today are between private sector companies and local communities. These companies either evade policy, or seek to influence it in clandestine ways, as we discuss in Section 5.

Communities

People at local level may have considerable capacity to influence forest management, or even to completely determine the fate of forests in practice. Physical proximity to the forest resources is their most obvious characteristic, but local people may also influence forest management through: exercise of *traditional or legal rights*, indigenous *knowledge* and systems affecting forest land and resources; *dependence* on forest goods and services for livelihoods; and *cultural integration* with the forest environment (Colfer 1995, 1998). Local communities may have considerable *adaptability*, being able to adjust local rules of forest use in response to changing circumstances – many traditional systems have evolved through learning how to respond to disequilibrium environments, as in the pastoralists of the Sahel (Scoones, 1998). Local-level mechanisms may also provide considerable *conflict-resolution capability*, at intra-community and interpersonal levels.

However, communities are rarely 'ready-to-use' units of social organisation for forest management. 'Groups', rather than 'communities', are usually the best unit for understanding motivations and responses to policy, as well as for inputs to policy processes (Scoones, 1998). Whilst institutional/ authority structures amongst community groups are often weak – precluding collective action – there are usually many subgroups with widely different interests, rights, claims, and aspirations for forests. Simplistic descriptions of community cohesion and natural predisposition to environmental care should be treated with caution. For example:

> ...*the much-vaunted populist view of Melanesian communities living in harmony with each other and the environment is a myth. In reality Melanesian communities have always been on the verge of disintegration and it has taken special qualities of leadership in each succeeding generation to hold them together. An egalitarian ethos may however explain why some communities are willing to degrade their land in an effort to keep up with their neighbours.*
> (Filer with Sekhran, 1998)

Local people are often marginalised from policy, but may well be central to its practice. For example, various studies have focused on local people's 'resistance' to international capital in forestry (e.g. Guha, 1983, 1990; Peluso, 1992) and their small acts of defiance in everyday life – the 'weapons of the weak' (Scott, 1985). In PNG local people can pose a considerable threat, or 'menace', to the operations of the loggers – converting this, if need be, into acts of sabotage, intimidation of company personnel or production stoppages. However, a focus on resistance in other contexts may romanticise the creativity of the human spirit in refusing to be dominated by large systems of power. It may also underplay the role of local power differences. Conversely, a focus on local resistance may also overshadow evidence of large social movements around forest issues such as the expanding global networks on indigenous peoples' rights and against the WTO. Many are working in 'lobbying' mode rather, than 'resistance' mode.

Farmers

Farmers are often cast as the villains of deforestation. Indeed farmers may be the *direct agents* of much forest clearance, although often not the *ultimate cause* of that clearance. But farmers and communities are also growing trees and regenerating and tending forests all over the world. Their efforts may not make it into most forest area statistics, as in Pakistan. Yet, if the regenerated areas and trees grown on small farms in Africa are added together, it represents a quite massive level of reforestation. In Kenya there is an estimated 90 per cent increase in trees in the landscape compared with 20 years ago. In Nepal, communities are reforesting landscapes with little help from government, whilst in India, joint forest management programmes between forest departments and community committees have successfully regenerated over five million hectares of productive forest. Rarely are these farmers organised to influence forest policy – although this can be done, as IIED's work with JUNAFORCA has shown in Costa Rica (Section 4.5).

2.3 Global changes and uncertainties in the forest-people relationship

The forest actors are having to change in efforts to respond to wider global and societal trends.

The rise and fall of state dominance

For developing countries, state control in forestry stems from the colonial period. It was bolstered by a post-colonial belief in state direction of the economy; in state provision and production of services; in large governments as major employers; and in political ideologies favouring public ownership and control of productive resources. By the 1980s, however, there was a widespread perception, and considerable experience, of government failure in the forest sector. Overextended forest departments had few resources and little expertise to manage or control forest land efficiently; and public subsidies for logging operations were extensive, but did not result in public benefits, and were challenged on ethical grounds in contexts where local people were denied access to forest land. Almost nowhere is the state's direct involvement in forestry increasing (Landell-Mills and Ford, 1999).

Let the market decide?
The widespread trend of pointing to the limitations of government has not been peculiar to the forest sector. Experience of inefficient enterprise management by governments, coupled with poor delivery of services, increasing public sector debt, corruption and rent-seeking, and the lack of accountability to the citizenry, led to the new mantra of privatisation, deregulation, and decentralisation. This doctrine became manifest in the transition of former centrally-planned economies to market-based economies; in structural adjustment programmes promoted by international lending institutions, which led many national governments to reduce public sector expenditures and price distortions; and in the increasing globalisation of the world's economy, with trade and capital liberalisation and currency reform.

In the forest sector there have been varying degrees of privatisation: from merely exposing state-owned forestry bodies to commercial pressures to encouraging an enterprise culture in these bodies; to corporatising government forestry bodies, so as to form e.g. parastatals, which are freer than government bodies to act in the ways they deem suitable; to complete transfer to the private sector. As a result:

- in some countries, much forest land ownership has been transferred to private individuals, corporations and some communities;
- the management of some state-owned forest enterprises has been transferred to the private sector;
- the production of forest management services has been increasingly contracted out to non-governmental bodies;
- more and more timber is produced by the private sector;
- market-based instruments are increasingly common; and
- private sector institutions are more influential in forest policy.

Problems with the market mantra
The market is the main way in which wood products are distributed. But

regulation and common social norms such as trust are required for the most basic operation of markets. Recently, there has been a growing recognition, or rediscovery, that the market suffers from three key failings:

- *Key forest goods and services do not enter the market.* Markets for non-timber products, biodiversity, and carbon storage, are often non-existent or ineffective, so the private sector does not recognise them.

- *Environmental costs are largely ignored.* Markets do not automatically internalise environmental costs and often shift these costs on to others. Without policy checks and balances, the private sector's wood production activities often degrade the production base for non-wood benefits.

- *Distribution of wealth through the market is rarely fair.* The pattern of private investment is often very patchy and does not address the needs and priorities of the weakest members of society. In contrast, the 'needs' which it does address are sometimes those which are 'created', through the power of advertising, amongst richer consumers. These may be temporary fashions. But they might also become integral parts of consumer life-styles (e.g. constant supplies of fine paper). *Bigger players* tend to be favoured whenever markets are developed, although smaller players may find temporary niches in the process of market development. This is the case of the rapidly-developing market for certified wood, where small players were originally dominant, but now big companies in the North are taking the lead. It is also the case with NTFPs and ecotourism.

Rising inequality
Global production and consumption patterns increasingly determine what is available locally in the way of consumer goods, production inputs, capital goods and technologies. Even remote rural areas are increasingly linked to the national and global economies. Poor people linked to forest environments are no longer necessarily isolated or 'traditional'. And they don't necessarily get what they *need* in order to make a good living from the forest and to be effective stewards of public forest benefits.

Poor communities deep inside forest areas in many countries are radio listeners; they have their souls contested by evangelists, and they have their communities organised by organisers of various kinds. In some cases they may be better informed and better poised to challenge their conditions. But in many cases this is little more than 'informed bewilderment'. And often their poverty is worsening. The global economy will certainly expand in the twenty-first century, using substantial increases in the power of telecommunications and information processing.

[It] will penetrate all countries, all territories, all cultures, all communication flows and all financial networks, relentlessly scanning the planet for new opportunities for profit-making. But it will do so selectively, linking valuable segments and discarding used up, or irrelevant, locales and people. The territorial unevenness of production will result in a new geography of differential value-making that will sharply contrast [different areas] (Castells, 1997)

Thus there are different abilities for taking advantage of globalisation – of the movement of capital and technology – and for bearing the costs.

Rising inequality is a potential powder keg in many societies. This is sometimes directly linked to forests. In Indonesia for example, the polarisation of wealth – one of the triggers for resentment against the current power brokers – is directly linked to control of the 70 per cent of national territory which is under forest reserve. Most of this land is apportioned out to forest industry. Ten companies control the $20 billion-a-year Indonesian forestry industry, and at least five of these have been owned by Mohamad 'Bob' Hasan. This story is explored further in Section 5.4.

2.4 Changing the forestry plot – the SFM 'power play'

The global trends described above have been reflected in some unexpected changes in the international environment and development drama over the last forty years. The major thrust, or paradigm, of forestry has taken some twists and turns over this period. Table 2.2 summarises some of the main perceptions of problems and of generally accepted solutions in the environment and development debate on the world stage, and the main forest policy responses.

Table 2.2 suggests that there has been something of an epidemic of policy 'weasel words' – 'slippery' concepts which can be redefined to suit the speaker. Indeed, the language used in policy discourse and statements can be as important as the substance – for specific forms of language are associated with certain moral positions, degrees of openness or closure to different views, and they therefore support the involvement of some groups and alienate others (see Annex 1 for further discussion of policy language).

Table 2.2 Global trends and forest policy responses

Decade	Problem trend	Solution trend	Forest policy response
1960s	'Under-development'	'Modernisation'	Industrial forestry as the 'engine' for development
1970s	Energy crisis and deforestation	'Basic needs'	Afforestation to avert the 'woodfuel crisis'
1980s	'Eco-disaster'	Sustainable development	National forest plans and social forestry projects
1990s	International muddle	'Free-trade', inter-governmental and civil society agreements	International quest for sustainable forest management
2000s	Globalisation with increasing inequality	Globalisation with initiative, rethinking the state,increasing taming the market, increasing formal local controls	Positive approaches appear to be: • equity and accountability • the state as facilitator • payments for global services • partnerships between government, civil society and the private sector

'Sustainability', 'cooperation', 'community' and 'participation' are sure-fire winning words – living blameless lives of their own in language, policy and analysis of every kind. Nobody, from any position, will want to say they are against these 'good things'. They are goals and aspirations, and they appear to provide room for everyone's principles and needs.

Yet we should be suspicious of terms agreed by so many people, which everybody likes – if they never become defined in terms which are meaningful locally. Such vague 'consensus' can mask important differences and imply that nothing is wrong. (Patently there is something wrong, or the debates in which these terms are wielded would not be raging in the first place.) Agreement over real-life meaning is only possible if negotiated by specific (usually small) groups of people.

But there are also underlying philosophical arguments which need to be explored and resolved, processes which will neither play out quickly nor be confined to the forest sector. If language constrains agreement, this is often because the underlying philosophies both clash and are unexplored. For example, the notion that forestry goals should accommodate *efficiency* is widely accepted. The goal of *equity* is gaining ground. But a conundrum is presented by the emergence of *sustainability* as a goal:

- Do we consider sustainability to be nothing more than a *technical constraint* to developmental goals, related to environmental limits – with the implication that this is primarily a matter of science? (Marcuse, 1998).

- Or do we consider sustainability, like liberty or justice, to be a *social goal* to which we aspire – but which has to be articulated and agreed locally before it can be achieved in practice – and therefore is primarily a matter of participation? (Holmberg *et al*, 1991)

If the second is the case, then the predominant political culture – and notably the way in which it encourages or discourages debate, participation and partnerships – will fundamentally condition the interpretation of 'sustainability'.

This brings us to the rather obvious but important point that policies are based on assumptions. In forest policy, typically, these assumptions have remained untested and unchallenged until quite recently (Box 2.3).

Box 2.3 Sticking to the story – the power of prevalent assumptions in forest policy

Many forest policies have been based on the assumption that once upon a time people lived in harmony with nature, that this broke down as people became numerous, needy and greedy, and that all kinds of environmental and social calamity will result unless the wielder of the policy (usually government) takes dramatic action by creating forest reserves, controlling use of forest products and generally ordering people around – and out of forests. Variations on this story, often spiced up with different sorts of 'crisis' at different times, have served policy-makers well and enabled forest departments and the other agencies given life by the policy to justify their existence and expand the territory and resources under their control (including donor cash in recent times).

Leaving aside the validity of such a narrative, it clearly overrides the huge diversity of perspectives on forests in different places and times. The language and 'labels' used in the narrative prop it up over time, and offer a rationale for management actions, projects and data collection methods which further strengthen the foundations created by the original assumptions (see Annex 1 for further discussion).

For example, local people get labelled as 'slash and burn farmers' and the forest department declares the forest area they live in as 'protected', so the farmers are then 'illegal squatters'. The farmers don't disappear, so the forest department collects figures on their illegal actions and the damage they are causing to the forest. The figures are used to justify further action, with the farmers now re-labelled as 'target groups' to be variously fought against – or now increasingly worked with – perhaps with donor support.

A few key *scientific* ideas would also likely underpin the foresters' justifications, these might include: the notion of climax vegetation (that in the absence of disturbance, closed canopy forest would be the natural vegetation); the supposed causal linkage between removal of forest cover and declining rainfall or increasing flood; and the idea of carrying capacity (that every set of ecological conditions can support a given number of people or livestock which, if exceeded, will result in rapid ecological degradation). All of these notions require very careful investigation in any particular context, and in many have been shown to be false (Leach and Mearns, 1996). Yet the narrative often lives on, despite the evidence mounting against it.

People on the 'receiving end' of policy may even join in to keep the same narrative alive. For example in Papua New Guinea, there have been cases where community groups have themselves dreamed up descriptions of the disastrous effects they are having on their forest environment, so that they can secure their own donor-funded environment rehabilitation or 'awareness' project.

Prevailing narratives also define people's priorities and ideals for forests.[1] NGOs, politicians, large companies and an ever more pervasive and persuasive media are adept at generating and changing dominant perspectives (with their attendant assumptions). WWF and their like have told us for twenty-five years that the forests are in trouble; politicians wave various flags of economic convenience whilst declaring forests to be national assets; companies tell us jobs depend on production forestry and there is no such thing as global warming; and the media, if it covers forest stories at all, sticks to simple sensationalism.

In forest policy processes, assumptions generated at international level are increasingly translated into national and sub-national initiatives, with mixed results. In Papua New Guinea the World Bank has used its considerable economic muscle in recent years to push policy change through a new forest law and forest revenue system. These are both very blunt instruments based on the assumptions that industrial regulation and national revenue generation are the top priorities. These assumptions had little in common with those held by local actors. The result has been rather minimal compliance with the new instruments – with much tension. The possibility of making progress with less heavy-handed policy instruments, building on an understanding of local narratives and priorities, has also been undermined in this process.

In Section 5, we examine the evolving policy language which has developed to suit the niceties of international environmental diplomacy. Whatever its drawbacks in terms of vagueness and obfuscation, the international forest debate has allowed actors with otherwise very different values and preoccupations to come to the same table. It has produced a general agreement on the main ingredients of good forestry whilst, for the most part, acknowledging that how these ingredients should be structured depends very much upon the local context. Box 2.4 summarises the elements of 'SFM' that are common in most of the intergovernmental, national and civil society initiatives.

---

1. For example, professional foresters and ecologists "have conventionally sought to maintain closed-canopy or gallery forest – practically defining 'forest' in these terms – so that any conversion of such a vegetation community is seen to constitute degradation. Yet such conversion may be viewed positively by local inhabitants, for whom the resulting bush fallow vegetation provides a greater range of gathered plant products and more productive agricultural land. Thus, what is degraded and degrading for some may for others be merely transformed or even improved" (Leach and Mearns, 1996).

Box 2.4 Common elements of SFM standards

Framework conditions:
- Compliance with relevant legislation and regulation
- Secure/ transferrable tenure and use rights
- Transparency and accountability
- Dealing with extra-sectoral pressures
- Clear roles of authorities
- Policy commitment to SFM

Sustained and optimal production of forest products:
- Sustained yield of forest products
- Management planning
- Monitoring the effects of management
- Protection of the forest from illegal activities
- Optimising benefits from the forest

Well-being of the environment:
- Environmental impact assessment
- Conservation of biodiversity
- Valuation and protection of ecosystem services
- Hazard (waste and chemicals) management

Well-being of people:
- Regular consultation and participation processes
- Social impact assessment
- Recognition of rights and culture
- Relations with forestry employees
- Contribution to socio-economic development

Thus, international preoccupations, and interpretations of what matters, have changed considerably over the last forty years. As new notions percolate through to national policy agendas, some concepts get inflated while others fall out of favour. But most countries and actors are now at a stage where:

- multiple actors, values, and objectives are widely acknowledged as important, but there is resistance to change;
- different forest needs from local to global level are beginning to be distinguished;
- the integration of all of these concerns is accepted in principle through some – as yet unclear – notion of sustainability planning.

But we lack the integrating institutions and policy processes to define this more clearly, or to put it into practice. This suggests that there are some elements of a common plot to the play in many countries and contexts. This report teases out those elements that have had a useful impact to date. It also looks at how they can be introduced to the plot where their absence has resulted, so far, in tragedy. The next section explains how we have conducted this work.

3 Policy That Works for Forests and People – a collaborative project approach

3.1 Aims and approach

The origins of the project, *Policy that works for forests and people*, lie in IIED's work over many years on information systems, field programmes and institutional capacity in forestry, on participatory methodologies in various fields, and on analysing various forms of national strategies and plans including land-use plans, conservation strategies, and forestry action plans.

All of this work pointed to the importance of the dynamics of actors, their decisions, and their experiments in the policy domain. In contrast, normative exhortations by a few people to introduce policy 'x' appeared insignificant. It became apparent that the real need is to improve understanding of what policy actually is in practice, how it is formed, what its impact is on forests and people and the relationship between them, how policy can change, and how it can be harnessed to improve forest management and human well-being.

The Forestry and Land Use Programme of IIED embarked on the project in January 1995, with the following goal:

to improve the understanding and practice of policy processes, so that they improve the sustainability of forest management and optimise stakeholder benefits.

The emphasis is thus on policy processes and the conditions conducive to good policy, rather than on specific policies or instruments.

The project was supported by the UK Department for International Development (the Overseas Development Administration until May 1997) and the Netherlands Ministry of Foreign Affairs (DGIS).

Collaborative policy research was at the heart of the project's approach. This can be described as a participatory process involving much engagement with the current 'holders' of policy (we aim to deconstruct the simplistic idea of 'policy-makers'). This treads the fine line between 'reasonably objective, disinterested' analysis of policy, and being 'drawn in' to policy processes. It aims to be strongly engaged in policy – to make the most of opportunities, to create political space, and to push the debate forward – but always on the basis of sound analysis that captures many actors' points of view, rather than promoting one.

Box 3.1 shows the basic form of the project.

Box 3.1 Policy That Works for Forests and People – project outline

Why?	Policy underpins the biggest forest problems, yet existing forest policy reforms often have little impact in practice, or benefit only a few people. What constitutes good policy for forests and people, and what makes it work?
What?	Understanding the actors and the forces at play in policy: the winners and losers and the factors that affect policy outcomes. Identifying policy that works, and pressing home the findings
Where?	Global review and detailed studies in six focal countries: Zimbabwe, Ghana, India, Pakistan, Costa Rica, Papua New Guinea
Who?	Country teams from a mix of local institutions and disciplines, and IIED
How?	Field work, extensive consultations, analysis both in the 'corridors of power' and in the forest, strengthening of local policy research capacity, and dissemination of results in focal countries and internationally
When?	1995 to 1999

The project began with IIED reviewing literature and consulting with various policy research organisations and forestry practitioners. The review aimed to understand the state of the debate on forest-related policy – both processes and contents – in as wide a range as possible of contexts and countries, North and South, within the constraints of time, money and information accessibility. A draft review document was produced in June 1995.[2] This wide-ranging analysis continued through to 1999 with a number of IIED-led and commissioned thematic and country studies.[3]

---

2 The initial literature review (IIED, 1995a) pulled together written experience from diverse international sources available to IIED. Lessons were drawn – about prevailing views on policy and process and apparent gaps in knowledge – and these were used to set the scope for the subsequent project. The review provided a source of reference material and a stimulant of ideas for country teams and collaborating organisations.

3 Early review work showed a clear gap in the analysis and tracking of developments in two main areas, which the project sought to fill: international forest policy processes; and the interactions of the private sector in policy processes. In addition, studies in 'northern' contexts were undertaken to investigate some of the key themes which emerged from the focal countries in the south, and to provide complementary insights from particularly interesting policy processes. These were: Australia – a developed institutional climate organising a process of policy change; China – much policy experimentation following major political and market change; Scotland – the changing importance of social objectives for forestry and its linkage to the land reform debate; Sweden – an example of an effective facilitatory, rather than managing, role of government; and Portugal – which is trying to reactivate the forest commons.

Project funding was available to cover six detailed studies within developing countries, and these countries were selected during this period. Although six studies is a fairly arbitrary number, it allowed a reasonable geographical spread and a range of contexts to be covered. In addition to representation from three continents, the aim was to select countries which together provide a range of forest endowments and policy-institutional environments. Additional criteria for country selection were:

- significant importance of the forest resource at both national and local livelihood levels;
- evidence that forest policy has some effect in the country, and/ or that policies affecting forests are under debate or reform;
- particular interest in the country from UK DFID and/ or Netherlands Ministry of Foreign Affairs, the project sponsors; and
- existence of previous fruitful collaborative research links between IIED and one or more local institutions interested in the project.[4]

This last criterion was felt to be important for methodological reasons: to ensure some working knowledge of the country context and forestry issues on the part of IIED; and to ensure mutual understanding about the role of the study with respect to the state of local debate and politics. It was also important in relation to the project's secondary objective – to utilise local policy research capacity, and strengthen it through the exercise – which is best served through a collaborative relationship. Finally, it was necessary to 'hit the ground running' by working with institutions of known credibility and reputation, because the studies were principally aiming to contribute to the improvement of policy in the countries concerned.

A planning workshop was held in June 1995, with potential coordinators of the country teams and key resource people. Core issues for investigation in all the country studies were identified, and a set of generic tasks for all the country teams to carry out was agreed. In addition, the specific policy dilemmas and opportunities facing each country were set out. Outline plans for the country studies, including who would need to be involved, were then developed, so that both generic and country-specific tasks could be achieved. In the following months, the country coordinators put the teams together and held in-country workshops and discussions to agree specific objectives and work plans.

Each of the teams had a base in the coordinator's institution, but also comprised a mix of experience and institutional affiliations appropriate to

---

4 Three further regions (and countries) were selected on the basis of these criteria: Latin America (Bolivia), Francophone Africa (Senegal or Mali), and South East Asia (Vietnam). However, funding restrictions prevented initiation of these additional country studies.

the thrust of the particular study. Each team also drew on the experience of a wider pool of people with an interest in policy issues, constituted in various advisory groups, discussion groups and interview schedules. These pools became vital both for generating information and 'feeding back' ideas and findings from the team as the studies progressed.

During the process of analysis the country teams attempted to enhance the level of debate and openness concerning policy strengths and weaknesses in each of the study countries. Through engagement with many stakeholders the teams themselves became involved in installing some of the attributes of adaptive policy processes, which in turn led to key opportunities to improve policies. For example, in Ghana the team's consultations with policy stakeholders led to a request from government to investigate options for forest certification. The team went on to prepare the ground for a promising national certification programme. In Zimbabwe, the team's approach to generating policy debate amongst groups from various institutions catalysed the formation of a policy unit in the Forestry Commission with a specific mandate to generate cross-institutional collaboration on policy issues affecting forests.

The process of preparing products from the work was itself used as a research tool – with drafts of reports and briefings being exchanged between IIED and the country teams (including exchanges between the teams). This allowed lessons or ideas from one country or context to be posed as questions for further investigation in another. For example, the Costa Rica team's approach to showing how the relative power of stakeholders has changed over time (see Section 4.1), was adopted and adapted by the Zimbabwe team. A preliminary results-discussion workshop, involving all team coordinators and key resource people, also helped this process. The country study drafts were consequently enriched and better focused.

Each of the six country study reports was launched at an event with a 'hook' appropriate to the country concerned. For example, in Costa Rica the formation of a new government provided the opportunity for a book-launch and well-attended presentation of findings to parliamentarians. In Papua New Guinea, the team's book provided the basis for a three-day national forest forum, which attracted much media attention and led to a considerable broadening and energising of the policy community. The country teams also prepared short policy briefings for key stakeholders, sometimes in local languages, in one case a video, and background papers. Detailed analytical papers were prepared by the India, Pakistan, Zimbabwe and Ghana teams, whilst the Papua New Guinea team published a monograph on *The political economy of forest management in Papua New Guinea*.

Further activities to maximise the impact of the country studies included: specific stakeholder workshops; incorporation of results in media and educational programmes (newspaper columns, radio transmissions, teaching and curriculum development); and, supporting the emergence of multi-stakeholder policy negotiation processes.

At the international level, project findings have been shared in a number of fora such as the Intergovernmental Forum on Forests, the World Commission on Forests and Sustainable Development, the Forestry Advisers Group, and a range of other networks, as well as being used for teaching in several international policy courses in the UK, the Netherlands and Indonesia. IIED and its collaborators are also utilising the findings in work with Forestry Departments and others in an increasing number of countries. Hence, use of this work, and engagement by the country teams and IIED in policy processes, continues.

3.2 The project's perspectives on policy

From the outset, it was clear that even within the group of IIED staff and country team members in the project, there were multiple perspectives on policy and on how the issues could best be treated. Indeed, perspectives amongst policy researchers and policy-makers always vary (see box 3.2)

Box 3.2 What 'policy wonks' say about policy

Forest policy commentators

- *"A forest policy specifies certain principles regarding the use of a society's forest resources which it is felt will contribute to the achievement of some of the objectives of that society".* Worrell (1970)

- *"In practice, normative guides (indicating what should be done) and positive reality (what is actually done) diverge. More often we find (quoting Charles Lindblom, 1961) 'coordination through partisan mutual adjustment'... Rationality is limited by the sheer complexity of interactions among economic and social forces".* Johnston et al. (1967)

- *"The starting point [for forest policy formulation] must be social objectives. It must provide for specified goods and services to go to specified groups by specified dates. That means finding out what people want....Thus the creation of a forest policy is a process which should involve all groups and institutions with a direct or indirect say in the forest or with responsibility for implementing policy".* Westoby (1989)

- *"It is easy to urge that forest policy should be treated as a whole, and that forest policies need to be strengthened. But it is very doubtful whether a comprehensive and consistent forest policy can ever be formulated, let alone prescriptions converted into reality."* Mather (1990)

- *"It can be argued that the more sophisticated a policy is, the poorer are the results in terms of implementation....The insistence on certain 'politically correct' issues such as 'country-led processes, partnership, participation, holistic approaches, inter-sectoral linkages, harmonisation, multidisciplinarity, sustainability' is to a certain extent misleading. All these aspects are certainly very important...but something is missing. It is policy tools which are essential in determining the course of action of a particular forest policy."* Merlo and Paveri (1998)

Public policy analysts
- *"Policy is subjectively defined by an observer as being such, and is usually perceived as comprising a series of patterns of related decisions to which many circumstances and personal, group and organisational influences have contributed... accidental or deliberate inaction may contribute to a policy outcome".* Hogwood and Gunn (1984)

- *"Public policy is commonly presented as a rational, linear activity. Policy-making in this view proceeds from the identification of a problem, usually in the form of a deficit, to a sectoralised statement involving data and 'research' and a short list of alternatives for a selection procedure known as policy-making, or a conclusion that there is no alternative, (quoting Schaffer, 1984) 'thereafter other and different (non-policy) things, known as implementation, occur'."* Harris (1991)

Natural resources policy analysts
- *"Policy is dynamic and is formed by actions (including the maintenance of the status quo) at all levels in the decision hierarchy from the central legislature down to the individual resource user. This means that there is no standing policy target against which the effectiveness of administrative structures, regulatory techniques or outcomes can be judged."* Rees (1990)

- *"[It is necessary to] analyse public policy on development as the process of what governments actually do, to explain the linkages between intentions and outcomes....There are two problematic aspects [with the 'mainstream treatment' of policy]: First, the myth of decisionality. Policy is treated as verbal, voluntaristic and decisional, in contrast with actual practice which is concerned with decisions, agendas and establishments. Second, the policy implementation dichotomy. Policy is regarded as mere utterances separate from implementation, so that whole zones of policy practice are ignored with serious, even disastrous consequences."* Clay and Schaffer (1984)

Development policy analysts
- *"Hearing, seeing, saying no evil. A fourth wise monkey would write no evil. But little policy writing is the work of a fourth wise monkey. The plainer and clearer a policy is painted, the more it is driven by evasion and disguise."* Apthorpe (1997).

- *"Effective development policies and programmes (i.e. ones that succeed in mobilising funds, institutions and technology) depend on a set of more or less naïve, unproven, simplifying and optimistic assumptions about the problem to be addressed and the approach to be taken."* Hirschmann, (1968), quoted by Hoben (1996)

Social anthropologists

- *"Policies are inherently and unequivocally anthropological phenomena, they can be read as cultural texts, as classificatory devices with various meanings, as narratives that serve to justify or condemn the present, or as rhetorical devices and discursive formations that function to empower some people and silence others".* Shore and Wright (1997).

- *"Stories commonly used in describing and analysing policy issues are a force in themselves, and must be considered explicitly in assessing policy options. Further, these stories often resist change or modification even in the presence of contradicting empirical data, because they continue to underwrite and stabilise the assumptions for decision-making in the face of high uncertainty, complexity and polarisation".* Roe (1994).

Common to most of the above are different degrees of preoccupation with: assumptions and beliefs; power structures; peoples' language, interactions and their dynamics; intentions regarding specific forest values versus actual practice; and patterns of correlation between the above.

It became clear that a flexible notion of policy was needed, which could accommodate different views within the country teams and could also provide a basis for the teams to engage with the prevailing perspective and state of the debate in each country.

The concept of policy which emerged over the course of the project is outlined in Box 3.3.

Box 3.3 Defining policy

There are almost as many uses of the word 'policy' as there are trees in the forest, and it is just as easy to get lost in either. Confusion about what policy is can obviously lead to confusion about what to do about it.

For the purposes of this report we use a rule of thumb: ***"Policy is what organisations do"***. This more or less tallies with the Oxford English Dictionary definition: *"Policy. A course of action adopted by a government, party, ruler, statesman, etc; any course of action adopted as advantageous or expedient."* [5]

In stressing 'organisations' rather than 'government', we highlight the fact that non-governmental and private organisations 'adopt' policy too. Whilst public policy studies concern themselves with 'what governments do', in our conception policy is not only the business of government. We think of policy as having substance or content – in the form of policy *statements* and policy *instruments* – and also having *processes* - chief among them being policy *making* and policy *implementing*. Thus we are concerned with both *intentions* and actual *practice*. Part of the job of policy *analysis* is to work out why policy statements often seem to bear so little relation both to what people actually do, and to the eventual policy *outcomes*.

---

5 A participant in this project's work in Papua New Guinea described Policy That Works as: "doing the right thing and doing the thing right"

The project recognised that policy is more than just the business of government because:

a. policy is strongly influenced by stakeholders' respective powers, and government is neither necessarily all-powerful nor in a position to moderate power inequalities

b. in practice, government policy is formed by actions at many decision-making levels – from central to local institutions; in turn, these are under pressure from, or at least influenced by, civil society and the private sector; and

c. many of the most important actions affecting forests – which may still have very 'public' or 'societal' effects – have little to do with government; they might be regarded as 'non-government public policy' or '*soft policy*' (although it may have very 'hard' effects!).

Thus, whilst the outputs of the project are mostly targeted at improving public policy in the project's six case study countries, some of the findings relate to intergovernmental, private sector and civil society soft policy processes and mechanisms. In the country studies the net of analysis was spread widely over these groups.

'The field' for policy analysis – the power play
A focus on policy needs to understand the workings of multiple, intersecting and conflicting power structures which are local but are also tied to non-local systems. 'Field work' for policy analysis therefore cannot be restricted to discrete local communities or bounded geographical areas, nor to corporations, élites and their centres of power. We need to focus on the social and political space articulated through relations of power and systems of governance between and amongst these levels. As anthropologists Shore and Wright have put it, "analysing connections between levels and forms of social process and action, and exploring how those processes work in different sites – local, national and global...tracing ways in which power creates webs and relations between actors, institutions and discourses across time and space" (Shore and Wright, 1997). This means looking at interactions between different sites or levels in policy processes, and at policy connections between different organisational and everyday worlds.

3.3 Methods

With the project's broad definition of policy, and its actor-focused approach, a wide range of methodologies was considered applicable. Collaborative approaches were key, and operated at several levels:

- IIED with international resource people on policy processes in developed countries, and the impacts of international processes
- IIED with country teams on project planning, methodology development, information flow and synthesis, peer review, editing, and project coordination
- country teams with in-country and international resource people on information-sharing and development of policy options
- country teams with other country teams on common methodologies and findings
- country teams with in-country stakeholders on sharing information, opinions and perspectives

Country teams utilised a wide variety of methods, some of which were invented and/ or adapted during the two-year course of the work. Box 3.4 lists these methods. Those which were particularly effective in several countries are described further in Annex 1 which is presented as a guide for others tackling this kind of 'policy work' – a term used for approaches which combine methods to research policy with tactics to influence policy.

Box 3.4 Methods used by country teams in collaborative policy work

Before research
- Consultation with current holders of policy – so that the research process and its results are not a 'surprise' to them, and they are open to change
- Team formation – balance of disciplines, credibility
- Inception workshop at national level – decision on focus of analysis (national purposes being primary and overt, generic or global issues being secondary or less overt; complementing other current analysis/ policy initiatives)
- Advisory group formation – for a 'sounding board', opening doors, and gravitas/ credibility of results
- Work plan and milestones

Throughout research
- Reporting progress and spreading information on the project – to unearth new collaborators, build an audience, feed debate, and develop outlets for findings
- Continuing communication with many key stakeholders – for verification, and to elicit opinions (policy issues are complex, and stakeholders do not always make the most pertinent input to policy research on the first contact – they need to react to others' opinions and researchers' findings)

- Team discussions on key themes
- Advisory group meetings and individual consultations
- Stakeholder/ key informant interviews – for those with stories to tell, but who lack the time, resources or skills to put them across; and to ensure a uniform structure for inquiry
- Liaison with current policy holders – to ensure that information is up to date and benefits from the current debates, and to ensure focus and timeliness in presenting results
- Resource material usage – IIED global literature review, bespoke information on possible methods and policy experience elsewhere, drafts from other country teams - to inspire, challenge and inform team members
- Circulating ideas and report drafts – in-country and with IIED

Situation analysis
- Literature review, update, critique
- Secondary data review – unpublished forest resource and usage statistics, economic data, internal institutional reports, project reports
- Forest resource survey/ inventory interpretation (Ghana, PNG)
- Mapping of policy, institutions and tenure
- Compendium of commissioned/ volunteered written perspectives from observers/ analysts (PNG, Ghana, India, Pakistan, Zimbabwe)
- Cross-country comparison (PNG – a comparison with Irian Jaya)
- NGO alliance meetings (Zimbabwe)

Defining 'success'
- Analysis of policy statements and laws
- Priorities of current policy holders – interviews
- Focus groups on 'success' and indicators of it
- Stakeholder narrative interviews
- Survey of community attitudes through field-based anthropologists (PNG)
- Participatory Rural Appraisal with particular communities on vision and priorities (Costa Rica, Pakistan, Zimbabwe)
- Interviews with actors in local institutions (Ghana, Costa Rica)

Policy analysis
- Focus groups on policy influences
- Regional debating workshops (India)
- Thematic workshops
- Stakeholder analysis – interests, capacities, relationships, impacts over time
- Stakeholder narratives of policy contests – tracking themes through history (Ghana, PNG, Zimbabwe, India)
- Review of press cuttings and media features over time (PNG)
- 'Fieldwork in the corridors of power' – interviews, liaison with policy holders
- Ranking/ diagramming policy influences and power
- Power analysis – distinguishing types, sources, and exercise of stakeholder powers over time

- Case studies:
  - policy in forest projects: logging, 'alternatives', conservation (PNG)
  - institutional change: land reform, state forestry roles and devolved resource management (Zimbabwe)
  - policy instruments: forest reservation, timber royalties, off-reserve controls, forest certification (Ghana)
  - institutional and strategy initiatives: forest cooperatives, state enterprises, participatory rural development projects, national planning and policy initiatives (Pakistan)
  - key areas for forest management: forest production, fiscal and financial incentives, protected area management, decentralisation (Costa Rica)
  - joint forest management: local practice, economic externalities, national policy framework, stakeholder positions, institutional change (India)
- Conceptualising and visualising the policy process (see Box 3.5)

Developing conclusions and recommendations
- Feedback focus groups on preliminary findings
- Written commentaries by stakeholders on preliminary findings (PNG)
- Consultation with parliamentary committee (Zimbabwe)
- National debate/ validation workshop (Pakistan, Costa Rica)
- Six country teams meeting together – debating/ refining findings
- Peer review of drafts by selected international reviewers
- Iterative refining of recommendations – including extensive 'questioning' by IIED and others to clarify issues and fill gaps

Outputs and dissemination
- Seizing opportunities to contribute to/ influence policy through the project (Ghana, Costa Rica)
- Synthesis reports by country teams in English – average 150 pages with: executive summary, forewords by key policy players, design-work, colour photographs and diagrams
- Spanish version of synthesis report (Costa Rica)
- Monographs and detailed background reports (Ghana, India, Pakistan, PNG, Zimbabwe)
- Policy briefing papers – and workshops tailored for key stakeholders
- Book launches with national dissemination workshops
- Newspaper, TV and radio features
- Incorporation of findings in educational curricula

A core method for each of the country teams was to develop a conception of the policy process which made sense in the context of the country. The diversity of concepts of policy amongst the country teams is illustrated in Box 3.5.

Box 3.5 Conceptualising the policy process

The six country teams each developed their own ways of describing the policy dynamic:

Forest policy as a tug-of-war

The reform of forest policy in Papua New Guinea was conceived as a social drama, or tug-of-war, in which the logging industry is pitched against a group of international donors for the prize of forest held by the third major group – the customary resource-owning groups and local communities. These resource owners may be cheering or booing one or other side, or they may be paying no attention at all. How these contestants fare depends in part on the actions and relative strengths of the groups in three sectors - the public sector (politicians, public servants), private sector (enterprises and private organisations) and social sector (NGOs, church groups, etc). The contest is shown in the diagram:

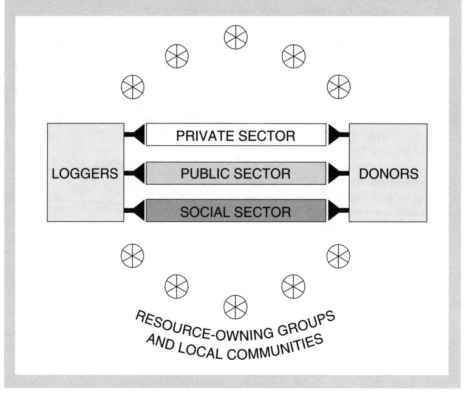

Factors shaping policy and practice

The Zimbabwe team was concerned with the 'gap' between policy pronouncements and practice. Practices which have impact on forests and people can be identified as emanating from a wide range of institutions. Within these institutions, trade-offs over capacities and resources are made and policies are interpreted (via key threads of reasoning, or 'stories') according to institutional actors' different motivations. Many policies can be identified and traced back to the broad strategies and themes informing governments' attempts to shape national development. These have themselves been shaped by resource endowments, historical factors, macro-economic pressures, and other contextual factors and pressures internal and external to Zimbabwe. These connections are illustrated in the figure below.

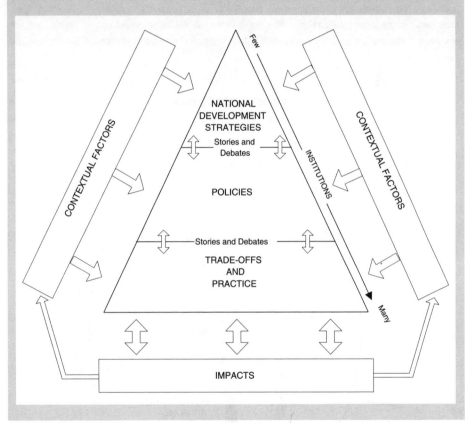

The policy 'water cycle'

For purposes of analysis, the Ghana team found it useful to think of policy as a cycle connecting the elements of policy-making, implementation, monitoring and evaluation, and review. However, these elements are not necessarily sequential and may well be overlapping. The team developed a policy 'water cycle' as a tongue-in-cheek analogy for the policy cycle. Policy rains down from above, having been manufactured by some invisible being. Some policies are absorbed and used by those below, sometimes gratefully, sometimes with much irritation. Other policies 'run off' with little impact.

Much public policy failure stems from a failure by the 'rainmakers' to see that a cycle is needed to maintain the process. Hence, community level participation which, like evapotranspiration from local surfaces, should throw water back into the atmosphere and drive the cycle, is kept weak and feeble, limited only to informal feedback processes. Furthermore, without key individuals ('condensation nuclei') on which water droplets begin to build, the rain will never fall!

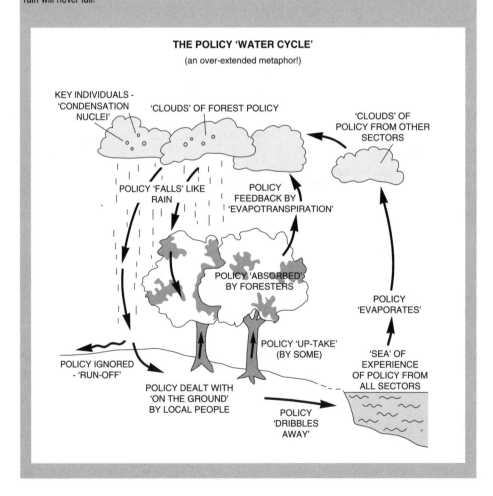

THE POLICY 'WATER CYCLE'

(an over-extended metaphor!)

Policy as an adaptive spiral

The Costa Rica team also used a 'stages' model of policy to picture an improved policy process. The interlinked stages identified are: goal-setting; negotiating roles and alliances; planning and developing mechanisms; putting plans into practice; and informing and monitoring. Although likely in practice to be out of sequence and unlikely to be self-contained, all of these stages are eventually needed for policy to be truly adaptive and to have the commitment of a wide range of people who are vital for sustainability to be achieved. Keeping the 'wheel' of policy turning is crucial – each of the notional stages of policy needs to be revisited regularly enough to permit learning, adaptation and improvement.

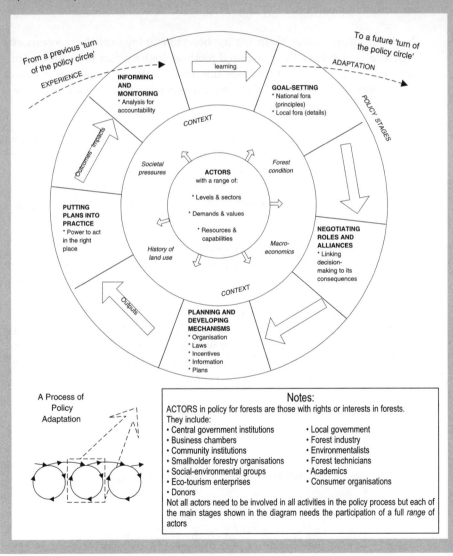

Methodological dilemmas were spotted early on. They applied in most of the countries but could only be dealt with on a case by case basis. A common concern was how to strike a balance between the project's generic objectives, country policy-holders' agendas, and country team ownership of the research. The project would have been perceived as meaningless if the country's priorities were swept aside. It was agreed that effective global comparisons entail a strong understanding of a country's cultural, institutional and economic context, and isolating what the 'success factors' are within that context.

A further dilemma was balancing the need to employ available professional expertise, which have their favoured methodologies *versus* introducing and developing new tools and methods amongst country teams that might reveal more information, but with which researchers were less familiar. We were selective with the latter.

It became apparent that there were limitations of time and money in relation to the needs for space, resources and processes to consult, to be innovative and responsive. This was dealt with primarily by allowing for more time. There is definitely an appropriate pace for 'action research' in policy, and this tends to have some resonance with the usual pace of decision-making within a country.

We were concerned about how to 'pitch' the work with respect to the political context. Key information that helps to decide this would be a good understanding of: who currently controls policy territory; how 'acceptable' it is to involve those who are currently marginalised by policy; what the immediately non-negotiable issues are; and how policy-holders learn and then undergo change (which helps to determine how to package the results of research). All teams planned their work on the basis that the end results would hopefully influence policy, but several were drawn into influencing policy processes in the course of the research.

4 Policy in the real world – themes in failure and success[6]

Policy positions, statements, practices, and even outcomes, are based fundamentally on value judgements. There are many different stakeholders' values and desired outcomes. Thus it could be said that there are no absolute, 'true stories' in policy. Instead, we have found it useful to identify the areas of congruence between different stakeholders' visions, and to explore what processes have helped this to come about. Thus our emphasis has been on:

- what appears to have worked for most stakeholders under known *conditions* – what contextual factors (e.g. cultural, power structures, institutional, information availability, market factors) are conducive to effective policies; and, given a context:

- what *processes* seem to have worked towards policy decisions that are agreed to be sound? (e.g. decentralisation, participation, change champions, intervention by international groups and donors, projects and other catalysts for change); and (although secondarily in this study),

- what policy *contents and instruments* have proven useful. Policy content and instruments are important concerns, but are also matters of efficiency and cost-effectiveness that become important once the process is in place to generate and review them.

In this section we present some of the key findings from the country studies. The section is structured as a series of linked themes, each theme illustrated by several substantive country examples. Thus, we lead into each theme through stories rather than giving a general discussion of each theme with multiple examples. This, we think, makes the themes in failure and success more 'real'. We should stress that the 'headline' theme in each case is usually only part of the story; many factors contribute to success or failure, and many of our stories could have had a partial 'billing' under other 'headlines'. Some of the richness of the country studies is revealed in this process, and we hope

---

6 Unless otherwise indicated, all the material in this section is drawn from the *Policy That Works* detailed studies of Costa Rica, Ghana, India, Pakistan, PNG, and Zimbabwe (which are referred to as 'PTW Costa Rica', 'PTW Ghana', etc); and the specially-commissioned papers on: Australia, China, Portugal, Scotland, and Sweden. The detailed country studies are listed inside the front cover of this report, while the commissioned country papers are footnoted in the relevant sections. All are also listed in the Bibliography.

that this will spur the reader on to get hold of these country studies! The themes covered are:

- **Changing power... over time**. Power may be defined, in large part, as the ability to control policy. Where policy is inert it is usually because weighty institutions are 'sitting on it'. But such institutions can and do change, given time. In any one context it is important to understand what actions have brought about change in the past – and sometimes these actions seemed rather insignificant at the time – such that appropriate strategies can be shaped for pushing policy improvement in the present or future. In Section 4.1 we investigate the historical context and extent of change in Zimbabwe, Ghana, Costa Rica and Scotland, and find that policy is often more susceptible to change than has been assumed.

- **Pushing formal policy reform.** A range of technocratic approaches has been used to try and bring about comprehensive policy change. The impact of these approaches has been a mixed blessing. Some have lasted only as long as donors prop them up, and many have benefited only a few. However, some approaches have kicked off considerable stakeholder engagement which has, in turn, generated forms of institution with real motivation for sustainable forest management. In Section 4.2 we examine the changes wrought in practice by attempts to push policy reform in Papua New Guinea, Ghana and Australia. We also look at the national experience of international models of policy reform that have been pushed, asking: are they planners' dreams or unifying frameworks?

- **Reinventing state roles.** The imperatives of financial belt-tightening, and the demands for more social and environmental benefits from forestry are putting pressure on government in many countries. In the past, government has often sought, to varying degrees, to be forestry player, manager, owner, referee and coach. Recent pressures tend to focus government – often reluctantly – on the last two of these roles, whilst private sector and civil society actors take over others. But this is often a painful process, and its results are not guaranteed. In Section 4.3 we examine experience in India, Zimbabwe, Scotland and Sweden to ask: how can the 'public interest' be secured as commercial activities are hived off to the private sector? How can the state ensure the (new) rules of the game are agreed and respected whilst also responding to the demands for information and extension? And how can this be paid for?

- **Linking the people who change things.** Many initiatives to change policy and institutions are premised on 'rational' arguments about objectives and roles which 'make sense'. But old institutional ways are found to persist because these initiatives fail to get to grips with people's real motivations. Even those fired up to change things often founder because of institutional cultures which reward only inertia. Yet innovative managers and other 'new foresters' of various kinds do sometimes 'break through'. With a view to finding out how this is done, in Section 4.4 we look at experience in Ghana and Zimbabwe and find that 'change agents' have emerged from government and NGO backgrounds. They tend to be characterised by their ability to: see the big picture, take on tactical battles, use a mix of 'insider' and 'outsider' traits in their institutions, make alliances, and use these alliances to tackle bigger issues.

- **Looking beyond the forest reserves.** Forestry has focused, traditionally, on a reserved forest estate. As a result, forestry institutions are increasingly missing the real action – on farms and mixed farm-forest landscapes – where a wide range of forest goods and services are being used, nurtured or abused. Managing the farm-forest landscape is the new challenge, and in Section 4.5 we examine how stakeholders are rising to it in Pakistan, Ghana, Costa Rica, Australia and Portugal. There is ample evidence that farmers will grow trees and take responsibility for private forests and woodlands, but government's enabling role is key. This often means paying more attention to smallholder forestry and the contexts in which public interest objectives can be guaranteed through shared private property – common property regimes.

- **Improving learning about policy.** One of the key elements of a policy process that 'stays alive' is its ability to link directly to initiatives that are experimenting with new ways of making things work on the ground. Local projects allowing stakeholders enough slack to investigate new alliances and roles can be vital learning grounds – but they only really become useful on a significant scale if they seize the attention of at least some of the current power-brokers or 'policy-holders'. In Section 4.6 we focus on Pakistan and Ghana and find that participatory projects and strategies can breach entrenched policy when this connection is made and can go on to catalyse better policy processes and capacity development.

- **Dealing with tensions in devolution.** Decentralisation is the proclaimed way forward for forestry in many countries. However, this often involves confused or conflicting objectives, sometimes from the same stakeholders. Saving money for the centre or empowering the people? Decentralise to promote forest industries or farm foresters? A Pandora's box may be the result. Whilst there may be much to be said for the centre strengthening its effectiveness through deconcentration, to do so at the expense of the periphery's forest management capabilities is a step backwards. There are worries that just this may be happening in some decentralisation programmes. Section 4.7 describes experience in West, Central and Southern Africa, India and China. Again it appears that experimentation is generally the best way forward – trying through experience to come up with spreadable models.

- **Building policy communities.** Those engaged with a policy process on a regular basis constitute a policy community. Such a community needs to be able to channel ideas of all those who are important to the prospects for sustainable forest management – the stakeholders – into the policy ring, and channel the outputs out again. A vague call for everyone to participate will not produce this. Mechanisms are needed which can recognise who has power (to help or hurt the cause of good forestry) and capability (actual or potential), and engage with them. If the process is too broad-ranging it will be unworkable; too narrow and the ideas will be the wrong ones. In Section 4.8 we note that few countries have achieved a policy community likely to guarantee sustainable forest management – although Sweden may be one of them – but several, Papua New Guinea and India included, are trying to build such communities.

Country profile boxes are included in the following sections at the point where material from each country is first introduced. These boxes provide essential background information to enable the story to work.

4.1 Changing power... over time

The current predicament of forests and people has not come out of the blue. Just as the landscape is shaped by people's interactions over time, so is the policy context. To understand the landscape, the current policy context, and the degrees of freedom to change in the future, we need to trace the history of these interactions – to see who has argued what and when, who has had the power to make their views count, and with what consequences. Historical policy often has huge inertia, but space for change can open up. The story begins in pre-Independence Zimbabwe...

Challenging the historical legacy framing land and forestry options, Zimbabwe

Policy choices about forestry in Zimbabwe are conditioned by the history of land allocation and inequitable land and resource distribution. The choices that can be made about forests, woodlands and trees depend on who you are, what quality of land you have, the rights you have to it, and the pressures exerted by neighbouring people.

The translocation of indigenous Zimbabweans to marginally productive land was made law in 1930, and has not changed radically since Independence in 1980. Population densities are as high as 58 persons/km^2 in these 'communal lands' and as low as 3 persons/km^2 in the private commercial farm areas.

Zimbabwe	
Forest	• Woodlands cover 208,000 km^2, or about 53 per cent of Zimbabwe's total land area (386,850 km^2), whilst bushlands cover a further 13 per cent. • Commercial plantations cover about 1,150 km^2
People	• Population – about 11.7 million (67 per cent rural, 33 per cent urban) with an average density of 30.2 people per km^2 and annual growth rate of about 2.1 per cent • About 74 per cent of the rural population is confined to the 'communal lands' created under colonial rule, which constitute 42 per cent of the land, and this land is mostly of the lowest farming potential in the country • About 19 per cent of the rural population is found on commercial farms – the majority of the best farmland – which occupy 31 per cent of the land area, and are controlled by just 4,000 whites • Most of the remaining rural population, about 62,000 families, are found on the resettlement schemes
Tenure	• Over a quarter of the woodland area is contained in state lands – national parks, wildlife reserves and forest reserves • Over 30 per cent of woodland is on commercial land and about 43 per cent is on communal land
Forest economy	• Annual industrial production is currently about 1 million m^3 of plantation roundwood (indigenous hardwoods now being only about 1 per cent of production) • The small but well-developed forest industry contributes about 3 per cent to the GDP and employs about 25,000 people directly. Processing is dominated by three enterprises • Many local livelihoods are linked to woodlands in the communal areas (fuel, poles, fruit, medicine)
Pressure on forests and people	• High concentrations of people in communal areas, increasing under structural adjustment programmes, leading to increased rates of woodland clearance and pollarding and coppicing of trees scattered through the landscape
Key policy issues	• The dual economy and the land question (this section). Dilemmas in state roles (see Section 4.3) • Tensions in devolving authority (see Section 4.7)

Land policy dualism

Much of the legislation on use and management of forests and woodlands was inherited at Independence and reflects the dualistic nature of the colonial period, i.e. laws favouring voluntary self-policing and investment in the (white) commercial private lands, whilst in the (black) communal areas state enforcement and regulation was the rule, with very little investment. Large commercial farms on the good land were able to employ intensive farming methods, and were encouraged to set up Intensive Conservation Areas to manage woodlands. In the communal areas, on the other hand, high concentrations of people on land with poor soils and unreliable rainfall tended to progressively clear woodland, as extensive forms of farming were the only practical option. The remaining woodland resource was heavily coppiced and pollarded. Communal area populations are also liable to 'spill-over' as 'squatters' and 'resource poachers' move into neighbouring state forests, parks, commercial farms and resettlement areas.

Land-use policy originating in the 1930s and continuing up until the 1960s, sought to centralise villages. The centre-piece of the policy was a vision of the farming landscape based on linear settlements, dividing an area of individual (male-owned) plots on the upland area and paddocked grazing in the lowlands. Technical solutions aimed at good crop husbandry and land management were then introduced. Some of these solutions were based on highly questionable assumptions and failed in their apparent aims. Others led to rapid depletion of the natural resource base, and excessive subdivision of land into many small, uneconomic units.

Propping up assumptions to keep control

Colonial administrators had additional motives for their land-use prescriptions. These centred on the need for more effective control and collection of taxes in the rural areas (see Section 4.7). The creation of village settlements, under the authority of a headman enabled this control. Although many of these prescriptions were abandoned with the nationalist movement in the 1960s and 70s, some of the post-Independence land-use planning exercises in Zimbabwe bear remarkable resemblance to previous initiatives.

Latter day arguments for 'sustainable development' and 'poverty alleviation' have both been employed to avoid addressing the existing inequitable land distribution. Old assumptions about the environmentally destructive livelihood practices of people in communal areas have been used to justify a version of 'sustainable development' that involves continuing centralised regulatory control in those areas. This is despite increasing evidence showing these old assumptions to be unsuited to the complex and diverse nature of the ecologies and livelihoods associated with woodlands. Poverty alleviation activities, meanwhile, are focused on those who bear the main brunt of the 'temporary' negative effects of structural adjustment – the communal lands population.

Land redistribution – a political football

In general, land redistribution has been used by politicians as a bargaining chip with the rural population. Since , approximately 62,000 families have been resettled – on land from commercial farms – but this is well below the target, set at Independence, to resettle 180,000 families. When elections are imminent, some commercial farm areas are bought and redistributed. After that, redistribution activity subsides. However, strong political pressures for large-scale distribution of white-owned estates re-emerged to dominate the national agenda in the late 1990s. Prospects for large-scale land reallocation have raised many questions about the intended beneficiaries, and about the ability of government to support the settlers with adequate extension and capability-building programmes. Donors have been reluctant to commit themselves, preferring to rally round the call for transparency in any land programme.

Poorly supported land reallocation may exacerbate some of the main underlying causes of woodland depletion, including: inter-sectoral confusion over responsibilities; inadequate integration of trees in 'off-the-shelf' land-use planning; and inadequate community institutional mechanisms. Concern is growing that without fuller participation of all concerned sections of society, emerging land policy may reinforce inequalities by concentrating too much on support for a relatively small number of smallholder producers and too little on allocation of land to those who need it for survival.

Thus, concerns for woodland management in Zimbabwe are generally being subsumed by wider issues of land allocation and management. Clearly, the prospects for good forest management are intricately linked to resolution of the 'land question'. And, since old assumptions about local people's tree-destroying tendencies have been used to maintain the current inequitable *status quo*, installing into the policy process a new understanding of the considerable local capabilities for tree management is a key challenge for resolving the question.

From consultation to diktat to collaboration, Ghana

As in Zimbabwe, today's forestry options in Ghana are strongly circumscribed by policies that were born in its colonial past. However, this has not stopped changes from being made which have pushed and pulled today's stakeholders into examining their past, and into developing new ways to deal with present problems. The ways in which this has been done offer lessons for Ghana and elsewhere.

Ghana

Forest	• Tropical moist forest covers 15,000 km², or about 7 per cent of national land area, and is found in the south of the country • Considerable tropical forest resources are also found in small forest patches and trees on farms (e.g. about 50,000 km² of land is estimated to carry some economically valuable timber) • Savannah woodlands are found in patches, and a few large areas, in the north of the country • There are many small woodlots, and about 165 km² of state-owned plantation
People	• Population – 18.3 million (63 per cent rural, 37 per cent urban) with an average density of 80.4 people per km², and an annual growth rate of 2.8 per cent
Tenure	• Most forest land is owned by landholding communities (represented by chiefs), but government has assumed the right to manage the forest reserves and to control timber tree exploitation outside reserves • Almost all tropical forest is in the 214 government-managed forest reserves, in various categories of protection • Landholding communities and some other stakeholders have certain rights to NTFPs
Forest economy	• Annual log production is currently about 1 million m³, from which the national government collects about US$9 million in royalties and fees • Timber industry consists of some 400 concession holders, many of them small companies, and about 90 export-oriented milling companies, which capture most of the industry's value i.e. little local processing • A wide range of NTFPs play a major role in livelihoods in the forest zone
Pressure on forests and people	• Most forest land outside reserves has been converted to tree-farm mosaic, with varying farm-fallow characteristics influenced by changing economic margins in agricultural crops. Agricultural encroachment in some reserves • Excess capacity in timber industry and patchy supply-side controls leading to over-harvesting in some reserves and many off-reserve areas • Uncontrolled fire in northern areas, and mining in some reserves
Key policy issues	• Conservators and change agents (Section 4.4) • Managing the farm-forest landscape – the off-reserve challenge (Section 4.5) • Linking local learning to policy and capacity development (Section 4.6)

Reservation by persuasion

Early colonial forest reservation policy was influenced by the need to protect watersheds and maintain climatic and soil quality conducive to the production of cocoa, the main export crop, combined with a policy of 'liquidation without replacement' of the forest outside the forest reserves prior to its conversion into cocoa farms. However, forest policies were strongly conditioned by the policy of indirect rule, and the need to reserve forests had to be reconciled with the need to bolster the chiefs' powers, as traditional institutions such as the chieftaincy became the main vehicle of local government. But finding the chiefs and traditional landholders reluctant to place the land under reserve through their own by-laws, the government later used the threat of compulsory reservation. The rights of communities in forest reserves, including access to harvest non-timber forest products, were 'admitted'.

Seeing only the wood in the trees

With the Second World War, timber out-turn became the dominant concern in forestry policy. The influence of foresters and timber merchants grew steadily, while the landholding chiefs' influence on policy declined and local communities began to be marginalised. The colonial government and local politicians held sway with an authoritarian approach: the first formal forest policy established forestry as a purely technical exercise. Furthermore, the creation of elected district councils distanced the landholding chiefs from forestry and land-use decisions. The seeds of future trouble were being sown.

Following Independence in 1954, the government saw the state as the deliverer of development, whilst the chiefs and traditional authorities were seen as having sided with the opposition. In 1962 government took formal control of land and trees – 'in trust' for the chiefs and people. 'Indigenisation' policy in the mid 1960s turned the timber industry from a concern of a small number of foreign 'merchant princes' to a plethora of local companies. Many of the latter had considerable influence at policy levels, and fought hard to implant the notion that the timber industry could be a driving force for national development.

The 1980s saw macro-economic reform and structural adjustment, which eroded social services and tended to deepen rural poverty. Reform of the timber sector, which had much donor support, sought to inject new life into the industry, at a time when the resource was in a precarious situation and forest managers were ill-equipped to cope. The Forestry Department was under-resourced and stretched to the limit. Many reserves had become badly degraded; the annual allowable cut bore little relation to estimates of sustainable yield; and some important timber species were threatened with commercial extinction. The landholding authorities and local communities had become marginalised and alienated owners of the resource, with few rights and even fewer responsibilities. There was over-capacity and waste in the timber industry. In certain quarters, patronage and corruption were rife.

Crisis, and a new policy landmark

This situation catalysed a period of study, reflection and reappraisal. Policy changes included: a reduction in the annual allowable cut; temporary bans on the export of round logs; indexation and improved collection of timber royalties; and a strategy for forest protection based on both 'fine-grained' stand-level measures and 'coarse-grained' landscape-level measures. The Ministry of Lands and Forestry became active in fostering coordination amongst donors in the forest sector, notably the World Bank and the UK's DFID (then the ODA), turning them away from invigorating forest industry, and towards resuscitating forest management.

A major landmark was the 1994 Forest and Wildlife Policy, which called for creation of the conditions suitable for sustainable forest resource management throughout the country – reversing the policy of 'liquidation' in the off-reserve areas. This is vital because the landscape outside the system of forest reserves in Ghana has been transformed over the years, from a pattern of forest and farm areas to a predominantly agricultural landscape with small forest patches and trees on farms. These off-reserve resources are still considerable, and the timber in this landscape continues to be vital to maintain a viable timber industry and to meet projected domestic demand.

Working groups and consultative measures

However, the 1994 policy coincided with the emergence of new Far Eastern export markets for round logs of species found off-reserve. There was widespread speculative felling and trade malpractice, including illegal trading in property marks, and unauthorised subletting of concessions to illicit timber operators. The Ministry of Lands and Forestry and the Forestry Department (FD) set up a Working Group, which set a new precedent in its outreach with stakeholders.

Traditional authorities spoke about timbermen who failed to honour their agreed obligations. Farmers explained why they were reluctant to leave timber trees on their farms because of the damage concession holders cause during felling and log-extraction, and because the farmers were not receiving any share of the timber royalties. Timber millers began to shed light on their business concerns, why they were forced to cut corners and apply 'tricks of the trade' which sometimes caused problems for others. In a few cases, loggers complained about abuse of verbal agreements made with traditional authorities and communities, and harassment of their workers. The FD described the consequences of over-exploitation and how the lack of operational funds was affecting its efficiency.

The Working Group produced the 'Interim Measures' which introduced:

pre-felling inspections – involving farmers and the FD; pre-felling permits; post-felling inspections; and the issuing of conveyance certificates before logs can be moved. However the FD was unprepared for these major new responsibilities and huge logistical constraints to full implementation remain. Nevertheless, the impact on checking illegal felling has been considerable, and revenue collected by the FD has risen dramatically since the Measures were launched.

Forest management balance swinging towards farmers

Logging contractors have generally supported the Interim Measures, and even timber millers – some of whom previously benefited from cheap illegal logs – have cooperated. Local governments have begun to play a more productive part in forest resource management. Some farmers have reported improved recognition by concession holders, improved relations with the FD, reduced crop damage, better compensation and improved confidence generally to act. This latter effect may be the lasting legacy of the measures: the balance is swinging from the timber industry towards management by farmers and landholders.

However, although wider than for any previous forest policy instrument, participation in developing the Interim Measures was still limited. Poor farmers, unemployed rural youth, local 'strong men' marginalised by the state forestry system, and others benefiting from the previous 'chaotic' situation, were the least involved. Whilst participation in implementation thus far is encouraging, it remains to be seen whether this model of limited participation in policy formation leads to the long-term wider participation and commitment in implementation which the policy seeks.[7]

Meanwhile, new legislation passed in 1998 seeks to replace concessions with Timber Utilisation Contracts – requiring stronger environmental and social commitments – and will improve landholder and farmer rights over trees. As the rights and capabilities of landholders, farmers and local government increase, the FD's sole responsibility for sustainable resource management off-reserve has become challenged. The FD has begun to experiment with a range of collaborative arrangements with other stakeholders, whilst some of the more far-sighted timber companies are also looking to the possibility of more long-term partnership options with farmers and the FD.

The story of forestry's development in Ghana over the last century has much in common with that of other formerly colonised countries. But in contrast to some, forestry in Ghana has always been marked by the

---

7 Similarly, there has been fairly limited participation in the certification system development to date; the notion perhaps being that strong drivers in government are needed in order to develop something substantial enough to be constructively debated.

involvement of a range of actors – whether government has liked it or not. With some of today's policy processes and instruments having considerable precedent in the past, there is much which can be usefully learned from the ways in which change in forestry has both caused and been affected by past power changes.

The evolution of policy space, Costa Rica

The balance of power through the practice of policy has also changed greatly over the years in Costa Rica. Like Ghana, cash crop agriculture dominated the agenda for some time before forest industry took the front seat in policy. Unlike Ghana, conservation interests have made major gains and, more recently, smallholder forestry has made it onto the agenda. The exercise of policy instruments has been at the root of these changes; some instruments have succeeded in their original aims, whilst others have failed miserably but have caused stakeholders to become better organised to bring about change.

State-sponsored deforestation

In Costa Rica before the 1950s, forest area slowly declined as agricultural society emerged. Large coffee-producing landowners dominated, and a collection of laws had been passed which, on the one hand, tried to mitigate certain impacts of agriculture on the forest and, on the other hand, set the scene for one of the most dramatic periods of deforestation anywhere in the world (as a proportion of forest lost each year).

Wholesale conversion of forest was stimulated in the early 1950s, when a new government sought to build a power base through a policy of extending low-interest credit to cattle ranching. Colonists were able to secure lands outside the Central Valley by clearing the forest. Some of these colonists were displaced smallholders; others were wealthier groups seeking extensive lands for cattle ranching. Timber industries benefited through a surplus of low-cost timber, cut in converting forest land to grazing lands, while coffee growers in the Central Valley and the new plantation owners in the south profited from increased export prices.

Rise of the conservationists and forestry incentives

However, an educated élite, influenced by international conservation interests, began to realise that dramatic efforts were needed to protect the environment. From 1970 to 1990 the country's famous protected area system became firmly established – it now covers about 21 per cent of the country. Between 1987 and 1989, international organisations bought debt titles worth US$75 million at a huge discount, which the Costa Rican government redeemed in local currency worth about US$36 million for conservation purposes, in so-called 'debt-for-nature' swaps.

In this period also, the forest industry continued its rise. Financial incentives

Costa Rica

Forest
- Natural forest covers about 17,870 km^2 or 35 per cent of the nation's land (total area 51,060 km^2), while logged forest in good condition covers another 1,500 km^2 and secondary forest plus regenerating forests in patches on agricultural land covers another 4,500 km^2
- Plantations cover about 1,700 km^2
- High biodiversity – 500,000 species or 4 per cent of world total

People
- Population – 3.6 million (50 per cent rural, 50 per cent urban) at an average density of 70.5 people per km^2 and an annual growth rate of 2.1 per cent. The main population centres are in the Central Valley area

Tenure
- Government protected area forest covers 12,870 km^2 in various protection categories
- Remaining natural forest area is under private ownership: 2,500 km^2 is designated private reserve – owned by individuals, NGOs and companies; while 4,000 km^2 is potential production forest
- Patches of forest are found on: small-holdings (under 10 hectares) which cover 5 per cent of the land; medium size holdings (10-200 hectares) which cover 48 per cent of the land; and large estates (over 200 hectares) which cover 47 per cent of the land

Forest economy
- Forestry sector contributes 4.7 per cent of GDP and employs about 12,000 people, including almost 7,000 in the industrial sector. Value of forest exports was US$28 million in 1993
- Government financial incentives were available for tree planting and forest management in the 1980s and early 90s. They are now oriented towards payment for forest environmental services
- Tourism is the country's top revenue earner, much of it linked to forest protected areas
- Costa Rica is a world pioneer in developing programmes to secure international revenues – initially from debt-for-nature conservation swaps and more latterly from bioprospecting and carbon offsets

Pressure on forests and people
- State-promoted conversion of forest to agriculture has greatly decreased, but still occurs in various agro-industrial developments
- Private sector forest protection – sometimes with negligible benefits for local people – is increasing
- Existing large-scale forest industry is protected by policy to the detriment of small-scale forest enterprise, although smallholder tree growing and forest management is on the increase

Key policy issues
- Powers and policy space changing over time (this section)
- Policy first preventing and then promoting smallholder forestry (Section 4.5)

for reforestation became the government's main forest policy tool. These mostly benefited larger landowners, and were generally insensitive to people's motivations for forest management and conservation. The main losers were the smallholders, who collectively own about two-thirds of the country's land. However, the short-comings of the incentives system generated considerable debate, and stimulated the formation of small-holder forestry organisations (see Section 4.5).

Thus, the relative influence on policy of different actors – their ability to create 'policy space' – and the resulting pattern of linkages between them has changed greatly over time (see Annex 1 Section 4.10 for a visual representation).

Growing momentum for changes in land tenure and forestry ownership, Scotland[8]

High concentration of land ownership and forestry benefits

Scotland has a history of massive dispossession of rural people from the land. In the Highland Clearances, beginning about 1800, some half a million people were driven from their lands to make way for extensive sheep farming and, later, sporting estates. The once-extensive native woodlands, which by 1800 were already much reduced, were all but wiped out under management for grouse and deer shooting. Displaced farmers who did not emigrate were confined to smallholdings – crofts – usually in coastal townships on the worst land. A key early motive for establishment of such townships was to provide a work force for the kelp industry, involved in the manufacture of industrial alkali. Crofts were deliberately laid out to make it impossible for occupiers to earn a full-time income from farming.

There appear to be few other countries in the world which can match the concentrated private ownership still prevailing in Scotland. If the 12 per cent of Scottish land which is publicly owned, and the 3 per cent which is covered by the major cities and towns is taken out of the equation, over half of the privately-owned land in rural Scotland is held by fewer than 350 owners, each with estates of over 3,000 hectares. Moreover, this pattern has remained much the same for hundreds of years – fewer than 1500 owners have held the majority of Scotland's land since feudal tenure was introduced in the 11th century (Callander, 1998).

The landlords of these large estates have, over time, secured many social advantages allowing access to other powerful and aristocratic groups and

---

8 This section draws from a paper prepared for IIED's *Policy That Works* project by James Mayers (1999).

Scotland

Forest
- 'Forest' covers 11,890 km², or 15 per cent of the land surface, of which 9,810 km² is exotic conifer plantation, and the remainder is broadleaf woodland with some patches of native conifer woodland
- Scotland has 49 per cent of the GB total of 24,230 km² of forest and woodland, and 64 per cent of the total 15,320 km² conifer plantation

People
- Population 5.1 million people: 10 per cent of the population of GB

Tenure
- Land ownership is highly concentrated: some 1,560 private estates own between them almost 60 per cent of Scotland's land; a further 12 per cent is publicly owned, including 8.7 per cent owned by the Forestry Commission (FC)
- Forest ownership: 58 per cent of the total forest area is privately owned and the remaining 42 per cent is owned by the FC
- Trees can only belong to landowners, not tenants – an important point with so much of the land in large estates

Forest economy
- Wood production, for GB as a whole, was 8.6 million m³ in 1997 (4.7 million m³ of this by the FC's commercial arm – Forest Enterprise – for a return of £100 million), of which 90 per cent was coniferous softwood, mostly destined for the board products and pulp and paper industries. This volume supplies about 15% of the timber used in Britain, the rest is imported. Production is expected to rise to a peak of 17.5 million m³ by 2025 as more plantations reach harvestable age
- The FC estimates that 10,660 people are employed in forestry and primary wood processing industries in Scotland, of which 2,810 are employed or contracted by the FC itself. This employment level continues to decline (it was 15,000 in 1992)
- There has been reliance on the private sector for much of the afforestation from 1945, and especially since 1980: through income tax concessions up to 1988; and through grant schemes since 1988
- Grant schemes aided planting and re-stocking on 149 km² in 1997, of which 60 per cent was for conifers and 40 per cent was for broadleaves. Management grants covered an additional 100 km²

Pressure on forests and people
- Agriculture is dominant in land-use policy whilst most forestry is consigned to marginal uplands. European Common Agricultural Policy continues to dominate, keeping the price of most good land too high for forestry investment
- About 55 per cent of Scotland is suited to growing primarily conifers, while another 15 per cent is capable of growing a wide range of tree species. But despite many improvements, continued contradictions between sectoral policies creates a balance of incentives against management and expansion of native forest and woodlands
- There is declining state forest estate and the rise of private afforestation by absentee landowners
- There is no tradition of farm forestry in Scotland. Only recently has there been an increase in locally-based woodland owners – through government grant schemes

Key policy issues
- Growing momentum for changes in land tenure and forestry ownership (this section)
- The rise, fall and rise again of social objectives in state forestry (Section 4.3)

lucrative business deals. Furthermore, such landed estates have come to command considerable 'prestige' value in real estate markets which has little relationship to the income that they can generate. Many of these large estate landlords have a strong interest in the encouragement of private forestry – and in recent times have tended to secure a disproportionate share of the benefits of government support for forestry at the expense of tenant farmers and crofters. Until the late 1980s this was through tax advantages; since then it has been through planting and maintenance grants. In 1993, the top 20 recipients under the government's woodland grant scheme were trusts, large investors, industrial interests, the aristocracy and large estates (Wightman, 1996).

Tenure system defines social relations between people

Scotland is approaching the millennium with a tenure system which is essentially medieval. The legal basis of land ownership remains the feudal system, and the pattern of reciprocal power relations which this system sustains has enduring consequences for land use. The feudal structure is embodied in a hierarchy – a 'pyramid of power' – which is topped, in formal legal terms, by God. The Crown sits below God and is taken as the ultimate owner of all of Scotland by virtue of being the Paramount Superior.[9] All other land owners are known as vassals of the Crown. Certain rights are reserved by the Crown but any vassal, when disposing of land, can retain an interest in it by specifying terms in the title deeds. They then become the superior of the new owner, who becomes their vassal. There is no limit to the number of times this can happen.

But the tenure system is not just a matter of titling arrangements. It also defines how society as a whole relates to land owners. Land is a unique form of property, since it is held subject to the rights and interests of others in the form of tenants' rights, neighbours' rights and public rights.

The 'public interest' and the basis for reform

The 'public interest' occurs in the lands and rights retained in the Crown's name or by public bodies. It also occurs as the overall aim of the system of land tenure and as the authority of government to regulate land ownership through administrative law – the range of laws which interact with the laws of land tenure, such as environmental legislation or the laws setting up, and wielded by, government bodies like the Forestry Commission. The public interest is also assumed to be manifest in the system of regulations and incentives used to promote or control the use of rights over land (e.g. grants and licences in forestry).

---

9 Similar terms were used by British colonial authorities in codifying the land tenure system, based on ownership by Paramount Chiefs, into statutory law in some African countries, e.g. Ghana.

Feudal tenure, for all its problems, may also contain an important basis for moving to a more modern and democratic system: "the conditional nature of feudal tenure is a legacy that should enable a new system of land tenure in Scotland to more readily strike an appropriate balance between public and private interests, than the more absolute notions of ownership in most other countries" (Callander,1998). Thus, any land reform programme which seeks to abolish feudal tenure will need to ensure that the public interest is not also undermined.

The case for land reform being necessary for better forestry
Reformers argue that concentrated land ownership, as exemplified by the large sporting estates, prevents the emergence of vibrant rural economies and limits land management possibilities, including more locally-beneficial forestry. Sporting estates are a legacy of a bygone age and were never designed to make money. Today they are bought and sold (to the extent that this happens at all) as assets of an élite leisure industry whose owners do not depend on them for their livelihoods.

Most people associated with sporting estates do not have a proprietorial stake in the land, but tend to be tenants or employees. The few that have this stake come from a narrow background, they tend to treat the land value itself as an investment or speculation, and have few ties to the rural economy (Wightman, 1996).This stands in marked contrast to more dynamic and robust rural economies, in some other European and Southern countries, which tend to have many occupants, pursuing a diversity of activities, creating a diverse economy, making maximum use of indigenous knowledge, and – where the land can give some return – investing in land capability and businesses with a stake in the land.

From feudalism to sustainability in one leap?
British institutions – in government, civil society and the private sector – have made considerable efforts internationally and at home to put flesh on concepts of sustainability. Considerations of sustainability could now provide the framework for land tenure's overall purpose of balancing public and private interests in the management and use of land and natural resources. It has been argued that Scotland has the potential to move from the position of being "the last country with a feudal system to become the first with a progressive system effectively targeted to the requirements of the new millennium" (Callander, 1998).

Experience in the South suggests that land reform requires more than changes in the tenure system itself: it also requires a redistribution of land and property resources, and appropriate back-up or extension arrangements to implement and support the changes (*viz.* Zimbabwe, earlier in this section). After coming to power in 1997, Britain's Labour government issued two land reform consultation papers – which received unprecedented public feedback – and the Scottish Parliament which began life in 1999 is considering proposals for land reform legislation.

Proposals include reforms to feudal law, a legal public access right to wilderness areas, compulsory purchase powers for mismanaged estates, and a community right to buy estates when they are sold (Scottish Office, 1999). This latter proposal is important to enable Scottish communities to stand a chance of competing with wealthy individuals and companies attracted by the influence and prestige value which can accompany a landed estate.[10] The new proposals would give local communities both time and access to money – from a land fund created from National Lottery finance – to meet a price set by a government-appointed valuer. This price would be set according to what an estate will yield, rather than its speculative or prestige value (Financial Times, 1999).

Thus, forestry may well take a different course in Scotland depending on how land reform proceeds. In broader terms, land reform proposals are highly charged with political significance and national symbolism. For a country still haunted by the Highland Clearances, this is certainly a historic moment.

In conclusion, change in forestry is brought about through diverse ways in which forestry power changes over time, including: changes in the relative strength of political actors; new consultation processes; the rise of new issues; and changing economic conditions. But few of these changes are quick – policy and power change take time – especially where fundamental assumptions are challenged and political interests are affected. The implied challenge is to accept change as both inevitable and necessary in policy, and to commit to dealing with it through continuous and adaptive learning.

---

10 The crofting community on a large estate in Assynt and the islanders of Eigg, in 1993 and 1997 respectively, have shown that popular support and finance for such estate buy-outs can be generated, given massive effort on the part of the communities themselves.

4.2 Pushing formal policy reform approaches – a mixed blessing

Since the late 1980s, and throughout the 1990s, a common theme throughout the world of forestry, especially in the tropics, has been policy reform. A whole host of initiatives – usually brought on by an air of crisis in forests, stakeholders' welfare, finance or governance – have sought to rethink the policy and institutional architecture. Organised policy reform has usually been led by – and may often be vehicles for – powerful (outside) players. Many of them in the South and East have been backed by donors and development banks, and have incorporated the interests of these bodies at least to some extent. They have also invariably overestimated what can be achieved without fundamental power changes. Here we report on several organised policy reform attempts.

Attempted reform of forestry policy, Papua New Guinea

During the 1980s PNG's forest industry grew until by 1987 it was harvesting 2 million cubic metres of logs annually. By then, allegations were flying of impropriety in the timber industry and government. Two months before a national election, the Prime Minister appointed an Australian member of the PNG judiciary, Thomas Barnett, to lead an inquiry into these allegations. After two years, and in 20 volumes, Barnett described a 'forest industry out of control' – dominated by foreign investors in questionable 'partnership' with PNG's leaders – in which the volume of logs exported was maximised with no regard for environmental damage and to the detriment of local processing capacity. He called for a slow-down in timber harvesting, and advocated the reformulation of national policy, establishment of a nationally-integrated forestry service, consultation procedures in allocation of permits, and formalisation of detailed provisions for sustained-yield forestry.

Crisis – the rationale for big policy changes

Barnett's findings were premised on the assumption that national technocrats and landowners could make common cause against the corruption of their state and society by an unholy alliance of foreign loggers, domestic politicians and wayward public servants. Below we illustrate that, with hindsight, this can be seen as a flawed assumption.

Following Barnett's findings, a small number of politicians and public servants lost their jobs, and a new government sought international assistance under the global Tropical Forest Action Programme, coordinated

	Papua New Guinea
Forest	• Tropical forest – covers 280,000 km^2, or just over 60 per cent of national land area (total 452,860 km^2) – and is renowned for its rich biological diversity[11] • Plantation forests cover between 350 and 400 km^2 of land
People	• Population – about 4.5 million (83 per cent rural, 17 per cent urban) with an average density of 9.9 people per km^2, and a growth rate of approximately 2.2 per cent per annum
Tenure	• About 97 per cent of all land, and more than 99 per cent of all forested land, is held under customary title; thus, most Papua New Guineans are 'resource owners' • Less than 4 per cent of land area has formal protected area status
Forest economy	• Very little commercial logging occurred prior to Independence (1975), but by 1994 raw log exports reached a peak of nearly 3 million m^3, with a value of about US$410 million, representing just over 18 per cent of domestic export values, from which the national government collected about US$111 million in log export taxes • Up to 7,000 people are directly employed in the industrial forestry sector • About 50 per cent of log exports are under the control of a single Malaysian company, Rimbunan Hijau, and another 30-35 per cent are under the control of other Malaysian companies
Pressure on forests and people	• Japanese general trading companies dominate the log trade – by controlling distribution chains, financiers and wholesalers. They effectively control the regional market for logs through the sheer volume of their purchases • Farmers are responding to population growth through forms of agricultural intensification rather than clearance of primary forest. Only 3 per cent of current subsistence farming takes place on land cleared of forest that was not previously cleared for agricultural purposes over the last 20 years • Establishment of agricultural plantations was the most important single direct cause of deforestation this century until the 1997/98 drought brought forest fires to highland areas and caused more damage to the forest in a matter of weeks than the logging companies have managed in ten or twenty years
Key policy issues	• Reforming forest policy – what effects from ten years of effort? (this section) • Making policy mean something – building a new policy community (Section 4.7)

by FAO. Eventually, PNG's National Forestry and Conservation Action Programme (NFCAP) took shape as a collection of projects which ran from 1991 to 1995. The NFCAP had a major emphasis on restructuring the forestry institutions, developing conservation objectives, and working with NGOs to form a bridge with resource owners. A succession of changes in the formal instruments of policy was set in motion by the discussions of 1989 onwards. The following table highlights these changes, and their impacts in terms of actual practice.

---

11 FAO's 'State of the World's Forests 1999' cites PNG's tropical forest coverage as 369,390 km2, or nearly 82 per cent of the land area.

Table 4.1 Changes in policy instruments and actual practice, Papua New Guinea

CHANGES IN POLICY INSTRUMENTS	CHANGES IN ACTUAL PRACTICE
NATIONAL LEVEL	
New National Forest Policy (1990) • Lays the foundations for detailed reform of forestry legislation • Proposes various fiscal measures to encourage 'downstream processing', and subsequent policy statements constantly repeat the need for such measures • Proposes various measures to promote reforestation by both government agencies and private investors	The new Policy does not contain any targets or guidelines specific enough to influence the course of subsequent public debate on forest management issues Government unable to devise any system of incentives which can satisfy the World Bank and encourage new investment in the domestic processing sector National Forest Authority unwilling or unable to make effective use of reforestation levies charged on logging companies, while developers make few additional contributions to this objective
National moratorium on the issue of new Timber Permits for raw log export operations imposed by Cabinet (1990), until the new Forestry Act comes into effect	More than 20 Timber Permits issued during the period of the 'moratorium', while gazettal of the new Forestry Act is delayed by the Minister and his Departmental Secretary
New guidelines for Environmental Plans for large-scale forestry projects produced by Department of Environment and Conservation (1991)	No marked improvement in the Department's capacity to evaluate such plans, monitor compliance, or prosecute offenders
New Forestry Act (1992) • Replaces old Department of Forests with a National Forest Authority under the direction of a board representing a range of stakeholders – including NGOs – in the forestry sector, with a view to reducing the exercise of arbitrary powers by the Minister for Forests • Requires production of a National Forest Plan as precondition for the development of new forestry projects. The required form of the Plan is essentially that of a land-use map (rather than a land-use strategy) • Includes a commitment to establish a new forest revenue system designed to encourage sustainable forest management	Minister continues to influence the decisions of the National Forest Board and National Forest Service through control over the appointment of the Managing Director. Forest Authority divides into factions supporting and opposing the policies of successive Ministers Cabinet approves some development proposals before the Plan is developed. A Plan appears in 1996 but can never be more than a statement of government intent because it covers wide stretches of customary land, whose multiple owners have not yet made their own land-use decisions Government raises log export taxes in 1993 and 1994, but comprehensive reform of the old system not implemented until 1996, and then only under intense pressure from the World Bank
National Forestry Development Guidelines – a radical set produced by Ministry of Forests (1993) • One of the statutory requirements of the National Forest Plan • Provide for review of all agreements made under previous forestry legislation	Guidelines widely attacked by forest industry representatives, and largely ignored by subsequent Ministers and senior bureaucrats in the National Forest Service Determined opposition by the logging industry, and lack of capacity in the National Forest Service, combine to halt the review process
Contractor to monitor log exports engaged in 1994 by government in order to control the incidence of transfer pricing by logging companies	Widespread agreement that this system has proven effective, though still substantially dependent on donor support
Logging Code of Practice endorsed by government in 1996 under pressure from World Bank	Logging industry complains about lack of consultation, and Bank still doubts capacity of National Forest Service to ensure that companies follow the code

CHANGES IN POLICY INSTRUMENTS	CHANGES IN ACTUAL PRACTICE
PROVINCIAL LEVEL	
New Forestry Act • Requires production of Provincial Forest Plans by Provincial Forest Management Committees, in all 19 Provinces, as the building blocks of the National Forest Plan	Provincial planning process envisaged by the Act turns out to be unworkable because PFMCs do not have the information or capacity to produce integrated land-use plans for their provinces, and officers of the National Forest Service take over the process in order to meet the Minister's demand for a National Plan
• Returns provincial forestry offices to national government control, under an integrated National Forest Service	Some improvement in morale and efficiency of provincial officers, and greater immunity to political interference at provincial level
New Organic Law (1995) appears to grant new powers to provincial authorities to determine their own forest policies	Production of National Forest Plan delayed by confusion arising from apparent inconsistencies between the new Organic Law and the new Forestry Act
LOCAL LEVEL	
New Forestry Act • Strongly recommends that local landowning groups be incorporated under the Land Groups Incorporation Act as a precondition of new Forest Management Agreements between themselves and the State	Task of land group incorporation exceeds the capacity of government officials, to the point where it either functions as an obstacle to state acquisition of additional forest resources, or else has to be completed with assistance from the logging companies
• Requires forestry officials to conduct a Development Options Study in consultation with local landowners as a precondition of new Forest Management Agreements	Development Options Studies consist of little more than proposals for large-scale logging
New forest revenue system intended to increase the landowner's share of resource rent from large-scale logging operations	Industry opposition, bureaucratic obstruction, and local disorganisation delay the transfer of additional financial and material benefits to landowning communities

Limits to heavy regulation

National government policy-making has tended to be dominated by foreign consultants. Nevertheless, for a brief period in 1991 to 1994, the forest sector reform process contained a real sense of national ownership and unity of purpose. However, of the two major policy instruments introduced, it is now clear that the Forestry Act grossly overestimated the capacity of the state to regulate the use of customary land, whilst the new forest revenue system is yet to find a way to deal with the variation in the quality of timber resources, the relative costs and efficiency of different concession operators, and their differential willingness to honour the agreements made with other stakeholders.

So far, the most successful experiment in encouraging better practice in the forest industry is the log export surveillance system, initiated in 1994, which was 'out-sourced' to a private contractor. Log exporters welcomed the opportunity to prove that they were no longer guilty of the transfer pricing documented by the Barnett Inquiry. This raises the question of which other elements of the reform process might have been better expedited by a less heavy-handed approach to industrial regulation, and a more concerted effort to build a wider policy consensus.

Pipe dreams of conservation areas and model local organisations

PNG's official conservation policies are constructed around the establishment of a representative system of protected areas. But the legal and institutional mechanisms established to pursue this goal have proven to be unwieldy and ineffective. Terrestrial biodiversity values are almost entirely confined to customary land, and there is no prospect in sight of this land being alienated by the state for purposes of conservation.

The forest policy reform process also aimed to foster new forms of community organisation, stressing accountability and democracy in the forest resource acquisition process. However, rural communities have a long history of resistance to imposed values, and the National Forest Authority's failure to negotiate either with the forest industry or landowner groups in developing these measures has resulted in limited uptake. Meanwhile, the Forest Industries Association has come to believe that the whole reform process was designed to drive a bureaucratic wedge between the logging industry and the resource owners.

World Bank leverage – depends on continuing crisis

After much opposition from sections of government from 1995 to 1997, the World Bank was able to use its economic muscle to push through government commitments to forestry reforms, in return for a major economic rescue package. The structural adjustment programme (SAP) included commitments to the new forest revenue system, the adoption of a logging code of practice, the provision of additional funding to the National Forest Authority, and improvements in the supervision of PNG's log exports. However the programme also began to hit the pockets of wage-earners and to have negative impacts on rural living standards. There was also concern that if the programme enabled the national government to balance its books, it would then give politicians room to escape the straitjacket put on their relationships with logging companies. The programme, with its forestry conditions, would then disintegrate.

If the World Bank is able to ensure the maintenance of the current fiscal regime then, on top of the South East Asian economic crisis of 1998, the local log export industry will almost certainly continue to contract. Those companies which make some environmental effort, along with a serious effort to bring returns to other stakeholders, may be the first to feel the pinch. The conservationists might gain more opportunities to ply their wares, and the new market for carbon offset schemes could assume greater importance. But PNG's ability to profit from such schemes is dependent not only on the managerial capacities of the national government, which are slender enough, but also on a wholesale transformation of indigenous property relations, such that commitment to a particular land use over a long term can be better guaranteed.

No alternatives to learning the hard way?

Villagers do not often see logging of primary forest as a direct threat to their subsistence lifestyles. Most of the trees which supply their food, fuel, or raw materials for buildings or other local artefacts are either deliberately cultivated within the secondary forest zone, where shifting cultivation is practised, or harvested from garden fallows. Thus, efforts by NGOs and donors to develop and commercialise small-scale forestry or non-timber forest products may not be seen locally as real alternatives to logging which might pave the way to more sustainable development – or to more certain income, at least. For example, some of the most successful small sawmill operators are found in existing log export concessions where they can utilise reject logs and get technical help or equipment. Furthermore, a recent survey of anthropologists and other social scientists suggests that the rural constituency for more sustainable development forms not as a result of awareness programmes and regulated activities but through the process of 'resource development', led usually by outsiders such as loggers – with all its benefits and costs – as people learn from experience.

In 1999, government was still playing cat and mouse with the World Bank over adjustment financing, and villagers were still learning the hard way. Somewhere in between, the comprehensive programme to reshape the forest sector had broken down. But the use of various policy instruments – and the experience of their successes and failures – was having some effect. A wider pool of people was engaged with the issues than before. The potential for this group to make real progress is examined in Section 4.7.

Regional Forest 'Agreements' as attempts to manage escalating conflict in Australian forest policy[12]

Under a very different set of circumstances, Papua New Guinea's neighbour Australia has also been through a period of perceived forest sector crisis which eventually precipitated a comprehensive technocratic approach to reform in response. The innovations and problems in the Australian Regional Forest Agreement process offer many useful lessons.

	Australia
Forest	• Total forest and woodland covers 1,558,000 km^2, or about 20 per cent of national land area (7,682,300 km2). Of this: 36,000 km^2 is rainforest and another 404,000 km^2 is closed and open forest in the coastal fringe; the remainder 1,118,000 km^2 is woodland, stretching into the dry centre of the continent. Eucalyptus is the dominant genus • Plantations, mostly of exotic conifers, cover 10,000 km^2
People	• Population – 18.2 million (15 per cent rural, 85 per cent urban) with an average density of 2.4 people per km^2 and annual growth rate of about 1.1 per cent
Tenure	• Of total forests and woodland: 9 per cent is state forest; 11 per cent is park and conservation reserve; and 69 per cent is private and leasehold. Three quarters of the closed and open forest is public land, in state forest and conservation reserve, the other quarter is privately owned
Forest economy	• Industry is primarily geared towards woodchip exports • About one-quarter of Australia's consumption of sawn timber and 40 per cent of its paper is imported • There are trends towards increasing foreign investment in both plantations and industries • About 30 per cent of Australian farmers are involved in Landcare groups and about 44 per cent of these plant trees concertedly
Pressure on forests and people	• Policy since the 1930s to encourage the pulp and paper industry, and since the 1970s to promote woodchip exports, led to widespread clear-felling and regeneration of even-aged crops, with loss of biodiversity • Since National Forest Policy (1992), there has been emphasis on creation of new conservation reserves, and resource security for large industries. Opportunities for small industries are likely to be foregone • Landcare and Catchment management programmes of recent years, with extensive rural community involvement, allow government to avoid regulating deforestation, and enable re-vegetation of some areas, but extensive deforestation is continuing unabated in other areas
Key policy issues	• Regional Forestry Agreements for state forests as a way of managing stakeholder conflict (this section) • Government's role on private land motivating voluntarism or avoiding responsibility? (Section 4.5)

12 This section draws from a paper prepared for IIED's *Policy That Works* project by John Dargavel, Irene Guijt and Peter Kanowski (1998).

Australian forest policies are in a period of dynamic change. In the 1970s and 80s, the old order, dominated by state and industrial forestry dedicated to producing wood, became steadily engulfed in environmental conflicts which governments were unable to resolve. The polarisation of views became amongst the most extreme of any country in the world. Environmentalists and later loggers became adept at organising blockades and other direct actions, to gain media attention and advance their causes. The challenges entered the political arena and were reflected in conflicts between the Commonwealth and State governments and between Ministers and their bureaucracies. Box 4.1 outlines some of the main elements in this saga. By the early 1990s, forest policy had become an ungovernable political nightmare.

In 1992, Commonwealth and State governments finally agreed a national forest policy statement. The statement expressed the desire of governments to progress from an era of intergovernmental and community conflict to one of negotiation between interested parties in which, they hoped, the political heat over forest issues would diminish. This marked the beginning of a revolution in Australian forest policy. Its major expression the Regional Forest Agreement programme – aimed to resolve the contentious issues.

The Regional Forest Agreement programme, still under way in 1999, concerns the regions containing rainforests and tall eucalypt forests on public land. These are the areas over which conflict simmered, concerning whether to use them for wood production or to preserve them as national parks. Both Commonwealth and State Governments wanted to make these conflicts manageable and get them off the political agenda. The Commonwealth's carrot was to promise that once a Regional Forest Agreement had been signed, export controls would be lifted; the stick was the probably empty threat that woodchip exports would be banned unless Agreements were completed.

Structure and conduct of the process

The Commonwealth's Department of the Prime Minister and Cabinet oversaw the Regional Forest Agreement process and negotiated with the State agencies. For each region, it set up a *Commonwealth-State Steering Committee* of officials and *task forces of specialists* drawn from different Departments. Different forms of consultative mechanisms were established in different States. For example: public meetings in Tasmania; community representative groups called Forest Forums in New South Wales; and direct representation of environment and industry stakeholders on the Steering Committee in Queensland. The process began by developing a scoping

Box 4.1 The background to Regional Forest Agreements – environmental challenges, conflicts and policy responses in Australia

Year	Challenges and conflicts	Response
1968-70	Objections to agricultural clearing of Little Desert area of mallee woodland in Victoria.	Political recognition of nature conservation. Land Conservation council starts reviews of all public land in Victoria.
1967-72	Objections to flooding of Lake Pedder in Tasmania.	Political rejection.
1973-75	Rising concern over environment.	Commonwealth Government passes environmental and heritage protection legislation.
1973-	Objections to clearing native forests for pine plantations.	Numerous inquiries. Commonwealth ceases funding States for plantations. States cease clearing and buy already-cleared land.
1973-	Objections to clear-felling native forests for woodchip exports and some for domestic processing. Environmentalists conduct blockades in New South Wales and Victoria.	Numerous inquiries but industry expanded. Forest Practices Codes introduced.
1979-82	Objections to rainforest logging leads to blockade at Terania Creek in New South Wales.	Logging stopped. Area nominated as National Heritage.
1979-83	Objections to dam development in wilderness. Blockade on Franklin River in Tasmania to stop dam construction.	Public recognition of wilderness. Area nominated as World Heritage. Constitutional power of the Commonwealth challenged in High Court. Dam stopped.
1983-87	Objections to logging and road construction in the wet tropical rainforests of Queensland. Blockade of Daintree road.	Area nominated as World Heritage. Constitutional power of the Commonwealth challenged in High Court. Logging and road stopped.
1987-89	Objections to pollution from proposed pulp mill at Wesley Vale in Tasmania. Provoked Commonwealth-State conflicts. Cabinet Ministers disagree publicly.	Pulp mill stopped. Push for better process.
1990-91	Objections to logging high conservation value forests in northern New South Wales. Blockades at Chaelundi State Forest.	Public inquiries. Some state forest service operations controlled by wildlife service in order to preserve habitat for endangered animals.
1992		Commonwealth and States sign *National Forest Policy Statement*.
1994-95	Environmentalists object to extension of woodchip export licences. Loggers blockade Parliament House in Canberra objecting to reductions and delays. Cabinet Ministers disagree publicly.	Prime Minister's Department takes charge of Commonwealth position on forestry and oversees Regional Forest Agreement process.
1997-98		First Regional Forest Agreements negotiated.

Source: Dargavel 1995; Robin 1998

agreement. This was to be followed by three further stages:

- conducting comprehensive regional assessments,
- integrating the assessments and preparing options, and
- negotiating the regional agreement.

Comprehensive regional assessment

This had to cover all the issues on which the subsequent agreement would be negotiated, and required groups of assessments covering:

- *Ecological or biodiversity assessment.* The greatest part of the effort was directed to this. Whereas there was a good knowledge base on timber, there had been few systematic assessments of ecological values on a regional scale. Large geographic information systems were developed to store, assemble and display the voluminous layers of vegetation and faunal data. Topography, climate, geology and soil data were also collected and added. The systems provided the basis for models which were developed to estimate the likely extent of each vegetation type prior to European colonisation, and hence to calculate the area of each which was supposed to be reserved to meet a 15 per cent criterion. Areas of old-growth forests were delineated following systematic surveys of where logging, mining and other activities had taken place in the past.

- *Cultural heritage assessment.* Surveys and community workshops were conducted, in which local people were able to record places of importance to them. Historic places included bridges, buildings, bullock wagons, cemeteries, charcoal pits, defence/ war relics, ethnic settlement and workers' camps, explorer's routes, fences, fire towers, mines, railways and tramways.

- *Indigenous heritage assessment.* There were several significant difficulties in conducting assessments of the history of indigenous occupation and of contemporary Aboriginal associations with forested lands. These included the high level of indigenous peoples' mistrust of State Government intentions. The accelerated timetables set politically for the process could accommodate neither the extensive consultation required, nor the fact that knowledge of sacred places in the landscape is commonly held in confidence by particular people and is not therefore readily amenable to public survey.

- *Social assessment.* Far less effort was directed to social assessment than to biodiversity assessment. However, a social assessment unit was established within the Commonwealth's Department of Primary

Industries and Energy and coordinators were appointed for each State (Coakes, 1998). Four assessment projects were conducted for each region:

- *Post-impact studies analysis*. Literature on previous land-use decisions and their social consequences was reviewed.
- *Detailed regional social profile*. Demographic and socio-economic characteristics were compiled from census and other statistical sources. Interviews and focus groups were arranged with key stakeholders to identify issues of concern, and some surveys were conducted to ascertain community attitudes to forest management and aspirations for the region.
- *Forest-related industry assessment*. Detailed surveys were conducted by mail and interview to document the location and type of industry, employment and some economic and educational factors.
- *Social case studies*. Selected case studies were conducted in communities which were particularly vulnerable to change as a result of the agreement process.

- *Wood resources*. Existing information and processes used by the States' forest services were largely taken at their face value. Indeed, given all the effort directed to the biodiversity values, there was little energy left to do much else. In effect, the wood in the native forests had come to be seen almost as a residual value; what industry could use once the conservation reserves had been created.

Integration of information – through technical methods and public participation
assessments were brought together in a process which differed from State to State, depending on the attitude of the Governments and the adequacy of the assessment data.

Geographic information systems dominated the whole integration phase to a remarkable degree. They provided a means for storing and displaying the various layers of information and creating composites of them, and a basis for various modelling exercises. They were ideal for the ecological and biodiversity aspects, which were largely driving the process, and were also quite well suited for recording the various types of heritage sites.

'Decision support models' were also developed. They carried an impressive aura of technical sophistication. Although they were effectively opaque to non-specialists, or perhaps because of this, they endowed a high level of credibility to the content of the assessments and the technical aspects of integration. Importantly, they provided the 'language' in which the negotiations were conducted, which made the inherent conflicts amenable to bureaucratic processes and thus potentially governable. However, many factors were left out of these models, and major shortcomings emerged in the attempts to ensure 'full and comprehensive integration'.

Outcomes of 'negotiation'

Each agreement is between the State and Commonwealth. It covers a 20-year term which may be extended by mutual agreement. Reviews are to be conducted at five-yearly intervals to monitor progress. The main features of each agreement are:

- *establishing the conservation reserve system* – built up from existing reserves and added to through purchase or covenant from private land;
- *establishing ecologically sustainable forest management principles* in the production forests, with planning systems and codes of forest practice devised to put them into effect;
- *removing the woodchip export licensing controls for the region*; and
- *providing security of resource access for industry*, or rather, removal of the Commonwealth's powers to enforce provisions of its legislation on world heritage, national heritage or environmental protection against the wishes of the State concerned (Tribe, forthcoming).

Solving or governing the conflicts?

Australia has committed considerable professional, scientific, economic, bureaucratic and emotional resources to the Regional Forest Agreement programme, the largest and most expensive resource planning exercise ever undertaken in the country. Although the programme was essentially politically driven, it has been strongly influenced and accompanied by expectations of professional and scientific standards.

As of April 1999, only three of the ten regions covered by the process have reached the agreement stage. These agreements notably exclude the claims of indigenous people to rights in the forests. In the hurry to complete the 'comprehensive' assessments, it seems that the indigenous heritage values were given little recognition, although the agreements include promises to undertake further consultation. Also noteworthy is the exclusion of any provisions to guard the interests of communities and people who depend on the forests. The immediate beneficiaries are the woodchip export companies who are freed of export controls.

Although it is too early to tell, some believe that the accelerated timetable may mean that such standards will fall and decisions lose their legitimacy. Critical factors will include: whether the legal basis of the agreements is satisfactory and enforceable; and whether governments will fund and implement the provisions of the agreements and respect the interests of the various stakeholders (Tribe, forthcoming).

The Regional Forest Agreement programme is directed to making the public contests over the use of the public forests governable. It is notably not directed to mediating or resolving the source of the conflicts themselves. The public participation programme, outlined above, aimed to produce inter-governmental agreements which in essence only required the top level of officials to conduct negotiations. Community level activities were essentially limited to information gathering on heritage and other matters, whilst the middle level of the programme – the Forest Forums – provided a legitimising mechanism and a channel through which some of the information could flow to the top level. Difficulties arose because the consultations raised expectations that the views expressed and submissions made at 'lower' levels would influence decisions. Some may well have done so, but the opaque, 'top-down' nature of the decision process did not provide any assurance that they were.

Despite the availability of highly sophisticated models and data bases and attempts at all-inclusive public consultation, compromises have been difficult to reach, and the attempt to achieve a 'harmonious whole' through these mechanisms seems fanciful. Two of the important sources of conflict – indigenous land rights and local economic development – are by-passed by the present process. Wider strategies of influencing industrial structure, investment and ownership for increasing employment are virtually ignored. Unless the participatory mechanisms can be developed to handle genuine negotiation with community and 'middle' levels, it remains to be seen whether diffusion, or perhaps exhaustion, of some conflicts is sufficient to keep the others at bay.

TFAPs, FSMPs, NCSs, NEAPs, NSDSs, etc. – planners' dreams or unifying frameworks?

The PNG and Australia policy reform programmes give us a mixed picture. Their ultimate impact is as yet a bit obscure but, although they are unlikely to achieve their original planner's dreams, they have certainly catalysed change. We now turn to a more general consideration of the national effects of some international models of policy reform. In Section 5, we look at the genesis and evolution of the Tropical Forest Action Programme, an international framework to encourage reform to address deforestation and other – essentially international – concerns. But there are several other international master-planning approaches, which have largely been the product of international institutions' thinking and which principally focus on environmental aspects. They have been perhaps the most rigorously developed and applied of policy reform methodologies, but they have not often had significant impact on forests and people.

Indeed, given their genesis outside the countries in which they have been applied, a number of alternative purposes of such strategy models can be postulated (Dalal-Clayton *et al*, 1998):

- to initiate real change, to deal with times of tremendous flux and uncertainty
- to rationalise the *status quo*, perhaps repackaging policies in the language of sustainable development
- as a delaying, marginalising tactic
- to 'spin' money to professionals engaged in such processes

The national conservation strategy (NCS) process was encouraged by IUCN, following the 1980 launch of the World Conservation Strategy, which called for the integration of conservation and development activities. NCSs involved various styles of policy analysis and reform, and often addressed the forest sector. The national environmental action plan (NEAP) process, driven by the World Bank, is essentially similar to the NCS, but has tended to be heavily influenced by the Bank's own agenda – forcing fast-tracking in countries irrespective of the 'natural pace' of policy reform, with the lure of loans under IDA10 if done on time.

In general, these organised attempts at policy change – TFAPs, the similar Forest Sector Master Plans (FSMPs), NCSs, and NEAPs – have often had the standing and momentum to be able to (quietly) forget real constraints to change. Approaches to date have been dominated by environmental officials and experts preparing papers and drafting chapters of a strategy document or action plan, workshops (again often restricted to officials and experts) and weak inputs from across government, political parties, the private and business sectors, NGOs and other interests, or from the public. The emphasis has usually been placed on delivering a document (often in a limited time); this has meant both rather sketchy analysis, and an inadequate process of building consensus on the key issues and possible solutions or ways forward.

Furthermore, most policy processes have tended to base their policy recommendations on an assessment of past and current trends. Few have generated scenarios which consider environmental and developmental conditions in the future or develop policies that respond to challenges, blind spots and priorities identified in these scenarios. Climate change, technological advances, and market trends tend to be kept out of these 'planners' dreams'.

It is no surprise, therefore, that the TFAPs, NCSs, NEAPs and broader strategies for sustainable development tend still to be seen as internationally-generated precepts. And they have not usually exerted much influence on the key decision-making processes which relate most closely to the national planning, finance, and major line ministries, rather than the environment authorities. They have had still less influence on political and business development processes. Relatively little advance has been made in providing lessons for better and more effective approaches. Over the last few years, numerous conferences, workshops and reviews have assessed strategies and made numerous recommendations on improving these forms of comprehensive policy reform. Yet few of these recommendations have been addressed or implemented: the various reasons for this are listed in Box 4.2.

However, some strategic processes involving policy review have had considerable success. In the Pakistan case, described below, the impacts have been not inconsiderable on national forest sectors.

Pakistan's conservation strategies – from top-down to a reinvention of natural resource policy at district level

In Section 4.6, we describe how participatory forestry projects have very recently had a strong influence on the forest authorities. But it was the new, multi-sectoral policy processes involving many stakeholders, notably the National Conservation Strategy and the Sarhad Province Conservation Strategy, which have provided the policy-level framework to push provincial forest authorities into scaling up participatory forestry approaches, and to consider developing new, enabling roles. The National Conservation Strategy was the biggest-ever consultation exercise on sustainable development issues; it incorporated a major technical review, and was the first ever attempt to integrate broad economic, social and environmental objectives. These initiatives have not remained abstract, but pointed to projects which have successfully dealt with local realities and found ways to overcome the constraints imposed by the boundaries in policy/ institutional domains – by bringing stakeholders together in purposeful ways. An Institutional Transformation Cell has been created in North West Frontier Province (Sarhad) to support the transition of both the policy and the roles of the Forest Department. The Forest Act of 1927 is finally being revised; and a modification has already been made to accommodate project initiatives in Joint Forest Management.

In conclusion, good policy reform approaches have created the space for 'vertical' hierarchies, and for 'horizontal' institutional, sectoral and cultural boundaries, to be crossed. They have resulted in a tremendous sharing of information and perspectives. They have led to multi-disciplinary networks and creativity. Rarely yet are they linked to such creativity in the field.

Nonetheless, where 'ownership' was strongly local, rather than by e.g. FAO or IUCN as a vehicle for their own growth, or by the World Bank for securing its operating conditions, or by other donors, such strategies have opened up the debate and presented fresh ideas.

Box 4.2 Why recommendations of past formal forest policy reviews have not been addressed or implemented

IIED's work over six years, assessing formal policy review processes such as TFAPs, NCSs, NEAPs and broader strategies for sustainable development in many countries, reveals a number of reasons why they tend not to be fully addressed by policy-makers, or fully implemented:

Policy recommendations not addressed:
- the key players involved in developing policy reviews have had no 'handle' on the pros and cons of *market* issues/ forces as means of achieving sustainable development; or on the *politics* that surround decision-making;
- lack of *institutional memories* (within government departments and in donor agencies);
- *staff turnover* – with loss of valuable experience of individuals;
- policy reviews in developing countries were seldom, if ever, designed to be continuing (cycling) processes, and therefore mainly *ended with the completion of a strategy document*;
- policy reviews seldom fitted with the resources (financial, skills, etc.) available;
- *ownership* of most policy review processes was perceived to be, or was in practice, outside the country concerned;
- policy review *fragmentation*, particularly through identifying fundable individual projects which were often 'cherry-picked' by donors, leaving important strategy elements unfunded;
- policy review recommendations were often flawed due to *inadequate ground-truthing*;
- policy review often *set no priorities* or gave no guidance to assist prioritisation.

Policy recommendations not implemented:
- policy reviews have not matched the level of *institutional* capacity in individual countries – they have often been too comprehensive/ complex for the prevailing institutional climate;
- no clear targets for *communication and advocacy*;
- some review documents have been too *generic* for decision-makers (lacking a clear 'hook');
- some reviews have been very descriptive with *too little analysis*;
- reviews have been *too focused on environmental issues*, partly because of their protagonists;
- lack of *indicators and targets*;
- lack of *ownership* within countries/ agencies – and thus reviews are perceived as only the opinion of their authors;
- inadequate attention to how to incorporate forest-related recommendations into other *(extra-sectoral) policy* processes;
- reviews have not adequately examined *situations where people have been motivated to change* (i.e. determining the effective points of entry).

Source: based on Dalal-Clayton *et al*, 1998

4.3 Reinventing state roles

What are the real public forest benefits which need public protection, as opposed to private benefits? And can any of the public benefits be produced by the private sector? Whether benefits can be considered public or private will depend greatly on the level of institutional development in a country and may change over time. There are no magic bullets, but collaborative approaches offer promise (Box 4.3).

Box 4.3 Public and private benefits – economic distinctions

Forest resources have been owned and controlled by national and state governments, because of the belief that many forest goods and services would not be properly produced and allocated under a system of private ownership and market exchanges. Indeed forests produce:

- goods that are well suited for market allocation and private consumption, such as timber;
- services that cannot be rationed by a market system and tend to be considered public goods, such as forest recreation, carbon storage and biodiversity.

Two key concepts help to distinguish between goods that are best suited for market allocation and goods which, due to market failure, are often considered to be public goods.

- *Excludability* – where an individual can deny the use of any goods or service to another individual
- *Subtractability* – the amount that the consumption of a product or service subtracts from its sustainable consumption.

Most consumer goods, like timber, can only be consumed once – they are highly subtractable. And, since it is also easy to exclude other individuals from using consumer goods, these goods are best allocated by the market.

On the other hand, biodiversity is characterised by low excludability and low subtractability, and is then best treated as public goods, subject to governmental regulation. Since there is little incentive for an individual to invest in the provision of biodiversity, it will tend to be under-provided – or not provided – unless a government or an association accepts the responsibility for the provision for the public's benefit.

However, there is no clear-cut case for private or public control of specific goods and services. The public or private nature of goods and services is not an inherent property, but depends upon the level of institutional sophistication, communications and technology. And mixed public/ private goods can be very effectively managed under strong common property regimes. It is possible to change excludability and subtractability through e.g. zoning and management agreements. Hence there is potential to transfer what once had to be public goods to market or community systems with institutional improvements and appropriate safeguards.

Forest Goods and Services	Excludability	Subtractability	Externalities and Comments
Timber	High	High	Private Goods
Hunting	Medium	Medium	Private – Congestion Effects
Grazing	Medium	High	Mixed public-private
Fuelwood Collection	Medium	High	Mixed public-private
NTFP Collection	Medium	High	Mixed public-private
Recreation/ Amenity Uses	Medium	Medium	Public – Congestion Effects
Carbon Sequestration	Low	Low	Public Goods
Micro-climate Moderation	Low	Low	Public Goods
Watershed Protection	Low	Low	Public Goods
Biodiversity Conservation	Low	Low	Public Goods

Source: Bass and Hearne, 1997

The following case examples from India, Zimbabwe, Scotland and Sweden show a considerable rejigging of state roles over time. Sharpening the state's focus on securing public goods, in the above economic sense, is a common theme – but so is a whole range of political considerations...

Who defines the 'national interest'? India

Most forest land in India is under government ownership, and the government is obliged to manage it in the 'national interest'. However the perception of what the 'national interest' is has been interpreted in different ways since Independence in 1947. Initially, forests were required to be host to accelerated extraction for national development. Since the late eighties, forests are to be conserved, again in the name of the 'national interest'. But quite how the national interest is defined and assessed is not clear – are these mere changes in policymakers' fashions or valid representations of societal consensus?

In a mixed economy such as India's, it might be assumed that the role of the government in protecting the 'national interest' is to look after the infrastructural or welfare needs of the people, whereas the private sector addresses market needs. But in India's forest sector, the reverse was attempted under the farm forestry programme of the late 1970s and 1980s. Public forest lands were planned to meet the commercial needs of the economy and farm lands were supposed to produce 'fuelwood and fodder' under a notion of community welfare. This conceptual reversal was perhaps one of the main reasons for the failure of both approaches. In practice village lands produced commercial polewood or urban fuelwood, and did not meet the subsistence needs of the poor. Meanwhile, the poor were at times displaced from common 'wastelands' as well as the public forest lands which once provided biomass (Hobley *et al*, 1995). The main beneficiaries were the mills of forest industry and the coffers of both central and state governments.

However, along with the radical shift encapsulated in the latest forest policy of 1988, from the earlier revenue orientation to conservation as a priority, came an emphasis on meeting the subsistence requirements of forest-dependent people. The policy proclaimed that forests are not to be commercially exploited for industrial purposes and that forest-dependent people's domestic requirements for fuelwood, fodder, NTFPs and construction timber were to be prioritised. Through Joint Forest Management (JFM) – the principal expression of the policy – government roles have been redefined such that forests are protected and managed through partnerships between forest departments and communities (see Section 4.7).

In seeking to understand how such an apparent about-turn in policy was possible, it is important to clarify the different arms of government which have some bearing on the forest sector. Forestry is a concurrent subject (as is the police service and the administrative service), meaning that both the states (in other countries: provinces or regions) and the central government have regulatory powers but the powers of the latter override those of the former. Although central government's direct role in the forest sector is relatively recent, the concurrent status having been introduced in 1976, it enjoys considerable and increasing powers, especially at the level of formulating policy and promoting programmes.

Whilst central government – through the Ministry of Environment and Forests but also other ministries with an influence on forests – is currently pushing a strongly conservationist agenda, state governments have regional and local pressures to deal with. State forest departments are in charge of government forests (i.e. most forests), tree growth in other categories of government lands, regulation of felling (in some states) and movement of timber from private lands.

On top of all this, the Supreme Court has become an important player recently, with powers above and beyond those of the Ministry of Environment and Forests. For example, the Supreme Court has made judgements regarding felling bans in certain states, and has stipulated the need for working plans to be drawn up. This implies that central government (the Ministry) is not yet adapting well to the new roles defined for it by the 1988 policy – or at least it is not taking the national-level decisions required of it.

Explanation for the slow progress in implementation of the new policy seems to lie in the complex power structure of those involved in the forest sector. Full implementation of a pro-people forest policy would seriously upset the existing power balance and adversely affect those who have

India	
Forest	• Forests cover 633,400 km^2, or almost 19.27 per cent of national land area (2,973,190 km^2), although 765,200 km^2 (23.6 per cent land area) is classified as forest land.[13] However, about 42 per cent of total forest cover has crown cover of less than 40 per cent. • Plantation forests are estimated to cover 132,300 km^2
People	• Population – 960 million (73 per cent rural, 27 per cent urban) with an average density of 323 people per km^2, and a growth rate of approximately 1.6 per cent per annum • Around 300 million people live below the 'poverty line', and around 200 million of these are partially or wholly dependent on forest resources for their livelihoods. • There are strong correlations between the locations of tribal people, forests and India's concentrated poverty area
Tenure	• All forest land is state-owned, apart from small areas of community and privately owned forest land • Protected areas cover 20 per cent of the forest land area (more than 4.5 per cent of the national land area)
Forest economy	• Large-scale industry accounts for only about 10 per cent of timber used in India, the rest being used for fuelwood and other domestic needs • The current forest policy prohibits use of natural forests as timber supply by industry, and encourages farm-grown timber • Wood processing, mainly carried out by small sawmills and cottage industries, is currently in decline due to a scarcity of raw material – due in turn to restrictions on logging • Forestry is estimated to contribute 1.15 per cent to GDP (1994-5) but this does not include subsistence use and local market transactions
Pressure on forests and people	• High dependence by a large proportion of the rural population on forests for subsistence • Competing claims on forest land: from conservationists calling for protected areas, from social activists supporting forest-dependent people
Key policy issues	• Pioneered in India, Joint Forest Management arrangements for protection and regeneration of degraded forest land now cover around 70,000 km^2; some JFM-protected forest has matured to the point where it can be harvested. But restrictions on JFM persist and in some cases forest departments are using JFM to regain control (Section 4.7) • Industry's current claim to lease degraded forest land for raising plantations is being strongly resisted by the supporters of farm forestry, and of rural people dependent on such land • Policy has become more open to different actors, but there are continuing clashes between new policy, restrictive law and 'fortress forestry' institutions (Section 4.8)

13 FAO's 'State of the World's Forests 1999' cites India's forest coverage as 650,050 km^2, or almost 22 per cent of the national land area.

benefited most from past policies. The fact that the new policy and subsequent documents make no mention of the institutional changes necessary to achieve implementation suggests that the policy was more of a 'power play' than a practical way forward.

In conclusion, the new policy direction is well ahead of the capacity to implement it at this stage. The policy remains the aspiration of only a few which, whilst they may gradually gather converts, are yet to galvanise sufficient institutional motivation to win through. It is also clear that in these days of multiple demands for forest goods and services, simplistic calls to respect or represent the 'national interest' in forests will no longer suffice as the sole rationale for government's role.

Dilemmas concerning state roles in forestry, Zimbabwe

Dilemmas concerning state roles in forestry, and how to finance them in the face of declining budgets, have been central to debates since the 1980s in Zimbabwe. As in India, there are different interpretations of what the 'national interest' is, and how it might be safeguarded. Should the state's Forestry Commission be primarily concerned with production forestry for revenue generation, or with biodiversity and environmental services? Or is there a wider developmental role – service provision for rural development which is more important? Or are all of these priorities? These dilemmas, magnified in an era of structural adjustment where government has to make hard choices, have both stimulated and confused a process of institutional change in the Forestry Commission.

Pressures for commercialising forestry functions

Donors sought to push state role changes in the early 1990s. A Forest Sector Review produced recommendations on changes to the law, support for local organisations, and restructuring the Forestry Commission. Institutional options may have been more strongly geared towards securing a financial base than to a full consideration of public forest benefits. The options have included: full or partial privatisation of the Commission's commercial arm; user fees/ consultancy services for extension; contracting out management of forest reserves; leasing reserve land for grazing or tourism enterprises; charging for research; and commercialising training services. But internal institutional reasons – to do with pressures on individuals and their motivations – and external factors – to do with the economic climate, donor relations and conflicts over institutional territories – help to explain why the Review's recommendations have been adopted only partially. Today, the parastatal Forestry Commission is still chewing over these options (Table 4.2).

Table 4.2 Options for organisational reform of the state role in forestry, Zimbabawe

Organisational option	Status	Advantages	Disadvantages
Full or partial privatisation of commercial arm into forestry company	Partial privatisation to be effected (full privatisation to follow?)	• Allows development of 'good business culture' and competitive strength in the company • Enables quick decision-making in the company • Allows FC to concentrate on coherent mandate for developmental/ social forest functions	• Tensions amongst the current public sector work-force over better pay and conditions likely in the company • Removal of the potential for commercial activities to fund developmental/ social activities • May require staff redundancies to promote efficiency • Company could be taken over by private conglomerate
Charge user fees/ run consultancy service for project design and management	Under consideration	• Increased cost-recovery for extension • Plantation and commercial farm sectors can afford it	• Poor rural communities cannot afford it
Transform extension services into a central (civil service) department	Not being considered? (This option is incompatible with the other options)	• Emphasises extension as a state social responsibility (government plans reportedly to close the salary gap between civil servants and parastatal employees)	• Reduction in salaries in line with civil service – possible inefficiencies; not favoured by current staff
Donors and government support extension activities; cost-sharing with NGOs?	Current situation	• Allows the FC to out-source its services • Reduces dependence on state funding • Allows FC to learn from others	• Divergence of perspectives between FC management, donors and NGOs • Doubts about FC as the most effective extension delivery agency?
Merge extension functions with those of other government agencies	Partially adopted. Donors pushing for this	• Greater pooling and efficient use of resources for extension • More integrated 'messages'	• Visibility/ importance of forestry in public opinion may be lost (if subsumed under e.g. crop production)
Commercialise reserve management, e.g. out-source management, lease tourism facilities/ grazing	Partially adopted, eg. one forest reserve leased to tour operator	• Much scope for increased revenues from tourism • Allows revenue-sharing with communities currently in conflict with FC • More efficient and effective tourism operations	• Potential for over-exploitation of resources • Loss of FC control
Charge out research activities	Under consideration	• Private sector gets the research it wants	• Non-paying research reliant on donor funding
Charges higher fees and raise revenue from training services	Under consideration	• Reduce dependence on state funding • Clear demand for this from the large-scale private sector	• Facilities geared towards cash generation rather than forestry learning • Key skills may not attract funding

Will the state be able to rescue its social forest functions?

Whilst, after years of debate, the privatisation of commercial timber operations of the Forestry Commission was a *fait accompli*, the nature of the remaining government role remains uncertain. For those other functions of forestry for which there is a market – some aspects of research and training for example – it would seem to make sense for the FC to limit itself to the role of facilitator, recouping costs through revenue generation and user fees.

However, unless the economics of the communal lands drastically change (see Section 4.1), government surely has a responsibility to continue its role as service provider and primary investor in these areas. The assumption has tended to be that communities possess enough information and institutional strength to make decisions and express demands which are in their own best interests. This may not always hold: history and structural constraints have restricted the information and choices about land use that are available to more marginalised groups of people. In such circumstances, is it not the role of state extension agents to help develop local aspiration and organisation to make a greater range of land-use choices possible? This is, as yet, new ground for the FC in Zimbabwe.

The rise, fall and rise again of social objectives in state forestry, Scotland [14]

Expansion of the forest – and rise and fall of the 'social objective'
The British Forestry Commission (FC) was set up in 1919, with the job of establishing public forests and encouraging and subsidising private landowners to expand their wooded areas. National security of timber supplies was the primary rationale. There was also a firm social development agenda in the early years of public forestry. In the 1920s and 30s forestry workers were recruited from Glasgow slum areas, as well as areas neighbouring the sites of plantation establishment. The sites were chosen partly to bring employment to relatively deprived areas. 'Forest villages' were set up for the workers and various land settlement schemes grew into the FC's 'Forest Workers' Holdings' initiative to integrate forestry with agricultural employment.

Employment grew in this pre-mechanisation phase of plantation establishment. By the 1950s, employment-providing objectives were paramount in the FC's Scottish planting programme and continued to be so up to the mid-1970s. Yet by the late 1960s, they were in increasing conflict

---

14 This section draws from a paper prepared for IIED's *Policy That Works* project by James Mayers (1999).

with the job-shedding which was going on over the FC's whole estate, in response to mechanisation, altered forestry practice, and above all the pressure to show positive financial returns.

From this period, the trend was to move from direct employment by the FC, to the use of self-employed contractors who tender for work. This rarely resulted in locally-rooted employment since these contractors often led nomadic lives, typically living in caravans as they chased the work from place to place. Today, the peripatetic nature of this work-force, combined with highly machine-intensive forestry practices and the rarity of local forest ownership, has effectively delinked forestry from local development.

It has been argued, by FC staff, that other government agencies have taken up the social and economic development objectives (albeit with mixed, and disputed, degrees of success) which the FC used to hold, and the FC development role has withered as a consequence (Forestry Commission, personal communication, 1997). The current picture is typically of remote areas with extensive plantations, managed by contractors with variable connection to small local communities who remember the planting days when there was work for perhaps 20 or 30 local people. Commonly, the FC has even withdrawn its local office, and its remaining staff are irregular visitors. With the publicly-owned forest estate now no longer expanding (except in particular urban areas), the existing area is mostly planted up and generally quite mature. Thus the visible operations, and potential for even basic interaction between working foresters and communities, are minimal.

Upland afforestation – an activity in permanent pursuit of a rationale
British forestry has had something of an identity crisis since 1957, when a government inquiry pointed out that the justification for expanding plantations to create a strategic national timber reserve had been rendered meaningless by the existence of nuclear weapons[15]. Since then, successive reasons put forward for forestry expansion have made it appear to be an activity in permanent pursuit of a rationale.

There has never been a concerted and systematic attempt to base forest policy on a process of gauging public opinion in the UK. Recently this was recognised as a problem by the FC itself (Forestry Commission, 1998a)[16]. In contrast, the UK's development assistance agency DFID has been in the forefront amongst international agencies in advocating and developing

---

15 The inquiry, chaired by Sir Solly Zuckerman, was on the whole supportive of national investment in forestry but called for forestry and agriculture to be planned as an integrated whole (Mackay, 1995).
16 In 1998, as a supplementary exercise to the government's consultation process aiming to develop a new National Sustainable Development Strategy, the FC, in collaboration with DFID and the Department of Agriculture for Northern Ireland, produced a 9-page public consultation paper (Forestry Commission, 1998a). However, the paper was not printed until July and the deadline for responses was only a few weeks later on 11 September – giving little time for effective outreach and response.

Changing justifications for forestry expansion:
primary objectives of 20th century British forest policy

1919	Strategic reserve of timber to be created
1957	Rural regeneration and employment to be promoted
1967	Conservation of natural beauty and amenity to be integrated
1980	Timber import costs to be saved
1987	Agricultural surpluses to be cut by promoting forestry
1991	Multi-purpose sustainable management to be pursued
1998	Sustainable forestry to be certified against government and civil society standards

public consultation processes on forestry in developing countries. Information exchange between FC and DFID, and joint initiatives in international processes, are growing. Nevertheless, changes in British forest policy have generally been responses to particular interest groups' campaigns (Grundy, 1997)[17].

Questioning state forestry's environmental record...

A major challenge to prevailing policy emerged in the shape of environmental interest groups in the 1970s. The challenge centred on the inadequate environmental functions of exotic conifer afforestation and the concomitant decline of native lowland woodlands. It was believed that, in pursuit of quantity, the FC had cast aside the other benefits of forests. By 1989, a former FC forester – turned critic – judged 75 years of afforestation as having produced "second rate forests in the uplands whilst the lowland woodlands have been shamefully neglected and inadequately protected by national forest policies" (Tompkins, 1989). Criticism on environmental issues developed into criticism of the FC's relationships with other groups, and criticism of the FC itself as an institution.

...relationship with the private sector...

The public accountability of the FC's management of the national forest estate has also been questioned. Its commercial mandate enables FC officials to conduct negotiations with other landowners, estate agents, contractors and forest industry at arm's length from the government, elected representatives, and people living in the locality, by claiming that because these dealings are commercial they are 'confidential' (Inglis and Guy, 1996). However, the interests of the different components of the FC, and of the wood and forest industries, are far from uniform, and have generally developed in a state of considerable tension.

---

17 Pressure group influence on forestry is not a recent phenomenon. The very existence of the Forestry Commission and the policy of national afforestation owes much to the persistent pressure of the Royal Scottish Arboricultural Society – a group of landowners, foresters and interested members of the public who, from the 1880s onwards, persistently badgered government to purchase land for afforestation.

...and blinkered institutional vision

The FC was criticised for being over-centralised, and too preoccupied with a 'forestry mindset'. Some of its lack of openness would seem to be an inevitable result of the FC's history of struggling to gain and retain hold of large stretches of agricultural land. After reviewing the history and effectiveness of the FC up to the early 1990s, Mackay concluded that "over the years the FC developed a fear of openness and devolved decision-taking which was inconsistent with its custodial role" (*ibid*, 1995).

Such criticisms are strikingly reminiscent of those traditionally made by British foresters about developing country forestry contexts. Indeed, Pryor (1998) notes the tendency of British foresters – in the wider forest industry as well as in the FC – to claim that 'Britain has the best forestry in the world'. But he also doubts whether many of these foresters really have significant overseas experience (Pryor, 1998). In contrast, many professional foresters in Commonwealth countries gained their forestry qualifications at British universities – notably at Oxford, Edinburgh, Aberdeen and Bangor – or in courses developed in the British forestry tradition. In partial defence of the 'forestry mindset' which such experience may generate, foresters are no different from other professions in assuming that ever greater supplies of their product must unquestionably be a good thing. But experience in an increasing range of countries shows that foresters should and can accommodate the ideas and realities of others.

The FC's public reputation hit rock bottom in the late 1980s, at the peak of the private sector's ability to gain tax advantage through upland afforestation – with little regard for any of its wider costs and benefits. This led to rapid changes in land ownership. High-rate taxpayers became significant investors in forestry, but had little interest other than financial, and were almost exclusively absentee landowners.

Although these tax incentives were abolished in 1988, and replaced by an upgraded direct grants system, various FC procedures were still the subject of much criticism from interest groups. Opaque procedures do not inspire widespread confidence that the Forestry Commission's influence is being used in the best 'public interest'. The FC's insularity was again manifest in 1993, when the UK government began a review of options for ownership and management of the state forests. The government's review essentially recommended retention of the existing system. However, institutional changes have been made since then – principally through granting executive agency status to the Forest Enterprise under the FC, which conveys some financial planning autonomy in the management of state-owned forests.

By the mid-1990s, the FC's image had experienced something of a revival. This was largely due to the greater environmental and social sensitivity of the woodland grants schemes, and the fact that existing landowners, rather than absentee land speculators, were benefiting to a higher degree from the schemes. Paradoxically, the FC also benefited from public concern over perceived threats to its continued existence under the review of privatisation options, and over the forest disposals programme (see below).

Continuing dominance of agricultural policy over forest policy

The incentives structure within the EU Common Agricultural Policy (CAP) means that 9 million sheep represent the dominant agricultural land use over most of Scotland. The CAP incentive for sheep farming in marginal areas is aimed at food production and maintaining farmers' income, resulting in very heavy levels of grazing. Both forestry and agriculture are subsidy-driven, but agro-subsidies out-compete forestry grants by £3 billion to £30 million a year. However, ways in which funding packages can be put together to encourage more integrated agriculture and forestry have recently improved, and the land tenure consultation processes in 1998 and 1999 suggest that a wider set of changes in rural policy is occurring.

Return of the social agenda

Today, the Forestry Commission is strong in its policy commitments to sustainable forestry (Forestry Commission, 1994; Forestry Commission 1998). The balancing act between multiple interests, which adoption of sustainable forestry requires, implies dealing with stakeholder pressures. These pressures currently pull the FC in three main directions: the Treasury seeks a greater degree of financial viability; environmental activists and the FC's own economists point to the need to take on board environmental and, increasingly, social externalities; and rural development and land activists, with increasing numbers of local groups, argue for the Commission to reassert the social (rural development) remit.

Public dissatisfaction with the 'open-market' sale of FC disposals – to speculative investors, pension companies and foreign firms – has been a particular catalyst for the return of forest social issues. And the maturity of the forest estate which, with much felling in prospect, brings management concerns to the fore – means forestry is becoming publicly visible once more.

A key driving force in the emergence of the rural development agenda for Scottish forestry is the Scottish Rural Development Forestry Programme (SRDFP) (Forestry Commission, personal communication, 1997). This is a partnership between three NGOs which began life in 1994 (SRDFP, 1994).

The programme developed an approach it called Participatory Forestry Appraisal, which involved local people in the analysis of their current situation and in developing their ideas and plans into action. Steered by individuals with considerable international forestry experience, as well as participatory planning skills, the approach is to investigate options for forestry aimed at community development. This often involves the consideration of approaches based on a small but steady flow of forest benefits, rather than an infrequent large cut. To date, the FC appears to regard this agenda with scepticism, and has not sought any significant assistance from, or alliance with, the SRDFP. However, it is evident that some participatory methods are slowly being incorporated into the skills bank of certain sections of the FC.

More than just a place to walk the dog: challenges ahead

Through a wide range of initiatives and engagement with international processes, the FC has covered a lot of ground since the early 1990s in responding to the demands of other stakeholders. Cross-institutional consultation mechanisms, an experiment with a community partnership, and the recent piloting of 'community panels' have focused arguments and made considerable political space for new collaborations and innovative institutional practice.

However, parts of the Forestry Commission retain the view that social issues in Scottish forestry are simply matters of rights of access, recreation and education. The Commission has fallen behind public opinion in attempting to limit the debate to these areas, and is only now beginning to catch up.

The state role in Scottish forestry, the management of a public forest estate, and incentives for private forestry integrated into the landscape, are likely to remain crucial to the population of a devolved Scotland planning for its future. There is much to be gained in this planning process from fast-tracking efforts to foster new partnerships between government, private sector and local people. A sharpened focus on learning lessons from policy and practice in other countries will also be crucial.

The early years of the Scottish Parliament and the development of the land reform agenda (see Section 4.1) will bring increasing attention to some of the drawbacks of large forest landholdings, and to the advantages of rural development forestry in local economies. It would make sense for government foresters to gear up for engaging with this process, to wait less for pressure to build on public issues, and to develop concerted programmes to bring social objectives into the mainstream of Scottish forestry.

Government as coach – not player, manager, owner and referee, Sweden[18]

Swedish society, and its governance are remarkably stable, having been without war for two centuries. Standards of living and education are among the world's highest, and the small and relatively homogeneous society retains an egalitarian character. Government is quite democratic and comparatively transparent. Given such conditions, well-designed policy has a good chance of promoting progress. So far as there can be said to be a Swedish 'national psyche', forests are deeply engrained in it. This is partly because of forestry's dominance as a land use and as a cornerstone of the economy.

At the turn of this century, forests were in depleted and degraded condition as a result of clearance for agriculture, of charcoal burning for the iron and steel industries and timber mining. Over the following fifty years restoration of the forests was accepted as a patriotic objective, achieved largely by rural community groups, notably schools and churches. There are now 23.5 million hectares of productive forest land, and the reserve of timber continues to grow. The strong public interest in forests is manifested in liberal public access rights, and a lack of competition for land by non-forest uses supports this. There are 350,000 private forest owners in a population of only 8.8 million, many of whom live in towns and cities. A great many other town dwellers own forest summer houses, and outdoor leisure activities – hunting, orienteering, walking, fishing, skiing – remain very popular.

The Forestry Administration (FA) is the state agency and comprises the National Board of Forestry (NBoF) and the County Forestry Boards (CFBs). Over the years, the state role has focused more and more on research, extension and training. This has occurred over the course of the three main phases of policy since 1900: first, afforestation; secondly, mechanisation/rationalisation; and more recently, improved environmental standards. Since the 1993 Forestry Act, policy revolves around commercial production and protection of biological diversity.

Keeping policy information flowing

The FA is Sweden's main provider of information and advice to forest industry and to private owners on established and new forest policy. The FA is, however, also concerned to reach the general public, to promote a national understanding of, and consensus for, forest policy. A campaign will typically comprise five components:

---

18 This section draws from a paper prepared for IIED's *Policy That Works* project by Bjorn Roberts (1999).

Sweden

Forest	• Forest covers 279,000 km² million hectares or 63 per cent of total land area (411,620 km²)[19]; of this, 23.5 million hectares is reckoned to be productive forest • Species: Norway spruce, Scots pine, other coniferous 85 per cent; birch, other deciduous 15 per cent • Large scale afforestation during the twentieth century in response to the loss of forests during the nineteenth century
People	• Population 8.8 million (17 per cent rural, 83 per cent urban). The large majority live in central and southern Sweden. Average population density is 21 people per km² and growth rate about 0.2 per cent. • Forest land per person = 2.8 hectares (EU average = 0.12 hectares)
Tenure	• Tenure: consistent and clear rules and conditions for land ownership to encourage investment and long term management • Half of all productive forest land (11.5 million hectares) is owned by private individuals, comprising 250,000 holdings, with an average size of 50 hectares. These are owned by 350,000 individuals. 89,000 owners belong to a forest owners' association • In the far north, private owners account for 37 per cent of forest land, while in the south the figure rises to 77 per cent. Limited companies own 37 per cent (8.6 million hectares) of forest land. 8.2 million hectares are owned by five companies. Most of their estates are in central and northern Sweden. The large companies also own saw, paper and pulp mills, and some own large forest estates outside Sweden • 8 per cent (1.8 million hectares) = communal/ church land, 5 per cent (1.1 million hectares) = state owned
Forest economy	• The forest products sector is Sweden's largest net export earner (SEK 76 billion in1997) • Forestry and the forest industries generate substantial employment (100,000 directly employed), but levels have declined with mechanisation. 60 per cent of wood received by Sweden's mills is from small private holdings • Among small-scale private owners, a small and declining proportion depend entirely on their forest holding; others depend on incomes from forest and agricultural production, and a growing proportion live in urban areas
Pressure on forests and people	• Degradation: air pollution, leading to acidification and release of aluminium into the soil, is the greatest threat. Sulphates, nitrates, ammonium and hydrogen ions are particularly high in the south • Competition from other materials and from overseas forest production • Disagreement between forest owners associations and the forest industry companies over competition laws and restrictions on land acquisition by forest industry
Key policy issues	• Government focusing on making information useful and developing consensus (this section and Section 4.7) • Certification: FSC certification has been adopted by the forest industry companies but rejected by private forest owners' associations that supply forest industry (Section 5.3)

19 FAO's 'State of the World's Forests 1999' cites Sweden's forest coverage as 244,250 km², or 59 per cent of the national land area.

- improving general awareness of the theme underlying policy and policy guidelines, through newspapers, television and radio;
- improving awareness of detail among forest owners and managers, through pamphlets and articles in specialist magazines;
- dissemination of operational knowledge, through manuals and text books;
- training courses, site visits by District Officers, and extension through examples of good practice by neighbours; and
- further courses, possibly leading to a forest owner examination and certification of competence, which bestows some prestige.

The FA's most important partner in providing extension and training is the Federation of Forest Owners' Associations together with the Associations themselves. Like the FA, these have both a centralised office to coordinate campaigns and produce materials, and a large network of local offices and representation allowing personal contact with individual owners, site visits and training.

Sharing stakes in research

The forest policy process in Sweden has put a high premium on access to good information for those making, influencing, understanding and implementing policy. Government has stimulated continual subtle shifts in policy priority, by redirecting funds through a framework of research bodies, and with funds for *ad hoc* projects. The allocation of funds to particular types of research affects both the direction and the quality of decision-making. Research programmes and funding, both public and private, are therefore integral, and sometimes contentious, parts of forest policy.

Silvicultural research is largely carried out with government funding through a plethora of agencies (principally the NBoF and the Environmental Protection Agency), research foundations, EU programmes, research councils and the Committee for Joint Nordic Forest Research. A nagging disagreement exists between the forest industry companies and government over the current extent of plantation research. Forest industry requires good information on forest yields and stem quality, and argues that government should step up its funding for research in these areas. Government counters that industry should itself stump up the funding.

Government also gives substantial direct funding to core forest research programmes, in particular the Swedish University of Agricultural Sciences (SLU). About half of SLU's research is purely biological, whilst applied work also includes social and economic studies. SLU runs the annual

National Forest Inventory (NFI), upon which the NBoF depends for basic information such as forest health, age-class distribution, stocking levels, increment, and pest and disease damage, as well as ecological conditions and change.

Skog Forsk, the Forestry Research Institute of Sweden, was established to make SLU's academic research meaningful and useful to non-academics. The institute is structured as a business and receives funding from Government, forest industry companies, and the Swedish Federation of Forest Owners' Associations. All of these funders influence research priorities and, conversely, Skog Forsk helps to provide the material necessary for forest policy formulation and extension. Skog Forsk's membership also includes individual forest owners, contractors, nurseries, community forest administrators, NGOs, suppliers, buyers, academics and churches. Total membership accounts for approximately 75 per cent of Sweden's forest land. Skog Forsk's research programme is focused in four main areas:

- optimisation of product value, production efficiency, and utilisation efficiency;
- 'ecological forestry' including the development of 'environmentally sound management systems' and the ecological impacts of forestry practices;
- production of improved regeneration material/ selection of breeding material;
- organisational development (business management and skills enhancement).

Converting research into knowledge

Skog Forsk is mandated to relay information and advice to its members in a form that readily assists decision-making. Researchers are often dependent on large numbers of forest owners for data, while those responsible for extension face the problem of reaching 350,000 forest owners. Its strategy for effective extension is based on:

- easily accessible, systematised and relevant information;
- a detailed quality control system to ensure correct information is applied, with input by external advisors;
- up to date information and rapid publication;
- targeted information (different forms, and different channels of communication, for the different recipients – using the criteria of: a forest owner's age; how recently he or she took up forest management; degree of interest shown; and the area of forest owned); and

- self-financing: marketing is encouraged; and market forces help to find the optimum distribution level and demand for high quality product.

Access to, and active dissemination of, forest knowledge has thus been the driving theme in the Swedish government's approach to forest policy. This has achieved remarkable results. Whilst tensions remain, the 'democracy of information' approach seems to underlie the high degree of consensus which typifies policy processes in Sweden. This is taken up further in Section 4.7.

In conclusion, the move to 'leaner and fitter' state roles is a widespread but frequently painful process – and not just for the state agencies themselves. The objectives are often unclear and the process is generally incomplete. As the state's commercial roles are hived off to others, the need to focus on guaranteeing appropriate regulation and on the social objectives of forestry – as a vehicle for improved livelihoods and rural economies – come centre-stage. But where state institutions are paralysed with old procedure, and where money is tight, this is a tall order. Progress seems to have been greatest where government has focused on generating the information that people really need, and making it as useful as possible.

4.4 Linking the people who change things

How do ideas gain and lose currency in policy? What does it take for sudden 'leaps' to occur, successes to get replicated, and innovations to become mainstream? Frequently, policy change can be correlated – not with grand reform processes or bureaucratic process as discussed above – but by 'change champions' – people with vision, experience, or just the right connections. Motivated about particular objectives and able to make 'political space' for them, such people are products of their time and circumstances. They are often from 'outside the system' which they are seeking to change, and therefore relatively uninhibited by its traditions and mores. In this section we discuss the role of individuals in Ghana and Zimbabwe.

Conservators and change agents, Ghana

Informal networks have helped shape opinions influencing forest policy in Ghana. One peer group of similar age and education, and with professional roots in the pre-independence period, have dominated the key positions in forest operational institutions. Another important network, which covers upper and middle management levels, is the annual meetings of professional foresters of the Forestry Department. Over many years, both groupings have been successful at charting a fairly coherent course for policy – an essentially administrative approach with some incremental adaptation to change. Policy became equated with what the Forestry Department could actually do.

Picking tactical battles carefully

However, there are also ample examples of policy affecting forests being pushed along by an individual or much smaller group. In the last few years, the elaboration and implementation of key policy reforms in Ghana the Interim Measures, the Timber Resources Management Act, the reform of the Forestry Department, the consolidated Forestry Bill have in no small measure been the result of the dynamic leadership of a small core group of Ghanaians – 'outsiders who have become insiders' – at the Ministry of Lands and Forestry and the Forestry Department. They are quite a diverse, and not necessarily unified, group who have tended to demonstrate a common characteristic in the ability to identify policy issues which are susceptible to change and then to find a way to make progress on broader problems. Thus, by picking their 'tactical battles' and scoring some successes, they have been able to build momentum and alliances to take on bigger or further aspects of the problem.

NGOs and innovative managers, Zimbabwe

It is clear that strong government/ civil society relations and information flows are key to improving policy. Where such relations are not very strong at an institutional level, relations between individuals from these different backgrounds can begin to achieve the same progress. We noted in Section 4.2 how Pakistan's original NCS worked very much through relations between inspired individuals in government, the private sector and NGOs, rather than by institutional mechanisms. The type and rate of change within Zimbabwe's Forestry Commission, described in Section 4.3, and the health of the Commission's relationships with other stakeholders, owes much to key players at the senior management level in government and NGOs, and their movement between these groups.

Since the 1970s, NGOs have had significant influence on the Forestry Commission (FC) and the wider policy debate. In the 1970s and 1980s, NGOs popularised the 'fuelwood crisis' notion and promoted responses such as improved cookstoves, biogas technology and woodlots for enhancing fuelwood supply.

The extension programmes which the FC developed in the 1980s generally focused on technical packages for producing fuelwood and *Eucalyptus*, combined with the enforcement of existing rules. Retrospectively, the programme was widely criticised for: ignoring the shortage of land available for the establishment of woodlots; the prohibitive cost of nursery establishment; the inadequate numbers and inappropriate training of FC personnel; and the lack of attention to management of natural woodlands. Above all however, the focus on fuelwood was found to be misdirected since it was based on the false assumption of an impending fuelwood crisis and associated inflated assumptions about market prices of fuelwood and forestry's role in terms of coal energy substitution costs.

Building strong connections

After it became clear that these types of prescriptions were unworkable, NGOs pioneered the concept of natural woodland management, emphasising participation and empowerment. Because of the good connections between NGOs and the FC at senior level, these ideas were later incorporated by the FC in its own work. For example, a project focused on community management of indigenous woodlands, begun in 1987 and managed by an NGO called ENDA-Zimbabwe, was a very direct attempt to counter the dominant fuelwood crisis/ exotic trees approach at that time advocated by the FC. Ford Foundation funding was also of great importance, providing a connection between the ENDA-Zimbabwe project, the FC's social forestry unit and other natural resources management projects which Ford was also funding.

This network of people, supported by the General Manager of the FC – who also moved between ENDA-Zimbabwe and another policy-focused NGO called ZERO (now at IUCN) – marked out the parameters of a new debate which was to have much influence. For example, the network of people in this group of organisations became strongly engaged with the World Bank/ FC Forest Sector Review process in the early 1990s. Although there were failures during this period, notably those programmes which tried to simply 'switch on' participation following a history of its absence, and schemes foundering on the slowness of growth of indigenous trees, these should be seen alongside these other achievements which had much wider influence on thinking in Zimbabwe. By the late 1980s and early 1990s, this network of people had brought the FC ahead of the game in many respects, with FC personnel leading the thinking on extension for woodland resource management.

Pragmatism was crucial in this process – finding ways forward through the constraints bequeathed by the past. For example, recognising that a number of laws were a potential hindrance to the emergence and implementation of new approaches to forest extension and local woodland management, FC managers effectively chose not to enforce the provisions of outdated legislation, whilst trying to develop better models and proposals for reform. These managers were also adept at donor liaison and insisted on local counterparts working with any external consultants assigned by donors. However, divisions among senior managers were also apparent, particularly regarding the privatisation thrust and whether non-foresters should be allowed to hold top positions in the FC. Meanwhile, the FC's parent ministry, under new management, began pushing for a stronger role in the affairs of the FC, with a particular interest in capturing tourism revenue and 'indigenising' the dividends from privatisation.

Musical chairs played by the experts

Today, other government agencies, donors, community-based groups, and the NGO sector are asserting themselves more. Political discourse has changed since the 1980s and the emphasis on public accountability, public participation and transparency behoves state forestry interests to link up with these organised groupings within civil society. There are reasons to believe that the basis for this is quite strong. For example, whilst flux amongst the senior positions in state agencies and NGOs caused some incoherence in pursuing initiatives at the time, over the longer term this 'musical chairs' game – played by a group of people moving between various NGO-academic-government positions – has created a pool of widely-experienced people capable of forming a broad-based policy community.

Movement of information between state and civil groups has allowed better understanding of the range of practices employed by people in communal areas in planting trees, and in managing naturally-occurring trees and woodland. Knowledge is also improving about the adaptations people make to changes in supply of woodland resources. FC extension workers are beginning to recognise and engage with these practices. The pilot 'resource-sharing' programme faces many challenges, but one important emerging effect is the percolation of ideas within, and between, government agencies about the positive aspects of collaboration with other stakeholders.

In conclusion, the above examples of progressive change have been catalysed by individuals who can be characterised by their ability to: find a way to make their personal interests overlap with the demands of their position; foster others' enthusiasm and trust through regular contact; see the big picture; take on tactical battles; engineer a bit of space to 'work around the rules'; use a mix of 'insider' and 'outsider' traits in their institutions (including linkage to international networks); make alliances; and use these alliances to move from tactical to more strategic issues. These people might be described as the 'enlightened policy élite' – in that they are already part of the policy game at central levels. There are other levels, however, at which policy 'change agents' operate. These include those who have the ideas and experience to work on new ways forward in areas challenging forestry, who are currently 'out of the policy loop'. Some of these people are working at local level with forests and trees outside forest reserves, to which we turn in Section 4.5

4.5 Looking beyond the forest reserves

In some countries, forest policy is about managing large blocks of land under centralised control – by government or corporations – for timber, biodiversity or other goods and services. But in many other countries, it is not so simple an affair. There may not be large blocks of forest from which forest goods and services can be obtained; scattered trees on farms may be the most important source of timber or fuelwood. In other areas, forests and woodlands may not be reserved for strictly forest-related uses, because they are central for sustainable livelihoods in the area; this may be farmed in settled or rotational ways for food, as well as providing forest goods and services.

Figure 4.1 Land-use spectrum

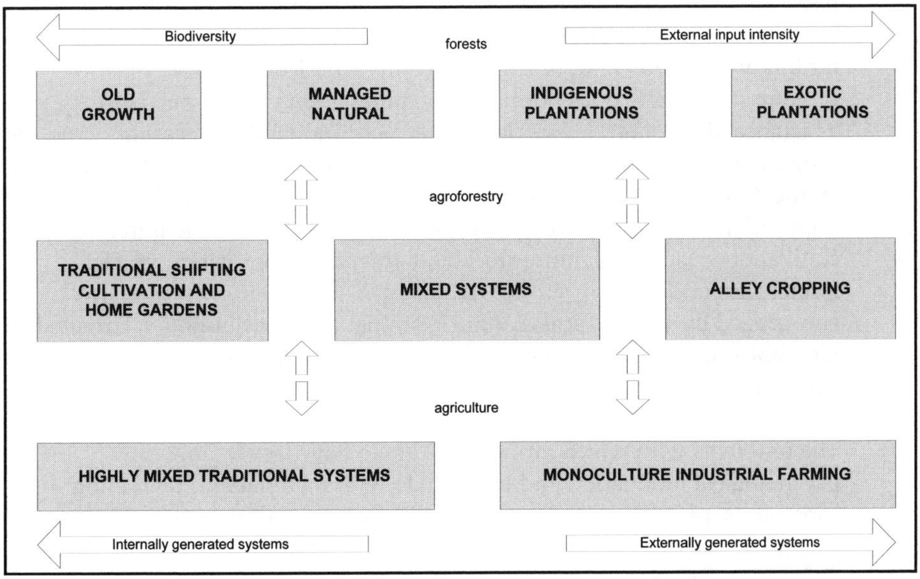

In other words, if forest goods and services can best be obtained from non-forest land (Figure 4.1), forest policy should reflect this. This entails the ability to assess the relative efficiency and sustainability of producing forest goods and services from different types of land, and coordination with agricultural and land-use policy.[20] And if forested areas are important for 'non-forest' benefits such as food production, forest policy should reflect this, too. One of the big policy challenges is thus to find the appropriate policy environment to enable forestry to take its place in broader land-use. In some countries, forest institutions are only just waking up to this.

Forestry institutions missing the real action – on farms, Pakistan

Pakistan's natural forest asset is very small; the legal categories cover 45,700 km^2 or about five per cent of the total land area, but the actual forest cover is less. Government forestry has traditionally been focused on these natural forests and, despite being better suited to producing the non-wood goods and services that are scarce and often cannot be substituted by imports, these forests are consequently under great stress from timber production. High local wood prices, in comparison with international prices, ensure that there are still incentives to cut wood from natural forests in Pakistan. Import tariffs keep the prices high.

Farmers are already motivated – they need support

Meanwhile, farm forestry has received rather little official attention. Although the government gives away seedlings to farmers on an annual basis, and there have been isolated projects in farm forestry, these attempts at promoting reforestation have enjoyed only limited success. The reasons for this are illustrative of broader problems with forest policy. The programme usually does not target the population most in need of its assistance: the poorer, less accessible farmers and landless peasants. And the programme's emphasis, with the limited farm population that it reaches, is on the delivery of tree seedlings and a motivational message. The implication is that the principal obstacle to farmer involvement in tree cultivation is lack of planting stock and absence of motivation. In fact, farmers are undertaking substantial on-farm afforestation on their own, encouraged by wood prices. So much so, that the contribution of irrigated farmlands to national timber and fuelwood production is estimated at 80 per cent.

The problems with which farmers need help have less to do with psychological motivation and more to do with material and marketing constraints, problems that, seedlings aside, are not addressed in the annual

---

20 Analysis of the relative efficiency of importing forest goods is also required.

Pakistan	
Forest	• Natural forest covers about 15,800 km^2 and total forest area is 17,480 km^2, or 2.3 per cent of national land area (770,880 km^2) (FAO, 1999) • Plantations cover about 3,800 km^2
People	• Population 144 million (35 per cent urban, 65 per cent rural) with an average density of 186.5 people per km^2 and a growth rate of 2.7 per cent per year
Tenure	• About 66 per cent of forests are under state control while 34 per cent are privately owned
Forest economy	• Annual production from forests under state control is about 0.5 million m^3 of timber and about 0.2 million m^3 of firewood. About 2.6 million m^3 is produced annually from private farmlands • About 80 per cent of commercial wood harvest comes from irrigated plantations • Consumption of firewood, at 51 million m^3 annually, exceeds that of industrial wood by over ten times, while projected sustainable supplies from forests are only 9 million m^3 • Imports of wood products were worth about 15 million rupees in 1994-95. Farm forestry is becoming more economically attractive with rising demand for wood and pulp
Pressure on forests and people	• Industrial growth at 6 per cent and population growth demand more construction wood, fuelwood, and water from forested watersheds • 'Timber mafia' are stripping timber from the few remaining natural forests, with some political collusion
Key policy issues	• Forestry institutions missing the real action – on farms (this section) • Lessons from participatory projects opening up the policy agenda (Section 4.6)

planting campaigns. Actions which would support the necessary change in forest policy towards farm and agro-forestry, and so meet the increasing needs of industry and subsistence use, include: census of tree stocks and growth on farm lands, as part of the agricultural census, and integration of this information into forest information systems serving policy review; integrated research into farm forestry, to optimise commercial tree and food yields; preparation of outreach materials that are specific to different agro-ecological zones; and, encouragement of farm forestry associations.

Meanwhile, high prices continue to serve as incentives for forest contractors and private forest owners to circumvent the controls of increasingly marginalised forest departments and to over-cut the remaining forests.

Trade liberalisation could breach the 'timber mafia'

The removal of the overt and covert barriers to imports would enable a reanalysis of the extent to which Pakistan's natural forests actually have a comparative advantage in wood production, *vis-à-vis* farm plantations and imports. This in turn should lead to a reassessment of just what types of forestry activities should be promoted. Liberalisation may also assist in breaching the 'timber mafia' domination of harvesting in remaining natural forests, by lowering their profit margin. Attempts by the 'mafia' to pass such cuts on to royalty holders would only serve to raise questions about the existing system, questions that they would prefer to leave unasked.

Installing farm forestry as a central focus in policy will require a general reorientation of policy away from forest area alone and towards securing the forest goods and services that people need. Given current forest institutional approaches and staff deployment patterns, this is a big challenge in Pakistan, and a precondition will be a system of forest fora at national, provincial and lower levels to work through the full implications. With good monitoring, review and information sharing, much can continue to be achieved through taking the above steps on a pilot basis, as suggested by the impacts already achieved by integrated development projects (see Section 4.6).

Managing the farm-forest landscape – the off-reserve challenge, Ghana

Ghana's Forestry Department (FD) cannot, on its own, guarantee sustainable forest management in the off-reserve areas, because other stakeholders, notably landowners, have property rights. Stimulated by the new Forestry and Wildlife Policy of 1994, the FD began seriously rethinking its off-reserve role in 1995. The FD's Collaborative Forest Management Unit (CFMU) posed the question, 'what kind of framework can sustainably develop a resource that cannot be permanently reserved for forestry, is under phenomenal pressure from the industry, and cannot be administered realistically by anyone other than the landholders and cultivators?' The CFMU went on to propose that a collaborative framework for sustaining the off-reserve resource needs to ensure that:

- a fair share of the benefits from utilisation of the resource for those who have tended the trees;
- wide consultation during planning for exploitation and management of the resource;
- people who can genuinely contribute to the achievement of operational objectives can do so, building on their comparative advantage;

- information on field activities is linked back to the forest policy; and
- the integrity of the resource is maintained, where that is the intention of the landholder.

Rethinking the forest concession

It was recognised that it would be very difficult to return to the pre-colonial position, whereby control over land and trees was held by the holders of cultivation rights (rather than by the chiefs, whose powers were built up during the colonial period). It was therefore concluded that rights to *planted* trees only should be 'divested' to the cultivator, and that the FD should continue to hold the naturally occurring trees 'in trust' – on behalf of the President – for the chiefs. However, it was also anticipated that in the long run, virtually all trees off-reserve will be planted or nursed, and full ownership should be with the cultivator.

Following a working group process, legislation passed in 1998 provides for the replacement of concessions with timber utilisation contracts (TUCs) – an instrument for area-based rights allocation. TUCs introduce the right for either the landowner or farmer to veto the harvesting of trees from their land. Social responsibility agreements, planned as part of the TUC, aim at ensuring more accountable relationships between timber contractors and land-owning communities, by partially formalising hitherto verbal agreements on community benefits. The main elements of the social responsibility agreement are a code of conduct for the TUC, agreed social responsibilities for the forest resource, and agreed contributions to infrastructural development in the area concerned.

It remains to be seen whether social responsibility agreements can overcome a major disincentive to protection of trees on farms: the fact that chiefs receive a share of revenue as owners, but farmers – who look after the trees – do not. Farmers argue that they provide a service and, like the FD, should get a share. The new agreements represent a possible mechanism by which the farmer's right to benefit can be legitimised.

Powers to landholders

A new consolidated Forest Act has also been produced, following the work of another FD working group and several consultancies. The proposed new Act provides comprehensive regulation of the forest sector as well as substantive aspects of institutional reform. Some innovations in the Act include:

- Abandonment of the ambiguous concept of 'vestiture' of forest lands and trees in the President, and replacement with a more accurate and

transparent concept: forests belong to the customary landholders and the management rights of the state are clearly spelled out.
- Introduction of the concept of 'dedicated' forest (see Section 4.6)
- Establishment of an administrative dispute settlement mechanism. Forest offence settlement committees are recommended in each region to deal with reported forest offences.

New roles and collaborative strategies

The proposed Act is understood to give a clear endorsement of forest management systems, based on collaboration and a recognition of landowners as the primary clients of a government forest service. The role of the proposed Forest Service is described as having two main elements: protecting, managing and developing in a sustainable manner, the forest reserves of Ghana in the national interest and for the benefit of the owners; and providing a management service on agreed terms to private individuals, traditional authorities, communities, companies, unit committees, District Assemblies and others outside the forest reserves.

Making tree management make sense

A major assumption in the FD's off-reserve planning is that people will start to plant valuable trees like teak in the very near future, which will produce a new harvest of planted timber off-reserve after the present stock of naturally occurring timber is used up. The Ministry of Lands and Forestry is preparing a private sector plantation programme to encourage farmers and companies to plant trees. In the current timber-poor areas, notably the dry forest zones, the government may need to institute measures such as tax concessions for forest development and regenerative activities with local communities, which take account of the additional costs involved in rehabilitating this zone.

Currently, trees on farms are actively nurtured, managed (coppicing, pollarding, canopy manipulation) or removed by farmers primarily for their effects on the farming system. Thus, as in other countries, improved tenure alone is not enough to promote and support tree management on farms. Progress in addressing the considerable technical problems of integrating timber and forest trees with agriculture has only just begun. However, opportunities exist in some agricultural sectors to develop more forest-friendly approaches. For example, in the key cocoa sector, increasing problems with pests have tended to encourage approaches to cocoa production which are more concerned with habitat management. Much will depend on finding ways forward based on the concrete experiences of farmers. Experience with collaboration by the FD thus far suggests that research and development agencies need to focus on experimentation and promotion of adaptive learning at both local and policy levels.

Getting smallholder forestry on the agenda, Costa Rica

If farmers are to commit themselves to forest stewardship or tree growing on the land which they work, they need confidence in their predictions about the returns they will get, often after quite long periods. Other uses of the land may bring larger or quicker returns, whilst the availability of labour and accessibility of markets for inputs and products will also enter the farmer's equations. Technical know-how and support may also be crucial factors. In Costa Rica, the potential for the smallholder to make these decisions in favour of forestry has considerably improved with the emergence of new forms of organisation stimulated by policy change.

Common interests and information needs

Smallholder farmers have, for many years, been members of union groups, formed at a community level to ensure benefits and fair relations among producers, and between producers and the government. These organisations are federated at regional and national levels. In 1989, certain smallholder organisations carrying out forestry projects, and others interested in doing so, met to establish their positions with regard to state forest policies. As a result, an informal national organisation – the National Smallholder Forestry Assembly (JUNAFORCA) – emerged to represent the smallholder sector at key decision-level meetings.

In 1991, JUNAFORCA was formally established, with the added role of coordinating medium- and small-scale farmer organisations involved in forestry activities. Some 56 organisations – cooperatives, agricultural centres and community development associations – have joined up, representing about 27,000 producers from every region. They are engaged in activities that range from forest plantations, to nurseries, agro-forestry systems, living fences, natural forest management systems, cottage industries and handicrafts.

One of JUNAFORCA's main achievements is its sharing of information and experience. Its 'bottom-to-top' integration has successfully linked those involved with viable local initiatives to regional organisations which in turn are supported by the national secretariat in installing their cases in national-level policy agendas. By doing this, JUNAFORCA has generated significant 'policy space' for smallholder forestry.

Reorienting policy and incentives with good organisation

JUNAFORCA became active in policy circles through clearly formulating proposals and commentary, which gained recognition by government organisations and opened the way to negotiation. Key factors in the success of this approach appear to have been the 'professional' and articulate lobbying by individuals who were able to open key doors and make government officers and others receptive to their argument. Through these means, JUNAFORCA has secured support for establishing several regional organisations. It has been able to precipitate modifications to forest legislation, gaining group access to government reforestation incentives. JUNAFORCA has worked on various national commissions, and now plays a part in management of the new National Forestry Office and the National Forestry Finance Fund. It played an active role in *Policy That Works*, and is now facilitating discussion on trade and forest certification.

When the government first introduced its incentive programme in the early 1980s, it was geared to creating a plantation resource. The programme was monopolised by larger landholders. However, a few smallholder organisations realised the potential for their members if government could be persuaded to reorient the programme to smaller parcels of land and appropriate financial schedules for such parcels. It took a considerable period of developing local, regional and national alliances, culminating in the establishment of JUNAFORCA and some other bodies, to secure these changes. Smallholders have since gained substantial benefits from the programmes in the 1990s. The programme has continued to evolve, and is today geared towards payments for environmental services, mostly on patches or blocks of primary forest or regenerating forest land.

But does smallholder forestry pay?

The technical aspects of smallholder forestry have also received considerable technical assistance and public research support since the 1980s. Together with the emergent strength of organisational support, smallholder forestry is on the verge of economic viability without subsidy from government or donor subsidy. However, vertical integration – smallholders investing in timber processing or sawmill owners investing in forest lands – is yet to occur to any significant degree. High real interest rates mean a high opportunity cost for both industrialists and smallholders, whilst stumpage values have not risen much in real terms, and are unlikely to increase in the future. In order to profitably operate their own sawmills the number of smallholders has to be large (and they need good business skills) whilst dividends remain low. In this context, smallholder timber production is likely to continue to find it tough to compete with larger-scale production and imported forest products. Until these economic factors change, continued support for smallholder forestry, and for timber processing capacity appropriate for smallholders, will need to be justified in social and environmental terms.

Deeper questions for political structures

Organisations thinking through the links between environmental protection and development are now relatively strong in Costa Rica, but locally-rooted political movements that integrate environment, forests, land and social concerns are weakly developed. However, several regional political parties have emerged, trying to develop a process of sound municipal natural resource management based on more equitable distribution of benefits for small producers. Some observers have noted the possibility that the impacts of globalisation, government economic difficulties, and privatisation may exacerbate local perceptions of insecurity, leading to more active and extensive questioning of existing political structures, and a more dynamic political process at community, municipality and regional levels.

Private forests and woodlands in Australia – farmers taking responsibility whilst governments avoid it?[21]

Traditionally, Australian governments were loathe to interfere with how farmers managed their land, while the state forest services kept themselves to their own forests and were not concerned with the vast areas of woodlands which could not produce wood commercially. Two factors have forced governments to broaden their concerns. One arose from land degradation and salinisation, which became widespread due to excessive deforestation and poor agricultural practices. The other arose from the historic dependence of Australia's wood supply on a handful of state forest services and large private companies.

Landcare – integrating trees and agriculture

Landcare started in the 1980s as a rural community response to increasing awareness of environmental degradation, and today about 30 per cent of Australian farmers are involved in roughly 3,000 groups scattered around the country. The groups, typically of 20-30 members, determine their own priorities, boundaries, and procedures. Activities range from changing tillage practices, to feral animal and weed control, planting trees, and anti-erosion earthworks.

Landcare is commonly seen as a social movement which reaffirms the importance of land stewardship for landholders. It encourages them to

---

21 This section draws from a paper prepared for IIED's *Policy That Works* project by John Dargavel, Irene Guijt and Peter Kanowski (1998).

resolve local or district level land degradation problems by group efforts. Its significance for forestry lies in the positive message it has spread about trees on farms, an essential step in integrating forest and agricultural policies.

Government chips in...

The Commonwealth government announced the 1990s as the Decade of Landcare, and allocated $340 million in support of two main programmes, *Save the Bush* and *One Billion Trees*. *Save the Bush* provides grants for community groups to protect remnant native vegetation. The symbolic 'billion' in the *One Billion Trees* programme raises the importance of reinstating native vegetation in land rehabilitation. Landcare thus became the first national natural resource policy to focus on community-level farmer groups, raising awareness, developing the planning skills of private landholders, and facilitating community-level solutions. More recently, the Landcare programme was upgraded and labelled as 'Bushcare', or the National Vegetation Initiative. The Commonwealth allocated $400 million over 4 years to encourage community revegetation plans, on a much larger scale than was previously the case. One of its goals is to exceed the rate of vegetation clearance by the year 2002.

...but should government do more?

However, criticism of Landcare is growing. Despite the generally acknowledged sea-change in rural people's attitudes towards natural resources, land degradation in Australia continues almost unabated. Views are growing that government has been able to use its encouragement and relatively minor funding of Landcare as a smokescreen for withdrawing other resources and services – effectively transferring responsibility for both land degradation and its solution to a community level, without allocating commensurate resources or decision-making authority (Campbell and Woodhill, 1997). Critics point to the many government policies, such as tax rebates for clearing land (in place well into the 1980s) or lease covenants stipulating livestock densities, that left farmers little choice but to clear all trees. Nor, they say, should present farmers be expected to pay all the costs of a process which has taken two centuries (Campbell, 1994; Martin and Woodhill, 1995).

Whilst Landcare gains ground, Australian policy is too weak to address the scale and severity of the continuing problems of soil and woodland degradation. At its core lies the unwillingness of governments to legislate and the considerable difficulties of regulating the use of farm and grazing land on a continental scale.

Wood industry also looking to the farmers

Despite Governments' traditional disinterest in private lands, in the late 1960s and 1970s some of the pulp and paper companies started to encourage farmers to plant trees and supply them with wood feedstocks. However, during the 1980s, the national policy of increasing the pine plantation resource, combined with environmental objections to clearing native forests, led State Governments to start programmes aimed specifically at increasing the commercial farm forest resource.

The 1992 *National Forest Policy Statement* set out measures to encourage plantation establishment and, in 1993, the Commonwealth Government launched its Farm Forestry Programme. This focuses mainly on regional projects for landholders growing trees with industry and government. Most State Governments have also developed farm forestry programmes which link the resources of landholders (e.g. labour, land, equipment) with government wood production in joint venture arrangements. In 1996, a new Commonwealth Government announced a national goal of trebling the nation's plantation estate by the year 2020. It was envisaged that this would be achieved largely through giving a greater impetus to farm forestry.

Farm forestry – rising to the risk-sharing challenge

Farm forestry has been enthusiastically promoted. It has the potential to reduce environmental costs by arresting land degradation in some areas, as well as to diversify farm incomes, and contribute to the development of the wood industries with their associated employment. In theory, it has the potential to be an important timber resource which could both reduce the pressure on Australian native forests and reduce imports.

As farm forestry is expensive, unproven, complicated and contrary to accepted farming ways, and is to be accomplished through voluntary adoption of new technologies, success might be slow. Where it has not been adopted, this has been associated with farmers' concerns about entering into long-term joint ventures, uncertain regional and farm benefits, and the complex nature of family-based farming. Addressing these underlying reasons of non-adoption is crucial for the success of farm forestry policies (Vanclay, 1992).

Landholders tend to be asset-rich yet cash-poor, requiring cost-sharing arrangements to enable them to adopt commercial farm forestry. Those arrangements which provide annuity payments to the landholder are likely to stimulate considerable farm forestry adoption. Joint ventures can spread the burden of establishment costs and commercial risk, while maintaining part-ownership for both parties. Some farmers who might otherwise have to leave the land might be enabled to stay. But many small-scale forest growers feel they are unable to negotiate with industry, and thus doubt that present markets will deliver fair returns for their investment. Improvements to the structural nature of forest product markets may require considerable investment by industry (e.g. offering 'lease' or 'index-linked' joint ventures); governments (e.g. improving access to more competitive markets through infrastructure developments to increase export opportunities or to support new industries); and growers (e.g. financing cooperatives).

Reactivating the commons, Portugal[22]

Despite the increasing shift in focus towards local capabilities for forest management, the re-establishment of forest commons – regulated out of existence in many countries – is a relatively rare event in modern history. The re-emergence of Portugal's *baldios*, or communal lands, is thus an informative case. Originating in the fifteenth and sixteenth centuries, the *baldios* covered more than 40,000 km^2 in 1875, but were all but wiped out under the dictatorship that prevailed in the middle part of the twentieth century. They have only made their reappearance in recent times in the northern part of Portugal.

Traditionally, the *baldios* have been important in traditional farming systems, by providing construction material, fuel, and bedding for stabled animals. They provided a 'social insurance' for landless poor who were permitted to pasture cattle and cultivate plots on a temporary basis. They were run by village councils which, on the basis of zoning and exclusion rules, determined the areas to graze, collect brush and reserve for regrowth.

Despite the importance of the *baldios* to local people, between 1928 and 1974 the government gradually took control of these lands through usurpation and government-backed privatisation. By the late 1960s the *baldios* had been reduced to some 4,500 km^2. Afforestation programmes were pursued by the forestry service on the remaining *baldios* areas (although, in legal terms, even these areas came under the private ownership of parishes or municipalities). Sixty per cent of the remaining *baldios* were planted up with pine under these programmes.

In the mid-1970s the formal restoration of the *baldios* system was the result of a battle between three main parties:

- The parishioners of the Vouga in the north of Portugal who decried mis-management by the forestry service and called for restoration of the *baldios* under the administration of the parish.

---

22 This section draws from a paper prepared for IIED's *Policy That Works* project by Olivier Dubois (1997).

	Portugal
Forest	• Forests and woodlands cover about 32,000 km², or 35 per cent of the nation's land (92,391 km²), of which 85 per cent is reckoned to be exploitable
People	• Population – 9.9 million (0.2 per cent growth rate), of which about 37 per cent is urban. Average population density is 109 per km² • Total forest and woodland per capita 0.3 hectares
Tenure	• 85 per cent of forest and woodland is under private ownership – of this 84 per cent is owned by farmers; 6 per cent is owned by industries • 12 per cent of forest and woodland is community land (no individual right to alienate) • 3 per cent of forest and woodland is state land
Forest economy	• Forest sector 3 to 4 per cent of gross national product • Total removals were estimated 11 million m³ in 1990, the share of industrial roundwood (majority for the pulp and paper industry) having risen from 30 to 93 per cent since 1950, with a concomitant fall in share of firewood from 70 to less than 7 per cent • Direct employment in the forest sector in 1987 was estimated at 10,000 permanent and 18,000 seasonal workers
Pressure on forests and people	• European Community and the World Bank have subsidised afforestation with eucalyptus for the last two decades creating some 300,000 hectares. Recent concerns about water table effects and tree diseases in these plantations led to a new law allowing no more than 25 per cent of an area to be planted with eucalyptus • Forest fires affect 50,000 hectares per year; 94 per cent of fires are caused by human agency e.g. to increase pasture land, facilitate access to hunting areas, and as an instrument for timber speculation
Key policy issues	• Pros and cons of communal forest land as a modern management system benefiting forests and people

- The communist party which wanted to use the *baldios* as a legal instrument for social and political transformation through collectives.
- The forestry service, which had enjoyed control of the *baldios*, favoured their transfer to 'the people', not the parishes.

The communist party's efforts foundered somewhat on the Northern peasantry's traditional opposition to revolutionary movements. This also showed that aspiration for common property regimes does not necessarily rhyme with a desire for collectivisation. The resulting Decree, published in 1976, was essentially a compromise – it devolved the *baldios*, from municipalities and the forestry service, to community-based users' groups. The law required that user groups, or commoners' assemblies, appoint a management council to be registered with the state, and suggested that staff of the forest service should be included on this council.

Some 637 management councils were created by the 1976 law. The councils were allowed to receive from 60 to 100 per cent of the proceeds of timber sales and licences for grazing. Indeed, by this time the contribution of the *baldios* to the local economy had shifted, from the provision of land and material for local farmers to the supply of timber to industries located elsewhere in the country.

However, by the mid-1980s, only about 137 management councils were still operational, largely due to ongoing uncertainty about their status. The forestry service also appears to have made a tactical mistake by not getting involved in the management of the *baldios* after the law of 1976. By failing to gear itself to assisting the commoners it missed the opportunity to gain support from the local population. Several bills were presented to parliament in the 1980s aiming to transfer the management of the *baldios* back to local municipalities. In 1993, another compromise law provided for NGOs to promote the creation of management councils, and for delegation of the administration of the *baldios* to other parties, including the forestry service, private sector operators (e.g. pulp and paper companies), or indeed, municipal councils. It also allowed for the termination of the *baldios* if unanimously agreed by the commoners themselves.

Whilst the 1993 law poses some threat to the *baldios'* existence in Portugal, and there is a high likelihood that the *baldios* land will be increasingly allocated for housing, there is strong evidence that local people still want the *baldios* to exist. Since 1974, their area has increased a little to some 5,100 km². Fuelwood has become more accessible in some *baldios* and is distributed by the management councils in ways similar to the traditional systems. Tree fellings are often used as sources of funds for the construction of local roads and other communal infrastructures, or the pine timber itself is used for the construction of footbridges, buildings and furniture. Pine resin and honey bring significant returns in some areas.

Although the forestry service is threatened by the 1993 law which allows the commoners to delegate the administration of afforested *baldios* to other third parties, demand for its assistance is growing. The challenge for the forestry service is to consider the commoners more as clients than landlords and provide technical assistance upon request. If current research succeeds in developing a variety of eucalyptus that can withstand the climatic conditions of the *baldios*, this may become of prime interest for the pulp and paper sector, which in turn may unleash a whole range of new prospects (and potential problems) for the commoners.

However, the survival of the *baldios* is likely to depend mostly on the successful development of other uses, such as mining, exploitation of mineral water, high-quality 'niche' agriculture (e.g. promotion of local, high quality meat) and ecotourism (including hunting). These strategies are beginning to gain political support from government, environmental groups, farmers' associations and the EU.

The *baldios* would appear to have an interesting, although uncertain, future. Whilst not providing an easily transferrable model, the example of Portugal's *baldios* may yet show that common property – or shared private property – can be an effective and adaptable modern system for optimising the use of land and for producing the forest goods and services that people actually want.

In conclusion, one of the biggest challenges for policy in many countries lies in the fact that the future of many forest goods and services is no longer to be found in large blocks of natural forest but in other parts of the land-use spectrum, notably in farm-forest landscapes. Traditionally, government forestry institutions have not been adept at encouraging production of forest benefits from farms and mixed farm-forest landscapes, but are rising to the challenge increasingly. This often means granting strengthened rights to farmers and brokering agreements about social responsibilities between farmers and the private sector. Indeed, partnerships of this kind demand a range of new institutional arrangements, production technologies and information systems. In encouraging the development of these, policy needs to adapt. Broader political challenges are likely to arise in this adaptation – since farm forest landscapes ultimately need to be sustained within rural economies which have united social and environmental objectives.

4.6 Improving learning about policy

In several of the above examples, we noted how a rooting of policy concerns in local realities has brought the right stakeholders together; and it has helped to make policy both more relevant and more implementable. The challenge is to make long-standing learning links between local actors and forest authorities, which continually experiment with policy-related change, monitor the impacts on the ground, and 'scale up' or adapt as necessary.

Linking learning processes to policy and capacity development, Ghana

The Ghanaian Forestry Department's Collaborative Forest Management Unit (see Section 4.5) has been successful in stimulating policy-level learning about local forestry potential. This started with the CFMU's review of NTFP resources and their use which involved surveys of people's attitudes in communities near forest reserves, under different categories of protection. It showed considerable local support for continued protection of the forest reserves, particularly for protection of drinking water supplies, rehabilitation of degraded forests and fire protection belts. Protection for biodiversity was reported to be a source of pride in some communities. The CFMU recommended the gradual abolition of the NTFP permit system and the development of specific NTFP management programmes and agreements with users.

The CFMU worked with various communities and District Forest Offices to recognise and find collaborative ways to manage 'dedicated forests'. These are mostly forest patches, set aside as ancestral groves and maintained by community consensus, taboos and low levels of exploitation, but now threatened by cash-crop agriculture and logging. Dedicated forests can now be legally recognised under provisions in the new timber resources management legislation of 1998. They are lands that have been committed to forest use for a period by the holder of land rights.

A key strength of the CFMU and its parent planning branch of the Forestry Department has been linking learning processes, such as the development of the dedicated forests concept, directly to capacity development and policy processes. The CFMU has been adept at linking pilot local collaborative initiatives to improved understanding at policy levels – effectively making stronger two-way links between government policy and local practice.

Participatory projects and strategies breaching entrenched policy, Pakistan

Experimenting with policy change through establishing national-local links is especially necessary if policy has hitherto remained 'stuck in the past', as was the case in Pakistan. Formal forest policies in Pakistan are principally a manifestation of the values and training of government foresters, who belong to a singular tradition which is largely based on pre-Independence objectives and procedures. They remain strongly focused on government reserve forests, and take little account of the fact that much of Pakistan's timber is produced on farms. Policy is thus, in many ways, considered the concern of the forest authorities – but its application in practice is distorted by political interference in gaining access to timber resources for a favoured few. Forest policy is also constrained by a peculiarly complex system of land and forest tenure, the result variously of colonial negotiations with different local groups.

However, a number of large field-based forestry projects, involving local populations to varying degrees, have tended to focus government and donor attention on forest policy's anomalies. The objectives and emphases of these projects have taken many twists and turns over the years, but in the process have successfully identified and dealt with many concerns and are now considered critical beacons for future policies. They have provided excellent case studies, and test cases for government staff to develop new approaches.

Tackling a local burning issue first

The Pakistan PTW team examined eight participatory projects which aimed principally at ensuring a security of wood supplies to meet local livelihood needs. These projects were all in areas which are out of reach of effective government intervention (or interference). As such, they have had to deal with local realities, but have not been able to count on government. This has often resulted in new approaches – exposing the lack of relevance of formal policy and procedure. The Kalam Integrated Development Project, for example, which started as a coniferous forest conservation project in the early 1980s, soon realised that it was not possible to achieve forest conservation without meeting the socio-economic and development needs of the local communities. Project interventions in social and economic infrastructure development helped to bridge the livelihood objectives of the communities with the forest conservation objectives of the project. Over time it emerged that one of the concerns of several communities was to get a better deal out of the forests over which they had rights.

The previous system of the sale of standing forest trees to contractors had been abolished by government in 1973 and replaced by harvesting through logging contractors and sale by state-owned corporations. The inability or reluctance of the FD to cut through overly-bureaucratic procedures meant that contractors became the favoured middlemen for dealings with communities, despite the fact that the communities were not obtaining their entitled income. In response, the KIDP project experimented with training teams of 3-4 local people in the use of key equipment such as skyline log carriage systems, and awarding them small logging contracts. This provided local employment and improved logging practices and their impacts. The approach has begun to be taken up in other areas where there is similar motivation.

Backing up rights with capabilities

Various projects have proven the thinking that private rights – if unregulated or unmodified by public sector laws and incentives – are unlikely to produce effective social outcomes, especially where government enforcement is weak. Similarly, community rights and regimes, although they may provide an element of law and order, have proven ineffective without equitable and well-resourced legal community organisations. The Aga Khan Rural Support Programme (AKRSP) has led the way in developing institutional arrangements that are key to the successful management of use rights – such as community management rules and regimes, partnerships, and the application of traditional knowledge.

Organisations that can deliver collective action

Collective action is often the prerequisite for managing public goods, especially in forests subject to much local demand. Although many question the nature and outcomes of decision-making in local organisations (both traditional and project-instigated), in the absence of local government, several key projects have shown ways forward in devolving decision-making to village-based entities. The AKRSP model of supporting Village Organisations for natural resource management has proved the efficacy of these institutions for collective action and local policy development at the grassroots level in Northern Areas. This has led to policy discussions about the village organisation model – if supported with adequate legal identities – as the basic block of organisation for good forest management, and more ambitiously as a possible form of village governments.

Pushing the institutional rethink

These participatory forest projects have all had some kind of forest authority involvement – even if it is only in terms of occasional review visits. Where they have involved forest officers more extensively – on the ground, or in joint training – they appear to be having a big impact on forest policy thinking in the authorities.

All of these initiatives, both area-based projects and wider consultative processes, have been quite specific in their results. The challenge remains the spread and replication of the institutional and resource management innovations generated by these initiatives – matters of institutional 'unfreezing' become critical (Bass *et al*, 1998). Whilst donors have been very influential through supporting participatory projects and the conservation strategies described in Section 4.2, which show unprecedented levels of participation in Pakistan, it is quite clear that the state has selectively ignored civil society institutions – especially for resource mobilisation and governance. This leads to alienation, which feeds on other sources of discord and fragmentation and breeds a relationship between the state and society which is largely acrimonious. Some form of wider constituency building and experimentation is sorely needed, and the better forms of consultation processes and local organisation-building projects will continue to need all the help they can get.

4.7 Dealing with tensions in devolution

Claims made about the benefits of decentralisation programmes generally include the greater sense of participation and ownership by civil society in public matters. The increased cost-effectiveness and responsiveness of government institutions is also often claimed, in terms of speed, quantity and quality of response, leading to increased satisfaction of local people and more congruence between government activities and locally-felt needs. Enhanced information flow between government institutions and citizens and faster learning of lessons from local realities are emphasised. However, decentralisation can take very different forms (see Box 4.4).

Furthermore, as the Pakistan PTW study argues, "decentralisation without participation is completely different from – and probably much less consequential for its impact on forests and people than – decentralisation with participation." Policies have often not worked because of institutional and managerial weaknesses, much of which are connected to failed attempts to decentralise.

Knowledge about local management practices, and community rules and institutions, have found their way in recent years into mainstream natural resources research. This knowledge is increasingly being made available in policy-making arenas affecting forestry. However this trend is somewhat confounded by the romantic rhetoric of many development practitioners about the powers and effectiveness of community-level management.

Community management now faces rather different conditions from the days when national and global forces were inconsequential. Decentralisation to communities cannot help much when community institutions have already broken down, or have been replaced by government structures that don't work. Institutions at local level are, after all, often vying for authority. Nevertheless, at least some of the decentralisation efforts of recent years have been spawned by the desire to foster local natural resource management. Here we examine experience from a variety of locations in Africa.

Box 4.4 Defining decentralisation – take your pick...

Decentralisation can refer to any of five different types of power transfer – deconcentration, delegation, deregulation, devolution or privatisation.

- *Deconcentration*: spreading authority from the central administration to its agencies closer to the 'grass roots'. A non-definitive transfer of decision-making and executive powers *within* the administrative or technical structure (e.g. from the Ministry of Interior to a governorship or from the national directorate of a service to the regional directorate). This takes the form of institutional modification from within an administration.

- *Delegation*: a non-definitive transfer of authority from an administrative service to a semi-public or private company.

- *Privatisation*: a type of delegation involving transfer of ownership and/ or management of (forest) resources, and/ or the transfer of the provision of (forest) services, from the public sector to private entities, either directly or through parastatal institutions (*corporatisation*).

- *Deregulation*: a transition in which a sector of activity previously regulated by a public authority ceases to be subject to such regulation.

- *Devolution*: a transfer of power from a larger to a smaller jurisdiction; this transfer may be total or partial (e.g. transfer to local communities of decision-making over renewable resources on their village lands).

Particular definitions are sometimes associated with particular voices. For example, 'decentralisation' in the language of government officials often really means *deconcentration*, whilst for local communities and NGOs it may mean *devolution*.

Sources: Thomson and Coulibaly, 1994; Banuri, 1996; Dubois, 1997; Bass and Hearne, 1997.

Power to the people, or just passing the buck? Examples from sub-Saharan Africa[23]

Failures in decentralisation programmes are legion

Many local government units are "neither local nor government" (Olowu, 1990). Common problems include:

Transferring problems and inflating bureaucracies. Often, central-level problems are merely displaced to local level, e.g. rent-seeking, lack of resources to pursue policies, confusion between public and private interests, and inadequate capacity. Moreover, decentralisation may bring a transfer of excessive bureaucracy from central to sub-national levels.

---

23 Much of this sub-section is drawn from a paper prepared for IIED's *Policy That Works* project by Olivier Dubois (1997). Key works on which his conclusions draw include: Olowu, 1990; Parren, 1994; Bonnet, 1995, Leroy, 1995; Ribot, 1995a; Gentil and Husson, 1995; Kruiter, 1996; Mortimore, 1996; Baland and Platteau, 1996; Buttoud, 1997.

Participatory burdens. Some programmes effectively create a 'participatory burden' – by imposing more responsibilities for 'participation' – without a concomitant increase of rights and income or other incentives. It must be remembered that formal means of participation are relatively new for many governments. Whilst Agenda 21 has been calling for 'the maximum possible participation' since 1992, it will take considerably longer to determine how – and sometimes if – this can be appropriately done. Actual experience remains much weaker than the rhetoric.

Inappropriate decentralisation models. Decentralisation approaches are often translated from European contexts, which are suited to nation-state hierarchies, but may be fundamentally incongruent with African contexts, and especially traditional forms of governance.

Weak state consultation and coordination skills. A lack of experience amongst civil servants in consulting with other stakeholders commonly goes hand-in-hand with a lack of coordination – both vertically and horizontally – between different levels of government. Participation with non-governmental bodies can hardly be expected to be efficient if it is weak within the government's own hierarchies.

Lack of accountability of local institutions. It is increasingly being acknowledged that accountability mechanisms are the most important element in successful decentralisation programmes. Three levels of accountability can be identified:

- accountability of civil servants to local leaders;[24]
- accountability of local leaders to local citizenry;
- accountability within decision-making bodies – both governmental and non-governmental.

Yet accountability has been notoriously difficult to ensure at all levels, often due to government's weak ability to regulate local affairs. Non-governmental mechanisms can assist in promoting accountability, e.g.:

- *Improvement of citizens' access to information*, thus enabling more informed participation in public debates. This can be achieved through the use of local media (e.g. through the hundreds of AM broadcasting stations in rural areas); and training in functional literacy.

- *Mechanisms to control daily operations which are based on shared responsibility*, e.g. the need for several signatures to approve financial expenditure.

---

24 Given the shortcomings of elections in many countries, we prefer to use the term local leaders rather than elected officials.

Uganda provides an interesting case under the current 'one party – several trends' system of government. Resistance Councils have been formed at almost all government administrative levels, in parallel to the local administration. This certainly provides some checks and balances regarding the use of local government funds, and an alternative means of recourse for some citizens.

- *Transparency* for reviewing and authorising contracts and verifying expenditures.

- *Formal redress procedures* which can be used against elected officials. This is essential for the mobilisation of local initiatives in the long term. But such mechanisms, where they exist, are often deliberately designed by governments and local élites to be cumbersome, so as to limit their use by local people.

- Better *representation* of local interests.

Lack of representation. Elected bodies may not be representative of local stakeholder interests, especially when elections are fought primarily amongst political parties, and do not involve independent candidates. Complementary mechanisms are often requested, as illustrated below with examples from Niger and Burkina Faso.

- In **Burkina Faso,** the *Tribunaux Départementaux de Conciliation* (District Conciliation Tribunals, TDC) were created in 1993 for settling land disputes. They are composed of four lay-assessors – 'honourable citizens' from the community – and chaired by a district officer. This blend of central government and community representation can be effective. But the TDCs often co-exist with customary rule systems, which tends to mean that the TDC either attempts to exert strong control, or that both the TDC and the customary system take a *laissez faire* approach, with anarchic results for land and resource conflict. However, in some cases a useful combination of the two systems has resulted in some innovative 'local laws' (Lund, 1996).

- In **Niger**, the *Commissions Foncières* (Land Tenure Commissions) (CFs) constitute a major instrument of the Rural Code of Niger. They are chaired by '*sous-préfets*' (county officers) and composed of non-elected representatives from the natural resources agencies, community groups and customary authorities. The CFs have a consultative role on use of land and a decision-making role on recognition and establishment of land rights; transformation of rural use rights into ownership rights;

determination of levels of compensation; and keeping of Land Registers. Three CFs have been operating since 1994 with support from donors, seven more started life in 1997, and eventually 57 are planned. The pilot CFs have been quite successful, although not without their teething problems: they have been criticised for over-representation of technical staff; they have had difficulty maintaining links with, but independence from, political, administrative and legal authorities; there is incompatibility with the Koranic oath, customarily used by traditional chiefs to settle land disputes; and there is a general resistance to formal land registration – "if the land already belongs to us, why should we register it?" (Gado, 1996; Yacouba, 1997)

Increase in local inequity. Since devolution tends to build on existing local power structures, often it is only the richer areas which are strengthened. Programmes may be dominated by local élite groups, or by élites within groups, including local government units. This may lead to the emergence of 'new feudalisms', and marginalisation of socially weak groups, e.g. in West Africa, migrant pastoralists. People living closest to the resource may be favoured, but they may or may not be less exploitative of the resource than more distant dwellers.

No environmental guarantees. Finally, even if devolution is successfully achieved, it is not necessarily synonymous with long-term environmental stewardship – *local decision-making may not prioritise environmental objectives* and especially those which are public benefits at national and global levels. In **Uganda**, there is little sign however that devolution has done much to conserve forests. Responsibility for national and local forests was devolved at an early stage, but recently recentralised in response to widespread conversion of forest.

Inadequate funding. Administrative decentralisation programmes are often imposed from above and distrusted from below. They are instigated in the belief that they will foster mobilisation of the people, even though many rural areas are characterised by strong scepticism of the abilities of government. Procedural changes cannot erase past coercive attitudes overnight on the part of government officers; nor can they erase local people's memories. Thus, alongside the common situation of *inadequate decentralisation of funding*, there is the related problem of *weak mobilisation of local-level finance*, due to people's lack of confidence in the system, or its objectives, or to the disinclination of local politicians to upset their constituencies.

Various *donor-supported projects have enabled experimentation* with ways to finance village activities under decentralisation programmes. For example *in 'Gestion de Terroir'*[25] projects in Burkina Faso and Mali, two types of funds have been developed: *village funds*, which are replenished by the villagers themselves, with project staff sometimes playing an advisory role; and *inter-village funds*, based on co-financing by projects and villagers, with local credit structures. Fund management committees have evolved different balances of representation between project staff, local technical agencies and villagers. Codes of financing involving criteria for choosing activities are being experimented with, and local NGOs and consulting firms are being involved in developing, training and overall advice, which further builds local development capacity (Fournier and Freudiger, 1995; David, 1995; Kabore, 1995).

Reluctance to relinquish control. Governments often claim control over forest resources on behalf of conservation concerns. Where decentralisation is in progress, this control is exerted through indirect means, i.e.

- In **Cameroon**, the state retains ownership over forest resources but, since the Forestry Law in 1994, management responsibility can be devolved to local communities upon submission of a simple management plan elaborated by the local forestry service, together with a *'cahier des charges'*.[26] However, such agreement can be cancelled by the forestry service if community duties are not fulfilled, without the possibility of appeal. In 1996, the government suspended the granting of community forest rights, after realising that unscrupulous forest concessionaires were paying local communities to obtain concessions under the guise of community forests (Egbe, 1996; Pénélon, 1996).

- In **Mali**, any land which is not privately titled must be first registered under the state's name and put into production (*'mise en valeur'*) before acquiring the status of legal goods. So far, in practice, this rule has only been used in urban areas and a few rural areas at their periphery. However, its broader enforcement through the establishment of new decentralised municipalities in 1997, constitutes a daunting challenge and is currently the object of much debate.

Why institutions can not, or will not, work with local realities

There are many understandable reasons why forest institutions, like other institutions, find it difficult to work with local realities. The inheritance or legacy of past assumptions, decisions and practices is a major reason.

---

25 For the purpose of this paper, the 'Gestion de Terroir' approach can be approximated to village-based land management.
26 A *'cahier des charges'* is a list of obligations for two parties involved in a deal. It is used for forest concessionaires in Central Africa.

Institutions are often fighting each other for control of land and money (fines and taxes), which tends to result in a siege mentality – a 'fortress forestry' tendency which reflects the needs and power games of central institutions more than local groups.

Institutions often try to apply uniform solutions across physical and social boundaries with very different complements of power and knowledge. Indeed, maintaining institutional coherence, mandate and turf (or holding on to project funding) may require the denial of these locally specific conditions of power and knowledge which in practice are largely unmanageable from a distant office. Forest agencies and development agencies often portray the local scene in ways which fit their organisational needs, such that the capabilities of these institutions, and the definitions of what needs to be done – through norms and rules – gradually become one and the same thing.

Field officers do not transmit to headquarters what they know of the realities of real villages and villagers, because headquarters is rarely capable of dealing with such complications. The information it requires is laid out in rules which mean that forest officers spend their time monitoring and adjusting to often irrelevant statistics (as in Pakistan at present). Local realities are covered only to the extent that field officers are told to find counterparts and local advocates who can 'get participation going' and make the central models work. These models are rarely what the field officer's local allies had in mind. Real devolved forest management therefore implies significant challenges to current power balances, which need to be teased apart.

Lessons on policy for devolved forest management
In summary, policies which set about transferring some level of power from a central to a more local level can be a godsend, a mixed blessing or a curse for local livelihoods and forest management. 'Devolution' is one of those aspirational weasel words, like 'sustainability' (see Section 2.4), which everyone will support – but the devil is in the detail. The required balance between granting local powers from the 'outside', and taking local powers from the 'inside', is unique to time, place and circumstance.

It is certainly possible to have too much 'political will', so often lamented for its scarcity, if it is unrealistic or over-powering. Those who 'hand over' responsibility to communities are themselves irresponsible if the communities' resources are poor, their rights are weak, and the institutions are not properly understood – leaving them ineffective as forest stewards.

Neither can sufficient autonomy to undertake development activities and modify local rules and institutions be achieved by forming local extensions of central administrations, i.e. deconcentration only. Accountable, self-governing bodies seem to need at least four basic ingredients: *local political leadership; local public interest; sufficient rights and resources*; and, *control mechanisms* based on peer pressure and transparency (e.g. in making contracts and verifying expenditures).

Review of experience suggests that key roles for the state, in achieving devolved natural resource management, include enabling:

- *Security of land and resource access rights* for local resource users and managers.

- *Information and guidance* on macro-level environmental changes, resource-conserving technologies, and legitimising local groups' claims to natural resources.

- *Subsidies/ economic incentives* for improving social and environmental values at local level, particularly where communities struggle to meet their basic needs or are at the mercy of powerful outside interests, and where these values are enjoyed at national or regional levels.

- *Protection against negative impacts of macro-level forces* (e.g. external corporations which aim at over-exploiting resources, resulting in pollution or social divisions) due to the better (potential) ability of the state to deal with externalities.

- *Formal conflict-resolution rules* where locally-derived rules do not suffice, e.g. where there are major conflicts between communities or with external entities.

A pragmatic *problem-solving* approach is needed. The goal might be to locate decision-making no higher than the level at which stakeholders know each other sufficiently to be 'tied' to each other, and are able to optimise the linkage between effort and outcome. It may be possible to develop autonomy at such levels on an *incremental basis*. Dubois (1997) has described this as the goal of achieving a 'wanted, planned and coordinated **subsidiarity'**. As the state develops these roles, it will be more effective if government officers are trained and subsequently work *with* local groups, and have the resources and incentives to do so.

Tensions in devolving resource management authority – the state retaining powers through language, budgets and rights, Zimbabwe

Decentralised governance is strongly promoted in the Zimbabwean government's rhetoric. The state no longer claims to have the capabilities to manage all natural resources at the local level and key state actors have ample evidence of the potential effectiveness of local control, through links with NGOs (see Section 4.4). Yet major tensions arise from the simple fact that, although the state is needed to create the conditions to enable local empowerment, the state reduces its capacity to exert its own local control by doing so.

Central government is likely to be wary of devolving authority to local levels for various reasons. The more that communities are able to control land and natural resource use, the less government can prevail in installing its approach to economic development in local contexts. Politically, central intervention through infrastructural development has been crucial for the government to maintain political control in potentially unstable areas. Intermittent interference by central government in the affairs of local authorities is also evident, apparently explained by government's wish to retain rural constituencies under its ambit. Manifestations of the underlying tensions in devolving authority to district and sub-district structures can be seen at various levels, described below. All of this is challenged by a focus on empowerment of communities.

Use of 'environment and development' language to justify centralised approaches.
Over the years, commercial farmers' representatives and some organs of the state have made repeated recourse to certain assumptions and studies which purport to show the environmentally destructive consequences of livelihood practices in communal areas. These arguments are marshalled against redistribution of commercial farm land, and to justify 'solutions' to perceived problems in resource use – through woodlots, destocking of livestock, etc.

More broadly, such arguments have been used to maintain the dualism established under colonial rule and still apparent in the overall framework of government policy on land use: state intervention, while benign, facilitatory and market-based in the commercial sectors, has been interventionist and restrictive in the communal and resettlement areas. Yet some of these arguments are based on studies now known to be deeply flawed or selectively interpreted (as discussed by Scoones and Matose, 1993; Scoones, 1996). In general, the technical assumptions formulated in the colonial period are unsuited to the complex and diverse nature of the ecologies and livelihoods associated with woodlands.

Budgets still channelled through line ministries

The Government departments with key mandates affecting forests and woodlands have, in principle, embraced decentralisation. But decentralisation of their budgeting and accounting mechanisms is yet to occur, and decentralisation of their policy-making functions appears a distant prospect. A chicken-and-egg situation prevails: resources are required for tackling the weak capacity in Rural District Councils (RDCs); yet sectoral agencies remain reluctant to transfer functions and resources, especially in the natural resource management fields, as long as capacity remains weak within RDCs.

RDCs themselves are also reluctant to devolve responsibilities and budgets to lower-tier structures. Since RDCs are constituted by elected representatives, councillors tend to maintain that they represent the wishes of the people, and in this they have the support of their parent ministry. Further sensitivity on this issue stems from the fact that central funding of RDCs has been declining under economic structural adjustment. This militates against further decentralisation of functions involving finance.

Policy equivocation about forest product rights of communities

The traditional reliance on restrictive legislation and the emasculation of local authority has meant that governance arrangements, particularly in the communal and resettlement areas, are unable to deal with matters of resource use. Local authorities – the Rural District Councils – which hold many responsibilities, are inadequately financed and tend to rely on the services of central government sectoral agencies with varying agendas, which further exacerbates overlap and contradiction at local level. Where commercially valuable natural resources are concerned, community interests are likely to be sacrificed by Councils starved of central government funding.

The future of woodland resources is crucially linked to the viability of local institutions. However, given the complex web of law that is disabling local management and control, and the conflicting signals from government agencies, it is not surprising that there have been few concerted initiatives in local management of woodlands. Thus, despite the considerable evidence and learning now available on the practices of individuals and small groups in woodland resource stewardship in Zimbabwe, the institutional support required to widen and deepen such practices remains elusive.

Forest policy experiments in a 'planned market economy', China[27]

On the afforestation campaign trail

Since the 1949 revolution, and particularly since 1978, China's forest cover has increased from 8 per cent to 13.9 per cent. Further afforestation campaigns are planned to raise national forest cover to 17.1 per cent by 2000.

	China
Forest	• Forest covers about 1,337,000 km^2 or just under 14 per cent of land area (total 9,326,410 km^2)[28]. Natural forest cover includes conifer forests in temperate north-east and south-west, and small amounts of tropical moist forest in far south. • Plantation forests cover 25 per cent of the total forest area: around 334,250 km^2.
People	• Population – about 1,244 million (68 per cent rural, 32 per cent urban), with an average density of about 133 people per km^2, and growth rate of about 0.9 per cent per annum.
Tenure	• All forest land is state owned, but responsibility for forests is split between forest enterprises, state forest farms and collectives. Recently, introduction of the household responsibility system has allowed families and small groups to lease forest land from state collectives. However, the right to private ownership of trees (in certain areas) was introduced in 1956. • 'Special purpose forests' cover 28 per cent of the natural forest area and include more than 900 nature reserves and 874 forest parks. However, capability for reserve management remains low.
Forest economy	• Forestry accounts for less than 1 per cent of GDP, but is an essential source of energy for 40 per cent of the rural population and supplies virtually all the timber for construction. • China is the second largest timber importer in the world. • Following devastating floods in 1998, a policy to reduce logging by 60 per cent was introduced; logging bans brought in under this policy are expected to lead to a reduction in harvesting of 10 million m^3 per year, the termination of 65 forest enterprises and retraining of 600,000 to 700,000 loggers. • The policy soon led to dramatic increases in timber imports; this was assisted by reductions in import tariffs.
Pressure on forests and people	• Huge demand for woodfuel – do large-scale afforestation programmes work? • Need to safeguard and extend environmental/ watershed services – new emphasis on environmental services of forests following devastating floods in 1998. • Pressures on natural forests from logging were suddenly reduced in 1998 following the introduction of logging bans.
Key policy issues	• Experimentation zones for forest policy and new tenure models in a complex web of regulations.

---

27 This section draws from a paper prepared for IIED's *Policy That Works* project by Liu Jinlong and Elaine Morrison (1998).
28 FAO's 'State of the World's Forests 1999' cites China's forest coverage as 1,333,230 km^2, or just under 14 per cent of the national land area.

According to Bruce *et al* (1995), afforestation efforts:

> *"are usually conceived in large part as reclamation of hill and mountain land, involving land-use models ranging from timber monocropping to household agroforestry as part of a mixed farming system. The preparation of this land is very labour-intensive and it is often organised by villages in a 'campaign' mode, with substantial support from provincial government agencies. However, responsibility for management, based in the use rights over the land, is often vested in smaller units or in households, and villages and projects have experimented with both individual and common property forestry".*

The 'Three-North' Shelterbelt Development Programme – also known as 'China's Green Great Wall' and claimed by the Ministry of Forestry in 1995 to be the biggest ecological programme in the world – stretches right across northern China. Started in 1978, its aim is to establish 35 million hectares of plantation by 2050 (i.e. equivalent to almost a quarter of the current forest area). By 1994, 13 million hectares of plantation, protecting 11 million hectares of farmland, had already been established. This Programme was established for the purposes of environmental protection, and it is claimed that completion of the first phase has effectively brought desertification under control in the region (Shi Kunshan *et al*, 1998).

Despite these aggregate-level afforestation successes, Chinese forestry is still unable to meet the needs of national economic development. This is particularly the case since the logging ban brought on by heavy country-wide flooding in 1998, and subsequent timber imports from south east Asia. Chinese forestry also seems unable to ensure the conservation of environmental services. Some observers have reflected on the enduring mindset, nurtured notably by Mao, that 'nature was there to be tamed'.

Experimenting with policy and forests

A strength of forest development in China is that there has been much experimentation for policy. In recent times this has been driven by the need to find appropriate models for adapting forestry to the imperatives of a 'planned market economy'. Since 1987, ten different types of *experimental zones* for forestry economic reform have been established in ten provinces, with the aim of finding suitable institutions, policies, management measures and information for making further forest laws and regulations. These ten experimental zones are located in different economic, social and environmental conditions with different objectives for policy reform. For example, some zones concentrate on land tenure reform, others on tree tenure reform, or forestry enterprise, marketing, tax and tariff reform. All zones aim to encourage producers to afforest land and manage their forests.

The great geographical and climatic range of conditions within China means that any central policy dealing with natural resources requires built-in flexibility for adaptation to local conditions. Local-level officials thus have considerable powers to adapt and detail policy appropriate to local conditions. In some places this leads to such extreme flexibility that there is an effective law vacuum. The unclear legal position of private entities "does not prevent but rather facilitates experimentation, and that experimentation appears to have been very valuable during this time of profound transition" (Bruce *et al*, 1995). For other experimenters however, expectations have been dashed by political vacillation and confusion over the permissible limits of their experiments (Zeng Hu, 1994, cited in Bruce *et al*, 1995).

Tenure trials and confusions

Large areas of mountainous land suitable for forest production, or orchards within village territories, tended to be neglected and denuded under the former commune control. With the break-up of the commune system, more than 60 per cent of rural forest land is now under the control of village economic cooperatives, rather than the state (Lei and Sheng, 1993, cited in Bruce *et al*, 1995). The well-defined legal framework of these cooperatives and their clear ownership of land have contributed to rapid development of forestry on some mountain lands.

Within communities, rigorously egalitarian distribution of small parcels of land has been carried out. There are also experiments with three types of larger management units for forestry, as alternatives to simple household management. All three involve common property management: direct management by the village economic cooperatives; private forestry by one or more 'specialised households'; and shareholder associations (Bruce *et al*, 1995).

Table 4.3 summarises some of the land and tree tenure types, the parties involved, the nature of the agreement, and the problems and benefits of the initiative. The forms of tenure available can be confusing to the various stakeholders in terms of their rights and responsibilities. In theory, a preferential bidding system for wasteland can favour local farmers, as they gain rights, responsibility and benefits. But they are frequently constrained by the lack of long-term credit and technical capacity needed to afforest such land. In general, when private ownership is clear and assumed to be long-term (e.g. in a 'four sides' plantation), private forest management has been shown to be successful. But when insecurity is perceived (e.g. in the 'returning land-use right' initiative), individual farmers are noticeably reluctant to invest in forestry.

Table 4.3 Examples of land and tree tenure systems in China

	Mechanism	Seller/ leaser	Buyer/ lessee	Length of lease/ tenure	
Land tenure					
Returnable land use right	Household contracts on barren hill land	Collective	1) Farmer 2) If returned, then another farmer or forest enterprise	Dependent on afforestation	
Leasing of forest land	Use rights of hill land leased by negotiation. Especially common in southern China	Farmer (use right only); collective or state (land ownership and land-use right)	Farmer, farmers' group, government agency, collective/ state owned forest farm, foreign investors	Fixed (4 -100 years)	
Bid for wasteland	Wasteland sold according to bid; buyer becomes both decision-maker and beneficiary, but according to the market. Common throughout China.	Collective; state	Farmer; forest enterprise	50 -100 years	
Tree tenure					
Transfer of ownership	Ownership of trees transferred at thicket and middle-growth stage	Collective; occasionally farmers	State enterprise; government forest department		
Bidding for harvesting in mature forest	Forest owner transfers mature forest to timber producers		Timber sale agencies/ middlemen	Period of harvest and possibly re-afforestation	
Transfer of fruit orchard	Management rights of fruit orchards are transferred once orchard ready for harvest	State/ collective forest farm			

Terms of agreement	Problems	Benefits
Land-use right returns to collective if not afforested	Insecure for local farmers; farmers may need land for other purposes; lack of technical and investment capacity; returned land tends to be recontracted to those from outside the local community	In theory, transfer of technology from outside the local area
By negotiation, but subject to relative power and influence of the negotiating parties	Potential 'under the table' negotiation: unlikely to benefit local farmers	Land-use rights stable for agreed period: a reasonably successful initiative
Bidding price tends to be kept low to be accessible to all	Farmers lack bargaining capacity and support: much responsibility falls to lessee. Wasteland requires significant inputs, and returns to the farmer depend on ability to provide inputs. Inequality can lead to local conflict. Potential conflict between social/ environmental interests and economic interests. Loss of common land as a community resource	System requires that each farmer should have equal opportunities to lease. An attempt to introduce the market mechanism to return responsibility, rights and benefits to farmers and enterprises, and to enable further afforestation by those with the means to do so. System is thought to have led to afforestation
By negotiation	Subject to market fluctuations. In areas of mature tree tenure market, taxes and tariffs are applied	Short-term income for seller – for immediate needs or investment; generally no taxes or tariffs. Quite commonly done
In some areas, legal only before trees are mature. Cutting quota is applied by forest owner; buyer/ lessee may or may not have to reafforest before returning forest land	May occur illegally and lead to over-harvesting; difficult to monitor extent of harvesting	Many timber harvesting units have good equipment and are efficient
Bidding or bargaining	Appears that state/ collective farms have difficulty managing fruit orchards for good economic return	Lessee able to provide degree of management required

Disabling regulations and healthy local scepticism

In contrast to the areas where freedom to experiment is the norm, in other areas regulations are so cumbersome and restrictive that they have become counterproductive. Menzies (1993) gives an example of villagers requiring three separate permits before they can harvest a single tree. Such overly cumbersome regulations are ignored and resources further depleted.

In a study of the role of controls and incentives in forest management in villages in Yunnan, Menzies and Peluso (1991) found that where incentives exist, they are more effective than controls on access, but they are themselves stifled by bureaucratic procedures. Thus the incentives also become another set of *de facto* controls imposed by the state. They conclude that, although the new contract and responsibility system are important steps towards improving villagers' access to forest resources, the labyrinthine rules allow few clear rights to villagers. As farmers say, 'it's all responsibility and no benefit'.

Frequent and sometimes drastic changes over the last fifty years in rural policy, and particularly rural land reform, have reportedly led to a belief amongst farmers that government policies may change again, taking away or further limiting the terms under which they manage forested land (Menzies, 1993). Currently, the emphasis of policy has, in effect, merely shifted from direct controls over land and people to controls over marketing, species choice and felling decisions (Tapp, 1996).

With the increasing influence of the market economy, the poorer members of local communities lose rights to, and returns from, land and trees, as wealthier members of the community or outsiders use their wealth, power and influence to gain access to such resources. What was formerly a broadly egalitarian system of land use, albeit with poor local control over management decisions, is becoming increasingly polarised.

In conclusion, the principle of continuous policy experimentation looks to be a useful one. But it can lead to conflicting signals and uncertainty, and may not help policy improvement unless the experiments are reviewed in the light of locally-agreed sustainable development indicators.

Regenerating both trees and problems through joint forest management, India

India's June 1990 resolution on joint forest management (JFM) forged a new path as, for the first time since Independence, it specified the rights of the 'protecting communities' over forest lands. The emergence of JFM is also in line with a national move to devolve considerable powers to village councils (*panchayats*). JFM has the potential to redistribute access to, and control over, forest resources between the state and actual forest user (although there is a need to move beyond abstract notions of the undifferentiated 'community'). However, the current imbalance in power and control that appears to be part of the institutional relationship between the Forest Department and local community is seen by some as being more geared to extending the Department's control over the community.

Since 1990 JFM has spread rapidly across India: it is estimated that around 7 million hectares of degraded forest lands are now under JFM, being managed by about 35,000 village forest committees. It appears that in some areas under JFM, forest cover has increased markedly: for example in south west Bengal, where a substantial area is under JFM, satellite images indicate that 4,100 hectares have moved from the category of degraded scrub (less than 10 per cent forest cover) to open forest (10-40 per cent cover). As such forests reach maturity, villagers managing such forests stand to receive considerable economic benefits.

However such devolution of responsibility for forest management has not been without its problems.

- *Only some areas are eligible.* Restrictions on areas eligible for JFM mean that barely 30 per cent of the country's total forest area is currently eligible; such restrictions are open to some interpretation by state forest departments, and in some states the eligible area is as low as 2 to 3 per cent.

- *Difficulties with the focus on timber.* Most state orders assure participating villagers a 25 to 50 per cent share of the net income from timber on 'final felling' of mature trees. This implicitly pre-defines JFM's primary management objective as the production of timber, diverting attention away from the diversity of existing forest usage and dependence. Even the villagers' share of timber is often offered to them in the form of monetary revenue after selling it, instead of making the timber itself available for meeting their own requirements – ironically villagers then have to buy back timber for their own needs.

- *Confusion over NTFPs.* Given the importance of NTFPs to the 200 million people estimated to be partially or wholly dependent on forests in India, and the restrictions on felling, tensions have arisen concerning access to NTFPs. Many states have vested monopoly rights over collection and marketing of NTFPs in forest departments, forest corporations or other agencies created for the purpose. Thus, these nationalised and other, high value NTFPs do not fall within the provisions of JFM agreements and the revenue they produce is not shared. None of the JFM orders promulgated by Indian states mention these existing institutional arrangements for the collection and disposal of many NTFPs from forest areas, which remain in force even when an area is brought under JFM.

Despite JFM representing a positive step towards devolved forest management, with the potential to empower and increase livelihood security for impoverished forest-dependent communities, it remains an institutionally fragile and inadequate intervention in relation to the 1988 forest policy mandate. This is particularly so because JFM is being implemented in a context of deeply entrenched institutions designed for achieving very different ends (see Section 4.3). These institutions continue to function at cross purposes with the new policy objectives, whilst concerted efforts to bring about institutional change are in their infancy.

4.8 Building policy communities

In the 'social drama' of policy, the characters may frequently speak loudly in their denunciations of each other and few are willing to admit that others act on any principle except self-interest (Box 4.5). Backstage, however, some actors get on well with each other, suggesting that it might be possible to play it differently in the next act. In other words, opportunities may emerge to establish some common ground upon which to develop a wider coalition of interests: to build a broader 'policy community'.

Finding common ground in Papua New Guinea

In Papua New Guinea (PNG), neither the donors, nor the logging companies, nor any other foreign stakeholders, have any proven capacity to close the gap between village politics and the public interest. The best hope lies in those institutions of civil society which derive their strength from both sides of the fence: i.e. building new constituencies from the building blocks of Melanesian society, using participatory structures, technical expertise and young leadership. It requires developing the quality and variety of negotiations between existing stakeholders, both national and expatriate, in their official and private capacities, in ways that create new partnerships and new stakeholders.

Box 4.5 The forest policy process as social drama, Papua New Guinea

A concerted and high profile attempt to reorient forest policy and institutions to support conservation and sustainable forest management has been underway in Papua New Guinea, since the early 1990s (see Section 4.2). The characters in Papua New Guinea's policy play are politicians, public servants, forest industry, NGOs, donors and local resource owners. However, some of these characters make more noise than others, and the national policy process is centred mostly on a struggle between the largely Malaysian-owned logging industry and a donor lobby for the hearts and minds of the resource owners. The former two characters have the most concerted voices, while the latter own the scenery (97 per cent of the country's land is under customary ownership). The weakness of the other three characters reflects the fact that nearly all Papua New Guineans are resource owners, and represent themselves in this light when flirting with the characters of politician, public servant and NGO.

The theme of the play, which these characters have been engaged with over the last eight years or so, is 'sustainable forest management', but the plot revolves around the relationship between the politics of the Melanesian village and the divergent interests of assorted foreigners.

A brief profile follows of the current powers to influence the forest policy process of the six characters. The profile shows why there is a sort of stalemate developing, but also shows that this may not last because the balance of power is constantly changing:

- *Resource owners* possess much *bargaining power* based on the simple fact of ownership of the resource. If need be, this can be converted into acts of sabotage, intimidation of company personnel or production stoppages. PNG citizens think of themselves as landlords and staunchly defend their territorial right to claim 'compensation' from the process of resource development which takes place on their land. The notion of 'landownership' as the foundation of national identity leads, for example, to denials that there is such a thing as poverty in PNG, and to popular resistance to anything which smacks of customary land registration, despite the fact that landowners in many parts of the country recognise the need to formalise their titles and their land-use options in some systematic way.

  Some communities are willing to degrade their land – through striking a deal with a logging company – perhaps in an effort to keep up with their neighbours. The numbers of highly educated members of local communities who might favour sustainable development are often smallest in those rural communities with the most forest of interest to other stakeholders.

- *The private sector* derives most power from its capacity to use *'divide-and-rule' tactics* with other stakeholders. However, politicians and NGOs derive much political capital from attacking resource developers, and the demonisation of the log export industry thrives on a mood of public disquiet about the 'Asianisation' of the national economy. Uncertainties about the physical and political environment constrain private sector players from manipulating the 'system' to their lasting advantage, but also reduce the potential for sustainable forest management to be pursued through corporate self-interest. The financial crisis in the national economies of Southeast Asia has contributed to the problems faced by log exporters. Loggers complain that they are being driven out of business by the combination of high costs, high taxes, and low market prices. A number have shut down operations.

- *Politicians* count themselves as resource owners, and many are, or have been, employed by government, NGOs and the private sector. So no clear distinction can be drawn between the country's political, bureaucratic and business élites. Political parties are little more than parliamentary factions. Governments have been formed by an unstable succession of coalitions of national politicians whose own electoral survival commonly depends on their ability to reward a very small local constituency with the maximum possible share of government resources. There is a vicious circle through which politicians justify the exercise of greater *personal executive power* by reference to the failings of a bureaucratic system whose own powers are diminished by the same exercise.

- *Public servants* compete to obtain the maximum benefit from each rearrangement of the institutional furniture under the various reform programmes. They keep their distance from each other by constantly mending the fences which the politicians like to break in their own search for additional executive power. Bureaucratic reforms in the National Forest Service encountered resistance from public servants as soon as they could no longer rely on the support of a reforming minister and an élite squad of donor-funded technical advisers.

- *NGOs* do not form a natural cartel and have very different degrees of leverage on forest policy. The relationship between NGOs and donors is a marriage of convenience which is forced upon both parties by their common disaffection with the powers of bureaucrats and politicians. The acid test for the power of many NGOs lies in their ability to simultaneously meet the needs of rural resource owners and satisfy the donors.

- *Donor agency* leverage over the national policy process revolves around a double-act between AusAID and the World Bank, where each relies on the other for specific actions, while other donor agencies play walk-on parts and keep their exit options open. The World Bank sells the prescriptions of a 'global' donor community, while AusAID gravitates towards the implementation of 'institutional strengthening' projects. Ultimately, *the power which donors exercise over the forest policy process*, through the Structural Adjustment Programme (SAP), *depends on the continued fiscal and governance crisis* which the donors are supposedly attempting to resolve.

The policy tools around which these characters have revolved and argued (described in Section 4.2) have caused a partial mutual understanding of different views to emerge. Through a combination of considerable tension related to forestry projects and policy tools in practice and several debating mechanisms – policy steering committees, sporadic multi-stakeholder workshops, and lively press coverage – an expanding policy community is forming...

Developing mechanisms for testing claims

There is a growing belief in PNG that stakeholders' policy positions need to be subjected to a set of trials or experiments which will reveal who speaks the truth, whose actions speak louder than their words, and what those actions actually achieve. For example, official representatives of the log export industry may claim that a specific large-scale logging operation should be regarded as a model of best practice under local conditions, or that a specific form of organisation exemplifies the best way for landowners to deal with logging contractors in the pursuit of sustainable development. Mechanisms are needed to simultaneously publicise and negotiate claims of this sort, by producing evidence rather than recycling assumptions. A 'route map' is needed which enables competing claims to be evaluated – to travel from the office to the village or a patch of forest, and back to the public media for all to see.

Only through negotiation can resource owners hope to achieve the capacity to protect their interests in the long term. Similarly, 'public interest' objectives such as environmental protection need to be balanced against conflicting private interests through location-specific negotiation, backed up by well thought out, practical market/ regulatory measures which can be applied in ways which respond to local realities. State agencies will have to take the lead, but will also need new partners, to:

- scrutinise the plans of developers
- publish model contract provisions
- legislate for court review of manifestly unfair contracts
- create finance arrangements, where landowners can borrow against future income to pay for preliminary investigations and professional advice
- enable non-government negotiation services for landowners

Generating private initiatives to pursue the public interest
In the absence of a massive upturn in public confidence and capacity in state institutions, policies will need to provide a clear mandate for 'out-sourcing' a wide variety of executive functions to an equally wide variety of 'non-government' organisations, in a manner which encourages these organisations to develop a common vision of their mutual responsibilities and separate specialities. Where government departments and resource developers agree to an arrangement, by which developers undertake integrated land-use strategies within an area of influence which is protected by government, the work of integration should then be relayed to a mixture of consulting companies and NGOs who share a common interest in breaking down the barriers between sectoral policy domains. The net result is to enlarge the size and influence of the 'non-government' policy community in each of these domains.

Working at the interface between village and state
A flexible method of working is also needed at the interface between the village and the state, between resource owners and 'policy-makers', which uses and empowers those groups of actors who specialise in adapting public

policy to the variable needs of rural communities, or in articulating these needs in ways which can transform policy. Some NGOs have already proven themselves quite effective in this interface, but the network needs widening to include church workers and social scientists, some of the staff and consultants employed by resource developers and donor agencies, as well as those government employees, such as primary school teachers, who work in immediate proximity to the 'grassroots'.

Finding the capacity for the job

If basic services are not delivered to rural communities in a manner which fosters their own self-reliance, then formal policy may become increasingly irrelevant to local practice. New contractual relationships or working partnerships between existing organisations are needed, which can match the needs of landowning communities with the specific capacities which different service-providers can offer. There is no shortage of previous experiments. For example, there are lessons to be learned from the experience of mining and oil palm companies in promoting local business development.

Formal policy stalemate and informal active negotiation, India

In India too, the evolution of forest policy and its changing orientation over time can best be understood in terms of the competing claims, and relative influence, of various interest groups. In particular, four broad interest groups seem to have had a major impact over the last hundred years: conservationists, foresters, industrialists, and social activists. From 1864 to 1988, forest management strategies were markedly biased in favour of commercial and industrial exploitation, with little attention paid to sustainability or to social justice. However as the forestry debate intensified in the late 1980s, the state increasingly responded to the claims of forest dependent communities as voiced by activists and NGOs.

Over the last eleven years, these groups have jostled for positions of influence, attempting variously to fulfil implementation of the pro-poor forest policy (see Section 4.3), or to subvert it. Each of these groups has played significant and changing roles in the ongoing policy dialogue: the proponents of JFM and pro-poor policies – the social activists; those in favour of expansion of protected areas for nature conservation but at the expense of the needs of forest-dependent people – the conservationists; those who want to lease so-called 'degraded' land for raising 'captive' plantations for raw material supply – large-scale industry; and foresters, who are charged with implementing the policy yet who find that their traditional roles and mandate ill-equip them to do so.

It is interesting to examine the emergence of the 1988 forest policy given the tensions between these groups. It appears that by the time the new policy resolution was tabled in Parliament, massive corruption scandals were dominating government business. Implementing a new forest policy became low priority. Even the subsequent 1990 Joint Forest Management Circular – which paved the way for JFM across India – may not have been issued but for the concerted efforts of a few individuals. The Circular was hurriedly drafted and was approved by the Minister of Environment and Forests, supported by a handful of officials and non-governmental individuals. Many of the initial state JFM orders were similarly pushed through by *ad hoc* initiatives taken by interested individuals without any open debate or discussion, and it is claimed that many of the subsequent state orders were issued under pressure from donor agencies. Hence the policy and subsequent Circular, whilst initially championed by certain lobbies, lacked the broad support to ensure its full implementation. Meanwhile the intended beneficiaries of the new policy – including hundreds of millions of forest-dependent people – lack sufficient political voice to influence national political processes.

Groups seeking to influence policy have used whatever political space is available to make themselves heard. For example the proposed – but subsequently unsuccessful – Forest Bill of 1994-95 was promoted by the conservationist lobby, which called for protection of the forest but at the expense of those dependent upon it. Had it been successful, the Bill would have reduced forest dwellers' rights and reasserted the control of the forest bureaucracy. It also attempted to alter the balance of decision-making authority between the centre and the states – towards the former. The Bill was almost submitted to Parliament – a reflection of the considerable support that the conservationists enjoy among the Indian Forest Service, and of their traditional access to some degree of political power. Only the intense lobbying of social activist groups, some of them acting on behalf of forest-dependent peoples, prevented the Bill from being enacted.

Tension between different policy interest groups is also manifest in the ongoing debate over the leasing of 'degraded' lands to industry for plantations. In this case, powerful industrial and commercial lobbies have used their considerable access to politicians to try to by-pass the 1988 forest policy by obtaining leases to thousands of hectares of 'degraded' land and getting the Forest Conservation Act of 1980 amended, to make such leasing legal. To some extent they have received an encouraging response from government. However, pro-poor social activist groups launched a massive campaign against the move in 1995 and continue to campaign on this issue. The social activists have successfully used a provision in the forest policy, that nationalised forests cannot be leased to any private agency, as the main basis for ensuring that government rejects industry's demands. The influence of the social activists is reflected in a recent Planning Commission Working Group, which concluded strongly in favour of farm forestry and agro-forestry, and against large-scale leasing.

Despite having been vigorously debated for years, this issue remains unresolved. It seems likely that this debate will resurface whenever there is perceived political space for change – such as each time there is a change of government. However, it appears that approval of industry's proposals would be too risky politically, and so far no government has acceded to industry's demands. Given the leading role currently played by the Supreme Court in deciding policy (see Section 4.3), this may be the only route to a potential resolution of this controversy.

India's forest sector is in a situation of semi-paralysis, whereby aspects of policy, such as the patchy implementation of JFM, are realised whilst the rest remains on paper only. The ongoing tug-of-war between different policy interest groups has ensured that no radical changes are made. At the same time, the major formal policy shift of 1988 has not been overturned, and the potential for its more concerted implementation grows by the day. Thus, the apparent paralysis masks sporadic but vigorous debate between interest groups which constitute a vibrant policy community – albeit a poorly connected one – which may eventually produce a strong drive for implementation as ideas spread and reach those who are motivated by them to take practical action.

One hundred and fifty years of consensus, Sweden[29]

Sweden has a broad-based community for forest policy, which has its origins in: a common perception that 'something must be done'; considerable disagreement over what to do; and a practical need for cooperation because of the disparate land ownership pattern.

Hard-won agreement in the 19th century

Sweden's forests were, by common consent, perilously degraded by the time the Royal Chancery began setting up independent ministerial departments in the 1840s, and a Central Forest Administration in 1859. But the existence of a range of strong rural institutions and other politically organised groups meant that no agreement could be reached for an effective and politically acceptable forest policy. Forests were still generally treated as an obstacle to more profitable land uses.

---

29 This section draws from a paper prepared for IIED's *Policy That Works* project by Bjorn Roberts (1999).

A process of widespread consultation and debate commenced, and continued for half a century, resulting in the National Forestry Act of 1903. The Act's main emphasis was the need to restock the nation's dwindling forest reserve. The Act proved effective and enduring. The preceding process of inquiry helped to establish an approach to forest policy by which interest groups are closely involved, and informed consensus among key actors is sought. Practical necessity for widespread cooperation played a large part, given that more than half of all forest land was owned in small private lots. The FA became an agency which prioritised the understanding and application of policy, with legal powers available as a last resort. This cooperative approach has characterised Swedish forest policy ever since.

Environmental argument and reactivated collaboration

However, during the 1970s and 1980s consensus began to break down. Conservationist groups drew particular attention to damaging forest industry operations in northern and central Sweden. Specific issues, including application of chemical pesticides and extensive clear felling, became the subjects of lively public debate. A more general challenge also arose, questioning the dominance of production over other forest objectives.

Government's response was to launch a Commission of Inquiry. Like other inquiry processes used in Sweden in the development of major laws, it was designed to allow different interest groups and opposition MPs to have a say, before government comes to a firm position for parliamentary debate. The objective was not, however, consensus at any price such that persistent and perhaps irreconcilable differences are obscured. Commission members who objected to the final report were free to append alternative proposals.[30]
In developing the new Forest Act, a general acceptance evolved concerning the need for greater environmental protection and recognition of forests' social importance. Forest industry and forest owner groups eventually became part of this general acceptance, when a compromise was reached that some environmental measures would only be introduced where they do not damage economic values significantly.

Deregulation and well-supported persuasion

The 1993 Act introduced a number of new regulations – on assessment of forestry operations' environmental impacts, and on the protection of key habitats. But it is largely deregulatory in character, and shifted responsibility for practical policy implementation to forest industry and private forest owners. Until 1993, the State Forest Enterprises of Sweden had owned extensive areas of forest land, 3.4 million hectares of which were then privatised along with the state's forest industry company.

---

30 The influence of interest groups on Parliament is not limited to lobbying and making submissions to Commissions of Inquiry; Sweden's proportional representation electoral system allows organisations such as the Swedish Federation of Private Forest Owners to field Parliamentary candidates – and they have done so successfully.

Even more than in previous years, the Forestry Administration now regards implementation more as a matter of persuasion and extension, than of regulation and coercion. In practice, legal actions and the imposition of fines to enforce implementation are rarely used, and never before extension campaigns are complete. However, it remains to be seen how far the Act's environmental and ecological priorities can be effectively achieved through voluntary action, and to what extent forest owners will expect compensation for production foregone. Implementation of environmental policy guidelines is tracked by environmental NGOs. Having done much to change their forest operations over the last two decades, the forest industries are thus far glad to point to (at least partial) endorsement by such NGOs.

In summary, participation of the main interest groups in developing and implementing major new policy, and in determining research priorities, has resulted in well-designed policy and good stakeholder cooperation in Sweden. A well-funded and professional Forestry Administration, with strong local presence, and the ability to adapt advice to local conditions has been a major factor in this success. A culture prevails amongst the full range of stakeholder organisations of involving representatives of other organisations in key aspects of each other's work. Although the balance of forest objectives has changed, and is complicated by organisations increasingly having interests and identities beyond Sweden's borders, strong mutual interests remain and the consensus approach continues to suit Sweden well. It must be stressed, however, that such a democratic approach has many precedents in Sweden. As we have said before, context is all-important for policy processes, and ideas do not necessarily transfer well to other countries, at least in their entirety.

Thriving policy communities – comprised of those with the power to help or hurt the cause of good forestry, and those with the actual or potential capability for good forestry – are the driving force for better policy processes. A policy community can channel ideas of all those who are important for the prospects for sustainable forest management – the stakeholders – into the policy ring, and channel the outputs out again. We have seen how various shifts in the relative powers of different stakeholders can enable the policy community to enlarge itself and make incremental or quantum policy changes. In the absence of such power changes, debate over particular policy tools can also prove to be an effective catalyst for the formation of policy communities. Few countries have yet developed a policy community likely to guarantee sustainable forest management – although Sweden may be close – but several, Papua New Guinea and India included, are making progress.

5 International policy trends and initiatives – their implications for forests and people

5.1 Setting the international scene

In previous sections, we have examined the interplay of actors in national policy processes. However, if national policy has historically been dominant in forestry, in recent years the nation has become a playing field on which global and local forces may clash, or interact positively.

On the one hand there are the many forces of globalisation noted in Table 5.1, and especially:

- those market forces that seek out comparative advantage for forest production in certain countries and which may bring capital and technology into the country; and
- the intergovernmental agreements which seek to secure global benefits from forests, such as biodiversity and climate moderation.

On the other hand, local forces of decentralisation are also increasing in significance. Trends towards democracy seek to improve local peoples' rights to have access to forests, and to use them for multiple local benefits. Other decentralisation forces arise because of pressures on government to cut costs and downsize – but not necessarily to reduce government rights and responsibilities – resulting in e.g. deconcentration (see section 4.7).

In this context – of what appears to be accelerating erosion of nation-state sovereignty over forests – it might superficially be concluded that national policy is becoming irrelevant. In contrast, we find it is even more imperative to improve national policy processes. Global policy initiatives to secure e.g. biodiversity, and biomass for carbon storage, will not work without local forest stewards being empowered and rewarded to produce these global benefits. Conversely, some local livelihood benefits from forests will not be secured unless some of the benefits of globalisation – access to markets, sources of finance and technology – are made available. At the very least, better communication is needed between levels in order to ensure compatibility in the production of local, national, and global values.

Table 5.1 Current international trends and initiatives – a framework for analysis[31]

International context	• Globalisation and increasing inequalities within and between nations, in relation to:
	  • capital and technology movement
	  • increasing communications and market links
	  • economic instability
	  • debt burdens
	  • different access to benefits of globalisation
	• Democratic movements and decentralisation
	• Increased demand for all forest goods and services
	• Trend towards intensively managed forests and plantations for fibre; economically inaccessible forest for environmental services
	• Net reduction in overall biodiversity

International actors and their policy interests	• Governments – distinguished as forest-rich/ poor; others income-rich/ poor:
	  • Want to protect sovereignty over forests
	  • Want access to finance, technology, trade
	  • Have multiple foreign policy agendas
	  • Some want SFM, others want status quo
	• Multinational corporations – often have more economic power than some governments
	  • Want forests viewed as economic assets and not solely as environmental assets
	  • Some want access to cheap forest assets and prefer weak policy/ chaos
	  • Others want long-term investment and prefer stable policy
	• NGOs (environmental and social) – may have more political power than some governments
	  • Want social and environmental services from forests
	  • Want binding commitments/ targets/ monitoring

Institutions which integrate actors	• Intergovernmental initiatives – tend to be bound by the desires and fears of member states. Varying degrees of participation but increasingly open to non-governmental input:
	  • Institutions – FAO, UNDP, ITTO, UNEP, UN Interagency Task Force on Forests
	  • Conventions on environment and trade – CBD, FCCC, CITES, GATT/WTO rules
	  • Development assistance initiatives: Tropical Forestry Action Programme; Forest Partnership Agreements
	• Regional governmental agreements
	• Professional networks

31 Table 2.2 in section 2.4 outlines the origins and evolution in recent history of these international trends and forest policy responses.

	• Private sector associations and initiatives:
	    • World Business Council for Sustainable Development
	    • International Forest Industries Round Table
	• NGO/ civil society initiatives/ policies:
	    • WWF/ World Bank Alliance
	    • FSC/ Buyers Groups
	    • World Commission on Forests and Sustainable Development
	• Mixed private sector/ civil society initiatives:
	    • fair trade and certification

Policy contents/ outputs	• International forest policies still focus mostly on forests, and are not engaged with 'extra-sectoral' policies – i.e. environment, peoples' rights, technology, trade, finance. Some address truly global issues (global forest services or global causes of forest problems), but others (merely) common national issues. Varying types:
	    • Fora for 'information-sharing' and discussion
	    • Non-legally binding principles and guidelines
	    • Legally-binding agreements
	    • Market instruments
	    • With/ without implementation programme
	    • With/ without standards or targets
	    • With/ without finance and sanctions

Impacts/ implications	• Impact of international initiatives depends partly on ability to interact and deal with national and local processes and power structures
	• A wide range of positive outcomes:
	    • improved understanding, language, or vision
	    • improved relationships
	    • improved political will to change
	    • legal clarity on SFM rights and responsibilities
	    • improved financial or technology flows
	• A wide range of negative/ ineffectual outcomes:
	    • Lowest common denominator
	    • Justification for inaction
	    • Inequitable coalitions and divisions
	    • Obfuscation or diversion from real problems
	    • Unrealistic 'dreams'
	    • Underlying causes not actually addressed

A viable and vital *national policy* process is therefore needed, now more than ever, to provide good communication between local and international policy processes. It is needed for finding the win-wins and sorting out the trade-offs between local and global, for ensuring that the benefits of globalisation are realised without unduly burdening specific nations and localities with the costs. It should support the kinds of institutional capacities and resources needed to implement the solution.

Without a strong national policy process, the surest phenomenon will be 'policy inflation'. In other words, scarcities of forest goods and services, and clashes between actors at whatever level – from local to global – will give rise to a whole multitude of policies, laws, or instruments which attempt to deal with the problem. But these new policies become uncoordinated and delinked from capabilities to do something about them – the *'policy inflation, capacity collapse'* which we introduced in Section 1.2.

Clearly, the precise form of international policy initiatives will determine whether they are conducive to improved policy and capacity at national and local level. At their best, international policy initiatives catalyse local initiative, build capacity, and offer common language and political momentum behind shared challenges and truly global needs. At their worst, international initiatives impose precepts which reflect the demands of a few countries only (or possibly none at all), remove national incentives and freedom to use forests for development and welfare, and destroy local institutions and other sources of resilience which had sustained forests and livelihoods. Caught somewhere between the best and worst, most international initiatives have spent much time and energy engaging with what we might call *'common national problems'* – which might be better addressed by national or local initiatives – and have made little progress with truly *global problems* (see below).

In this section, we examine a range of international policy processes, and observe where they have supported the pursuit of forestry and sustainable development at three levels:

- have they actively and sustainably dealt with global issues? These are of two main types:
  - o the cross-border activities that may either help or hinder good forestry, notably: trade, aid, foreign investment, forestry activities of foreign companies, trans-border forest protected areas, and pollution;
  - o the security of global goods and services i.e. globally-important biodiversity, carbon storage and climate moderation, and global natural heritage such as wilderness landscapes
- have they supported the development of viable national policy processes and institutions?
- have they recognised local realities and supported local (livelihood) rights, capabilities and needs?

We examine three very different types of international policy initiatives, all of which have had considerable impact: intergovernmental processes, civil society approaches, and private sector activities. Some of our findings are drawn from the six PTW country studies, others from specially-commissioned studies, and the remainder from IIED's own experience of fifteen years engaging with, and assessing the impacts of, international forestry initiatives.

5.2 Intergovernmental forest initiatives – global change and international games

The actors in intergovernmental initiatives and the processes they adopt

Governments

The key international actors are governments. But these are not at all homogeneous in their perception of 'global' forest benefits or causes of global problems. Their perceptions depend upon many factors, principally their status as forest-rich or -poor, and income-rich or -poor countries (Figure 5.1). These factors shape how they see their own forests, and those of other countries.

For example, income-poor countries with a wealth of forests (e.g. Guyana or the Democratic Republic of Congo, formerly Zaire) tend to treat forest capital as a source to be drawn down, in order to create other forms of capital – physical, financial, or social. This is especially the case if the forest asset is valuable, but its growth rate is poor, as with many tropical forests (most people, faced with a large bank balance earning low interest rates, will remove their money from that bank account). In contrast, an income-rich country with few forest assets (e.g. the UK or Saudi Arabia) will be concerned that better-forested countries should make timber and environmental services available to them at low cost.

Furthermore, these governmental perceptions are tempered by what they consider the 'national interest' to be. In less democratic societies, or those with few checks and balances on those in power, the national interest turns out, in practice, to be the interests merely of those in power and their associates. Very often this results in an elevation of timber production concerns, and a demotion of environmental and equity interests (see Sections 4.1 and 4.3).

Hence governments may have a number of goals in intergovernmental forest processes, ranging from:

- maintaining the *status quo* (including power structures and relationships), which may take the form of preventing new agreements, as much as forging specific new outcomes, to
- positive *goal-oriented* approaches, aimed at local, national and global sustainability[32]

---

32 Multi-stakeholder policy initiatives can be divided into two basic types: relationship-maintaining (where it is seen as important to keep the *status quo* and ward off external threats; these can often be influenced by covert or informal needs); and purpose-led (where there is a perception of the need to change, some idea of what is desirable, and commitment albeit to varying degrees).

Intergovernmental institutions

These bodies are somewhat secondary actors in intergovernmental forest policy. In practice, they serve the current order between nations. Those concerned with forestry (FAO), forest products trade (ITTO), environment (UNEP), development (UNDP) and forest-dependent groups (ILO) are usually late-comers or poor siblings in the intergovernmental family of institutions; with limited capabilities to correct intergovernmental inequities, as they derive their powers from governments which retain sovereign control over forests, and limited executive capabilities. Nonetheless, they do provide certain routine functions such as forest monitoring (but they have limited powers to call on for information and to extract the truth), aid disbursement (but they are obliged to spread this thinly, rather than focus it on where the real needs are), and debate, notably through the Intergovernmental Forum on Forests, IFF (although their function is more that of 'leveller' to common denominators than supporting progressive new approaches). Recently, they have come together, to reduce duplication and to coordinate functions, through the Inter-Agency Task Force on Forestry connected to the IFF. Their influence in terms of defining language and creating legitimate agendas is considerable, although they are not known for particularly incisive solutions that get to the heart of forest problems.

NGOs

Some tertiary players have recently moved centre-stage, and appear ready to change intergovernmental initiatives or to provide an alternative to them. NGOs and, to a lesser extent, private sector/ market actors (see section 5.4) have really begun to alter the mix of international policies. The generation of an international sense of crisis about deforestation by NGOs and some prominent forestry/ conservation professionals, spurred on by the media, helped to give rise to the TFAP and gave both narrative and language to the WCED and UNCED debates.

NGOs (and private sector associations albeit to a much more limited extent) have sought to work right within the intergovernmental system, as advisers to national delegations to the ITTO and IFF (such as WWF and IUCN) and as highly vocal advocates for certain policy positions in these fora (such as the Global Forest Policy Project, a multi-NGO activity).

William Mankin has catalogued the gradual improvement in facilities and rules that circumscribe an NGO's ability to affect intergovernmental outcomes[33]. These have progressed from demonstrations outside UN halls, to open-mike sessions within the halls, to actually taking part in the drafting of

---

33 He has coordinated the Global Forest Policy Project from the beginning. As such, he has probably taken part in more intergovernmental debates than most national delegation members – especially as delegations change.

policy text. ITTO provided a valuable training ground for NGOs in this respect. The CSD generally, and the IPF and IFF in particular, have helped the gradual progression from intergovernmental antipathy of NGOs, to partial partnership. These initiatives have offered learning about participation in policy – as well as representing the massive (yet inefficient) diplomats' training courses in sustainable forestry for which they are more frequently credited. However, whilst intergovernmental processes have opened up debate, have improved transparency by not allowing governments to duck issues, and have provided some new solutions, NGOs remain sceptical as to just how far they can go (Mankin, 1998).

The story: from 'top-down' TFAP to 'bottom-up' Criteria and Indicators

Since the 1980s, there has been a wide range of intergovernmental initiatives focused on forests. It is difficult to ascribe impacts on forests or stakeholders' positions directly to intergovernmental initiatives, as these initiatives only really operate once they are translated by government legislation or aid agency programmes, where other influences come into play, or through indirect influences on relationships and trust, debate and attitudes. Nevertheless, in this section, we attempt to contrast the earliest major initiative, the Tropical Forestry Action Programme, with the more recent initiatives to define and implement Criteria and Indicators (C&I) for sustainable forest management.

In Boxes 2.2 and 2.3 we noted that a sense of crisis has been a principal catalyst for policy. The *TFAP* could be characterised as a top-down, quick – but none the less comprehensive – fix to the perceived tropical forest crisis, the perception being promoted by NGO and media concern about 'deforestation'. The response was essentially a bureaucratic and technocratic one, led by professional foresters, and lubricated by development aid. The product of FAO, UNEP, World Bank and World Resources Institute thinking in the mid-1980s, TFAP set a 'standard' for a balanced forest sector for the next decade, and defined a new liturgy for forestry aid planning.

Whilst the TFAP set out a broad set of worthy areas for aid intervention, in practice it resulted in fewer improvements in forestry than had been hoped. Because it was closely associated with the government-to-government aid system, the TFAP was not able to challenge the inequities and perverse policies that underlay deforestation, and then to build the necessary trust between governments, NGOs, local people and the private sector.

Its very standardisation, within a global framework, and the exigencies of the aid system that supported it (which often installed expatriates to lead the in-country planning), meant that the TFAP did not adequately recognise diverse local perceptions, values, capacities and needs. Finally – and despite efforts to house TFAP exercises in powerful but 'neutral' bodies such as planning ministries, the TFAP failed to generate real extra political support to the broad range of forest values, and thus to appropriate aid and investment.

The Papua New Guinea PTW study describes that country's experience with the TFAP approach in the early 1990s – a classic case in which the international programme was developed, largely by outside experts, as the comprehensive solution to the crisis in the forest sector felt by many at the time. Wholesale policy and institutional reform was mapped out, and a range of major projects got off the ground – many stressing the importance of involvement of customary landowners and NGOs, yet few achieving it (Mayers and Peutalo, 1995). Ten years on, many of the anticipated direct impacts are hard to see – they were simply not based on genuine local motivation. Yet the programme did create a considerably wider pool of those engaged with forest issues than before – a policy community whose shared experience is now critical to making progress in the forest policy process.

The TFAP's master-plan approach did not offer the kind of national-led policy process which is required to trade off global, local and extra-sectoral national needs. However, in its later years, the TFAP protagonists, reacting to a perceived lack of progress and local ownership of the initiative, put considerable (if unfocused) attention on '*stakeholder participation*', the two words now forming a new mantra. Parallel changes in national conservation strategies (coordinated by IUCN) and to a lesser extent in national environmental action plans (coordinated by the World Bank) also put an emphasis on 'bottom-up' participation of notional stakeholders[34]. Much of this reflected genuine learning: for example, reviews of NCS noted how those which were considered to be successful were characterised by 'legitimacy', 'ownership', 'commitment', 'equity', and good 'networking' – all functions of participation (Bass *et al*, 1995).

Forests were among the politically hottest issues at the 1992 Rio Earth Summit. Calls both for and against a forests convention were strongly

---

34 The *stakeholders* in a forest are all the people and organisations who have a 'stake', or interest, in the forest and may be affected by any activity in it. The idea of stakeholders implies both that they have some legitimate claims to forest values, and that they are able to exercise this. In reality, many are not legitimate (because of how they may have appropriated others' rights). Others may not be able to exercise claims in the forest or in the policy process (because of inequities). Hence the term 'stakeholder' may imply a right or an ability to engage in policy processes that may not, in fact, exist.

polarised. At times, it seemed as though forests were being 'elevated' as a ground for fighting other (non-forest) battles – in the same way that national forest debates (over 'monoculture', for example) may only really be a metaphor for more fundamental arguments (over rights and power).

The stalemate on the need (or otherwise) for global regulation on forest issues continued through the subsequent Intergovernmental Panel on Forests (IPF) and indeed the following Intergovernmental Forum on Forests (IFF). Some countries stalled on implementing forest-related decisions under the CBD, for example, under the guise of promoting a global forest convention (the net effect, intended or otherwise, being substantial indecision and inaction).

One of the outcomes of the CSD/IPF/IFF series of discussions, and of the renegotiation of the International Tropical Timber Agreement, was the principle that future intergovernmental initiatives should not prejudice against tropical forests. All countries (and especially richer countries) should take on board social and environmental provisions – as *constraints* on the use of their forests. Thus, in contrast to the previous notions of top-down, uncompromising regulation (a convention) or centralised action plans (the TFAP), intergovernmental processes in latter years have come up with a subtler mix of 'bottom-up' initiatives which apply to all forests and respect sovereignty yet, on paper at least, offer bountiful scope for local adaptation and compromise (although compromise can include inaction).[35]

Both the varied strands of international debate, and the practical searching for 'SFM' at national and forest level, led inevitably to demands for greater clarity of definitions and goals. With a political and professional climate that was now more accommodating of local differences, this led in the mid-1990s to a major phase of defining *Criteria and Indicators (C&I)* for good forestry.

Sometimes this was a case of the governments in a region getting together to work out what a bottom-up approach to the CSD forestry commitments would mean for them (notably the pan-European, dry-zone Africa, and Amazon Pact processes). At other times, progressive governments from both North and South came together for mutual encouragement to define the key dimensions of SFM (notably the Montreal process). Both these types of intergovernmental approach were aimed at giving countries considerable degrees of freedom to adopt different approaches in their forests, whilst at the same time rightly being able to claim that all such responses amounted

---

35 Sovereignty may rightly be viewed as a legal fiction, and has *always* been compromised by international economic and social forces. Assertions of sovereignty serve to insulate states from the international effects of their own policies. Furthermore, they serve to consolidate the status of the government of the day and the economic interests with which they enter collaborative arrangements (Humphreys, 1996, p171, citing David Potter).

to SFM. A technical and political precedent for all of these was the earlier work by ITTO in defining C&I for natural forest management, and later for plantations.[36]

In addition, civil society and forest industry actors have come together to do the same, with an emphasis on managing forests to high social and (especially) environmental standards. The most influential of these has been the Forest Stewardship Council's principles and criteria (section 5.3). Again, these define the main dimensions of SFM, and encourage some local interpretation, rather than prescribing a universal set of actions (although it must be said that the FSC's P&C are more specific and clearly set high standards).

The three basic families of C&I are illustrated in Figure 5.1. There was much cross-fertilisation in their development, certain professional foresters and environmentalists working on several of them. As the figure illustrates, these international policies have already resulted in some national standards and indeed verification programmes. C&I provide an important picture of the many dimensions in which change is required if forestry is to provide more environmental and social benefits – even if they (deliberately) do not describe the means to make these changes, and if they leave a clear distinction between local and global benefits out of the picture for the time being. Nonetheless, C&I have laid out a new game where timber is not the only stake, which everyone can play, and for which there are few losers. They offer a 'politically correct' language which unites local, national and global actors but acknowledges their separate situations. And they are precise enough to allow rationalisation within these situations – as the UK government has done with all its varied legislation, regulations and incentives by matching them up with the Helsinki C&I and generating a new UK forest standard as a result. As such, C&I offer a potential currency which minimises 'policy inflation' and helps to focus capacity development and monitoring.

Parallel to, and connected with, the C&I initiatives has been a 'reinvention' of the TFAP concept – but this time broadened to all countries, North and South, and taken outside the purview of aid.

---

36 ITTO's C&I were meant to guide committed producer nations towards better practice, and later as a means by which progress towards Target 2000 (trade only in timber from sustainably-managed forests) could be assessed. However, once ITTO's C&I were fully developed and articulated for individual nations' forests (as they were for only a few countries) the enormous scope of changes necessary to reach sustainability became apparent, and rather reduced producer nations' commitment to change. Instead, there was some deviation of effort towards assessing the (huge) costs involved, and in getting the richer (consumer) countries to commit to similar improvements in their countries. ITTO officials will admit that no producer member country will meet Target (now Objective) 2000, even as defined by the accommodating C&I.

Figure 5.1 The main international forestry standards initiatives

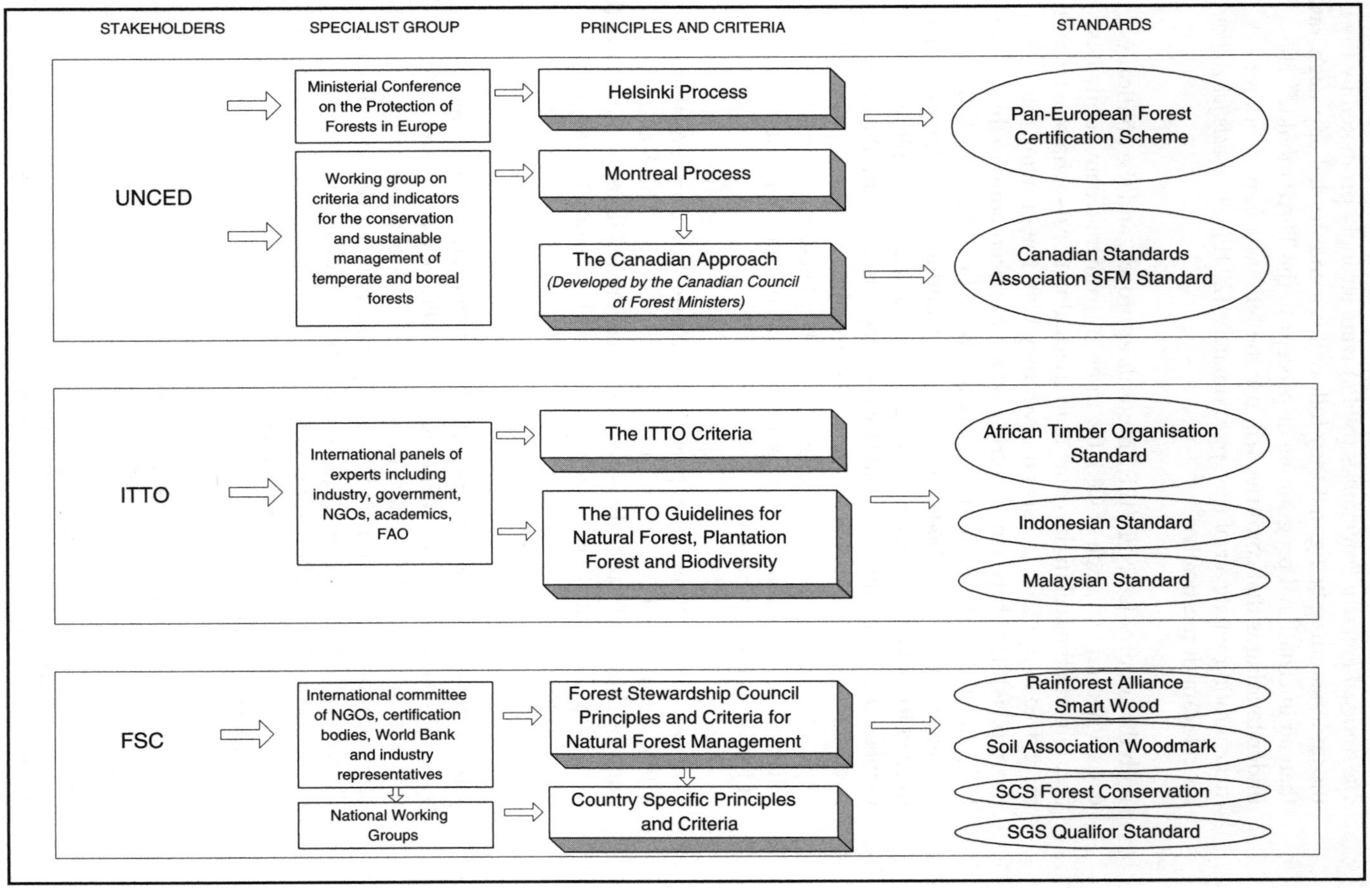

| STAKEHOLDERS | SPECIALIST GROUP | PRINCIPLES AND CRITERIA | STANDARDS |

UNCED

- Ministerial Conference on the Protection of Forests in Europe → Helsinki Process → Pan-European Forest Certification Scheme
- Working group on criteria and indicators for the conservation and sustainable management of temperate and boreal forests → Montreal Process
- The Canadian Approach (*Developed by the Canadian Council of Forest Ministers*) → Canadian Standards Association SFM Standard

ITTO

- International panels of experts including industry, government, NGOs, academics, FAO → The ITTO Criteria
- The ITTO Guidelines for Natural Forest, Plantation Forest and Biodiversity → African Timber Organisation Standard / Indonesian Standard / Malaysian Standard

FSC

- International committee of NGOs, certification bodies, World Bank and industry representatives → Forest Stewardship Council Principles and Criteria for Natural Forest Management
- National Working Groups → Country Specific Principles and Criteria → Rainforest Alliance Smart Wood / Soil Association Woodmark / SCS Forest Conservation / SGS Qualifor Standard

The National Forestry Programmes (NFPs) were strongly promoted by the IPF, the understanding being that they follow it is a country-led approach, rather than an international programme or precept in the TFAP mould.[37] The understanding is that countries will use the NFP as a framework for incorporating whatever of the approximately 120 IPF *'Proposals for Action'* are relevant domestically.[38]

With the country-led NFP/IPF Proposals approach, and the politically-correct neutral language of C&I, the intergovernmental community has securely put the ball in the court of individual nations – who remain pretty much free to use their forests however they like. The growing (but incomplete) trust between countries, which has been built up through the IFF process in particular, is necessary for this NFP/C&I approach to work at the international level. Better national policy and implementation capacity is needed for it to work locally. A weightier set of legal principles has been pioneered through international law connected to CSD and the multilateral environmental agreements (MEAs) – the precautionary, user pays, polluter-pays, and inter-generational equity principles. These have already been reflected in national laws, and could improve the national implementation of the NFP/C&I approach.

However, this 'bottom-up' approach puts a premium on national (and to an extent local) needs, and leaves out global dimensions. This leaves the question: how well are truly global issues covered? Figure 5.2 illustrates the range of global initiatives, which can be 'mapped' along two axes – from general/ political to focused/ technical, and from global-only to involvement in local conditions.

Global environmental services are partially covered under a number of MEAs – the Convention on Biodiversity (CBD), the Framework Convention on Climate Change (FCCC, which partially covers carbon reservoirs, sinks and sequestration), CITES (for traded rare species) and the World Heritage Convention (WHC, for managed landscapes of global heritage value). But few are well-informed by, for example, the C&I of good forestry, which means that they view forests in a very partial light, stressing only certain environmental functions.

37The notion of the NFP, as a generic programme (and hence the acronym should not really be capitalised), was developed by the Forestry Advisers Group, adopted by FAO, then endorsed by the IPF. Given that the FAG arose from the TFAP, the NFP represents an example of lesson learning. It can also be noted that the aid system may still have an important programmatic lever on NFPs through the mooted Forest Partnership Agreements, where countries will receive aid for keeping to certain of their NFP commitments. These agreements have also been proposed by the FAG.

38These Proposals for Action are the product of political negotiation, and are not written in operational language. Nonetheless, they form a useful 'checklist' of actions from which a country might choose in making the transition to SFM.

Figure 5.2 Global initiatives in forest policy

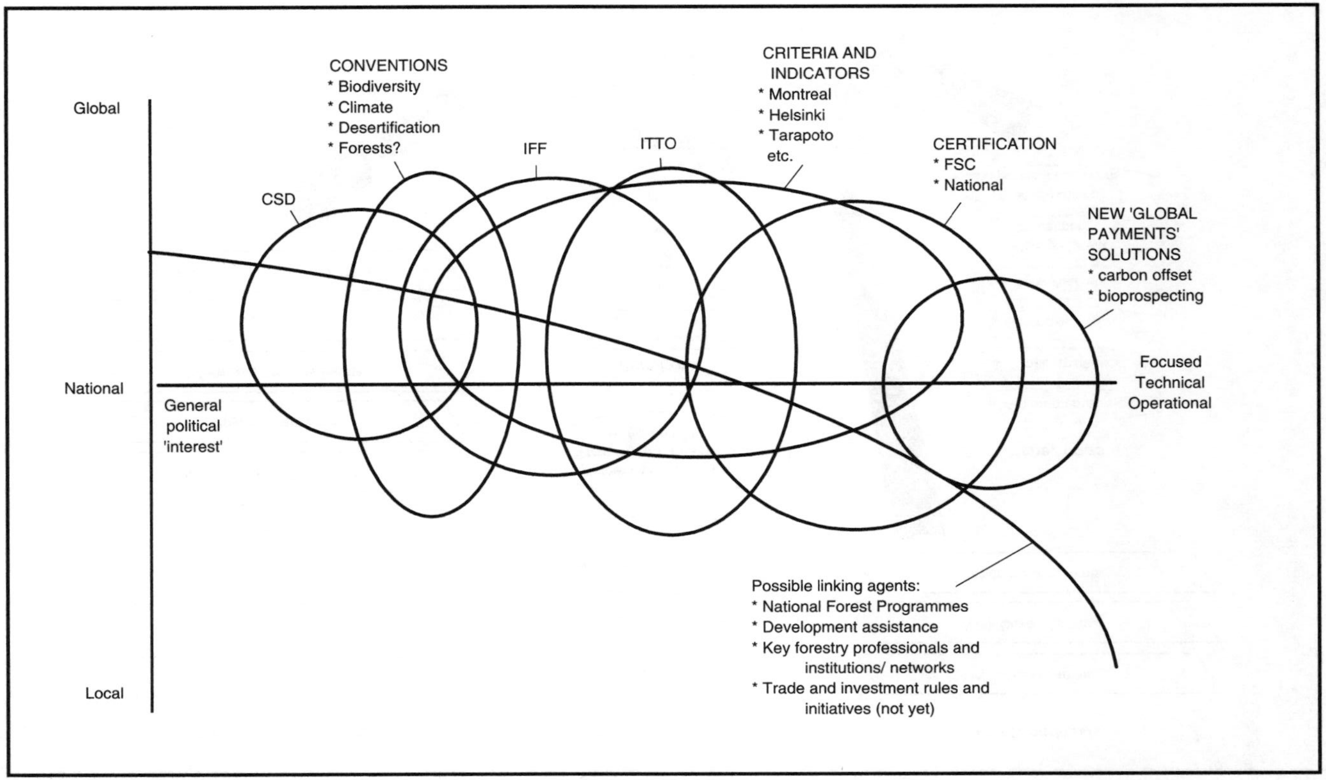

CONVENTIONS
* Biodiversity
* Climate
* Desertification
* Forests?

CRITERIA AND
INDICATORS
* Montreal
* Helsinki
* Tarapoto
etc.

CERTIFICATION
* FSC
* National

NEW 'GLOBAL
PAYMENTS'
SOLUTIONS
* carbon offset
* bioprospecting

CSD

IFF

ITTO

Focused
Technical
Operational

Global

National

General
political
'interest'

Local

Possible linking agents:
* National Forest Programmes
* Development assistance
* Key forestry professionals and
  institutions/ networks
* Trade and investment rules and
  initiatives (not yet)

Source: After Thompson, 1998

Figure 5.3 Countries are bombarded by international processes

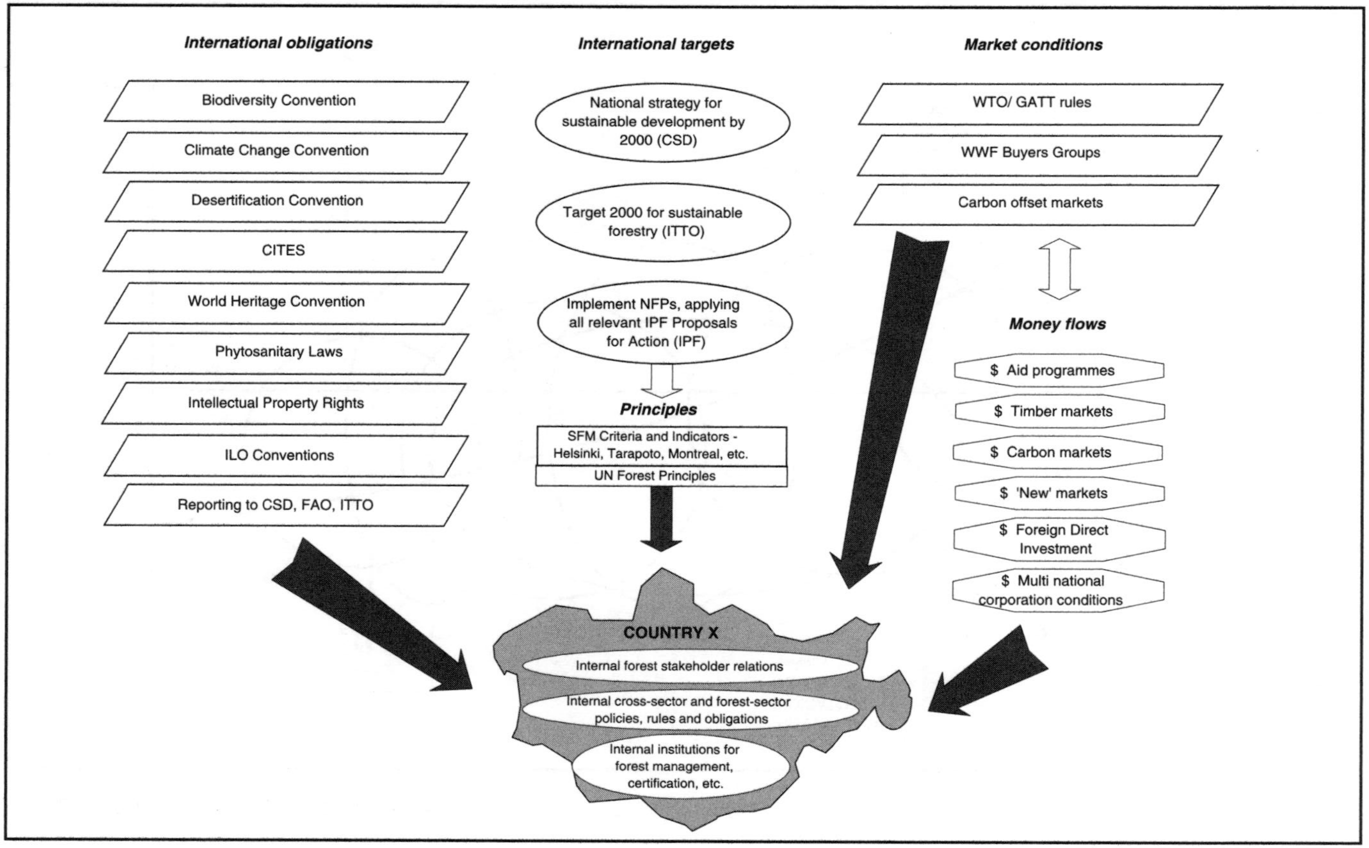

International obligations

- Biodiversity Convention
- Climate Change Convention
- Desertification Convention
- CITES
- World Heritage Convention
- Phytosanitary Laws
- Intellectual Property Rights
- ILO Conventions
- Reporting to CSD, FAO, ITTO

International targets

- National strategy for sustainable development by 2000 (CSD)
- Target 2000 for sustainable forestry (ITTO)
- Implement NFPs, applying all relevant IPF Proposals for Action (IPF)

Principles

- SFM Criteria and Indicators - Helsinki, Tarapoto, Montreal, etc.
- UN Forest Principles

Market conditions

- WTO/ GATT rules
- WWF Buyers Groups
- Carbon offset markets

Money flows

- $ Aid programmes
- $ Timber markets
- $ Carbon markets
- $ 'New' markets
- $ Foreign Direct Investment
- $ Multi national corporation conditions

COUNTRY X

- Internal forest stakeholder relations
- Internal cross-sector and forest-sector policies, rules and obligations
- Internal institutions for forest management, certification, etc.

The *global causes of forest problems* (most of which are 'extra-sectoral' to forestry) are not well-addressed at all. Only when we 'ratchet up' the forest debate into these circles will the most powerful underlying causes of forest problems – trade, investment and debt conditions – be addressed. There is precious little intergovernmental action to consider the developmental and environmental aspects of forests and forest-dependent people in such crucial negotiations as those connected to the WTO, a mooted multilateral agreement on investment, and international debt.

The net result is a bombardment of individual countries with a 'policy inflation' of often conflicting or overlapping international obligations, principles, and programmes, which somehow need reconciling and marrying with local needs. Figure 5.3 illustrates these. It will be appreciated that if all the possible links between these international forces were drawn, the diagram would be impossible to read. Yet, in practice, countries are expected to trace all such links. It is not surprising that many international links remain 'ghettoised' within individual government departments, who are perhaps unwilling to coordinate with others, or else the links are plainly ignored (Thomson, 1996). This is evidently the case in all six PTW case study countries.

The forest convention dilemma – identifying the limits to intergovernmental processes?

Confusion over what is a global issue, and the complexities of relations between nation-states and the intergovernmental community, is nowhere more evident than in the contention surrounding a legally-binding global forest instrument.

Some parties suggest that a *global forests convention* will improve attention to global forest issues, improve global governance, and/ or increase control over, and support to, national and local forest use. Some of them would wish to do this by 'hardening' the Forest Principles, Agenda 21, and the IPF Proposals for Action. But others may be pushing for a convention to justify delaying action on forests under existing conventions, notably the CBD.

A *legal* instrument will not guarantee the improved *political* attention that forests need at the global level. Rather, it may weaken other practical and political efforts to improve forests by being unnecessarily legalistic; or it may enshrine unsustainable or inequitable objectives – partly because the wrong people are involved in negotiating intergovernmental agreements. Indeed, some observers suggest that we may be reaching the limits of what can be achieved through intergovernmental initiatives, citing how the

Figure 5.4 Questions to be answered regarding further international regulation on forests

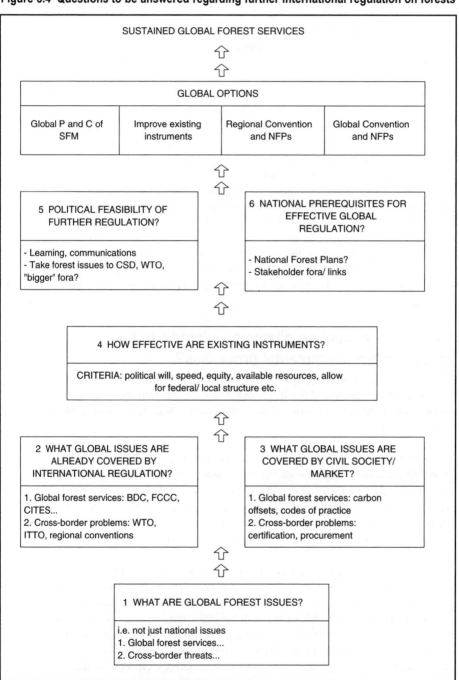

SUSTAINED GLOBAL FOREST SERVICES

⇧
⇧

GLOBAL OPTIONS			
Global P and C of SFM	Improve existing instruments	Regional Convention and NFPs	Global Convention and NFPs

⇧
⇧

5 POLITICAL FEASIBILITY OF FURTHER REGULATION?

- Learning, communications
- Take forest issues to CSD, WTO, "bigger" fora?

6 NATIONAL PREREQUISITES FOR EFFECTIVE GLOBAL REGULATION?

- National Forest Plans?
- Stakeholder fora/ links

⇧
⇧

4 HOW EFFECTIVE ARE EXISTING INSTRUMENTS?

CRITERIA: political will, speed, equity, available resources, allow for federal/ local structure etc.

⇧
⇧

2 WHAT GLOBAL ISSUES ARE ALREADY COVERED BY INTERNATIONAL REGULATION?

1. Global forest services: BDC, FCCC, CITES...
2. Cross-border problems: WTO, ITTO, regional conventions

3 WHAT GLOBAL ISSUES ARE COVERED BY CIVIL SOCIETY/ MARKET?

1. Global forest services: carbon offsets, codes of practice
2. Cross-border problems: certification, procurement

⇧
⇧

1 WHAT ARE GLOBAL FOREST ISSUES?

i.e. not just national issues
1. Global forest services...
2. Cross-border threats...

IPF/IFF have repeated many cycles of argument but appear to have evaded decisions, and noting their weak capacity to truly incorporate civil society and market players. Or they state simply that a convention will be toothless, as no-one is willing to offer funding to compensate for the implied relinquishing of sovereignty over some aspects of national forest land.

International regulation is time-consuming to negotiate and operate, there will be opportunity costs and diminishing returns, and possibly no real outcome – as Mankin (1998) points out. There has been too much discussion of form, and too little on purpose – delinking international debate from both scientific and local socio-economic realities. The priority at this stage is for nations to individually address the question: *what forest issues are best dealt with by international regulation and, for each, how can today's regulatory environment be improved?* Figure 5.4 suggests a process for this (Bass, 1998a).

5.3 'Soft policy' – international civil society initiatives and the case of certification

Civil society initiatives have been one of the defining features of the international forestry stage over the last decade. Certification is a key initiative – does it/ will it work? In trying to answer this question we will touch on the experience with certification in some of the PTW focal countries – Ghana, Costa Rica and Papua New Guinea – but our major focus is on the effectiveness, efficiency and equity of certification as an international policy.

Forest certification provides independent verification that the wood in a product originated from a forest managed in accordance with certain standards (Box 5.1). The 'ingredients' of the certification process may be described as 'soft policy', as they cover: a standard describing what many stakeholders agree is good forestry; a means to audit the achievement of the standard; a rewards and sanctions structure (through the market); and a multi-stakeholder governance structure. For most of these provisions, however, there are usually 'hard policy' parallels and precedents with slightly different aims and assumptions, winners and losers. In other words, the forest management and trade purposes of certification could also be achieved by regulations or other alternatives. This has led to much contention.[39]

In the last decade, certification has become a major vehicle through which the forest policy process has widened from a government-led affair, to one with

---

39 However, certification is synergistic with the government-led regime: a sound legal and policy framework is needed for certification and related markets to operate; and adherence to laws is a fundamental condition of certification.

Box 5.1 What is certification and how does it work?

Forest management certification is a relatively new type of procedure. A third party inspector (the certifier) gives a written assurance that the quality of forest management practised by a defined producer conforms to specific standards. It is conceived as a voluntary procedure, which buyers may choose to specify, and which producers may choose to employ. By providing information about the origins of a traded forest product, certification attempts to link market demands for products produced to high environmental standards with producers who can meet such demands. As such, it has the potential to act as a market incentive for better forest management. Forest certification has evolved since 1989, and is part of a general trend to define and monitor standards for environmental and social improvements in natural resource use.

The general practice of forest certification is as follows: At the request of the forest enterprise, the third party certifier conducts:
* an independent audit of forest management quality,
* in a specified forest area,
* under one management regime,
* against specified environmental, social and economic standards;
* by assessing documents which prescribe and record management, together with checks in the forest,
* followed by peer review of the assessment,
* resulting in a certificate for a period and/ or a schedule of improvements ('corrective action requests' or CARs)
* plus regular checks thereafter to maintain the certificate.

The three main approaches to forest certification are:

The Forest Stewardship Council (FSC) approach: this is currently the only established international system of forest management certification. The FSC was established precisely for the purpose of forest certification to promote high performance standards. The approach offers a global set of Principles and Criteria (P&C) for good forest stewardship; an international accreditation programme for certifiers; a trademark which can be used in labelling products from certified forests[40]; and a communication/ advocacy programme. At present the FSC-accredited schemes are dominant.

The International Organisation for Standardisation (ISO): offers a framework for certification of environmental management systems (EMSs) through its ISO 14000 series. This covers similar ground to forest management certification except that it does not specify forest management performance standards, and does not confer a label on products, severely limiting how products can be promoted in the market. It certifies the EMS rather than the forest. In some instances, companies are having their EMS certified in preparation for forest performance certification under FSC or a national scheme.

National certification programmes: some are developed under the aegis and following the procedures of the FSC. But others are independent e.g. in Indonesia, Malaysia, Finland, Norway, Canada and an emerging approach in Ghana. Many of these combine elements of the FSC performance-based approach and the ISO process-based approach.

---

40 'Chain of custody' also certifies the route of products from the forest through the processing chain and verifies that the product is indeed from a certified forest.

greater interaction of interest groups and forestry professionals. Indeed, government has been a rather marginal actor in the whole process.

How and why has this form of 'democratisation' of forest policy arisen, what are its impacts, and how does it fit with the regulatory regime?

The actors and their interests

Certification is very much a product of different civil society groups acting together, both influencing and using the power of the market. The key actors are:

- Civil society as concerned *citizens*, through their membership of *environmental organisations*. WWF has been the major protagonist for forest certification since the inception of the idea (by several small NGOs). Now many other NGOs also support the approach. There was considerable frustration with the slow pace of change in forestry that was being achieved through intergovernmental processes; and NGOs believed themselves to be having little impact on these processes. On the other hand, NGOs had been gaining experience in relations with the private sector. Independent certification was a logical step beyond NGOs 'selling the panda' to companies; instead of merely raising money from companies to be spent on environmental projects elsewhere, certification might actually directly influence that company's forestry practice. The NGOs' concern to expose bad forestry practice, and only secondarily to improve adequate practice, has had a significant influence on the way that certification has turned out.

- Civil society as *consumers*: In north western Europe and, to an extent, north America, consumers have increasingly been demanding products which do not contribute to environmental degradation and social disruption or inequity. They have been buying more labelled products. The evidence is that consumers are becoming more discriminating about the type of label and its origin. Independently-awarded labels have higher credibility than producer self-labelling. Sometimes certified products occupy only a small market niche; at other times a larger proportion of production has been affected, as with recycled paper (although this may require legislative support).[41]

- *Major forest product traders and retailers*: Whilst civil society as a whole has a latent interest in social and environmental conditions – which can be

---

41 A note on civil society as *investors*. This is perhaps the current 'sleeping giant', which may soon be more active in encouraging certification. Whilst the amount of 'ethical investment' is but a small fraction of one per cent of stock market capitalisation in the UK, for example (Grieg-Gran *et al*, 1998) there is increasing interest – in some *portfolio investment*, and in certain *stock markets* – in looking at social and environmental standards in addition to the normal financial parameters. WWF is now trying to actively drive fund managers toward certified forests through its 'Forest Finance' initiative.

sharpened by groups such as WWF – the real drivers of the certification process have been forest product buyers and retailers in Europe and north America. They have established the 'brand' of certified timber, using their market power and the influence of advertising to make it really take off. In many countries, they have grouped together into Buyers Groups.[42] Whilst these groups have in practice supported FSC-based certification only – since, in fact, this was the only forest certification system until recently, the ISO approach merely certifying the management system rather than forest management – the signs are that many buyers are open to certification approaches other than FSC i.e. ISO-based certification, national forest certification systems, or fair trade certification.

- *Forest producers*: Two types of producers have been significant in bringing certification into the mainstream. One is larger producers, usually with existing high forest management capacities and a desire to demonstrate this to the market (i.e. moving beyond their original defensive approaches to environmental criticism, and the subsequent promotional approaches, to transparency through independent verification). These companies have tended to be based in northern countries. The second type is at the other end of the spectrum, i.e. small, niche operations, often with a community level involvement, and development assistance or philanthropic funding. The key actors behind the latter's support of certification have often been the technical assistance personnel, and the donors, who viewed certification as both a means to demonstrate the (complex) multiple use management which they were trying to forge locally, and as a means to gain niche markets for the project's products.

- *The forest certifiers*: The 'front line' in certification – in terms of developing and refining standards and procedures, driving it into normal procedures of many companies, and taking the flak where things go wrong – are the certification bodies. These have been of two basic types: the environmental groups who saw certification as a way to achieve their missions, and possibly to be able to earn income (the Soil Association in the UK, and the Rainforest Alliance in the USA); and the professional audit companies who saw forest certification as an extension of business into a new sector (SGS in the UK, and SCS in the USA). Other certifiers have since joined the fray and, in addition, many forestry consultants and audit companies – including many in developing countries – have become qualified local assessors for the certifiers. Between them, certifiers and assessors include some highly experienced foresters, as well as networks of specialists such as sociologists. The development costs of certification, the changing rules of the game (e.g. the certifiers' own standards having to give way to FSC's

---

42 Whether or not these buyers' groups amount to cartels is an issue worth exploring. For example, WTO rules proscribe cartels.

Principles and Criteria), the intensive participation required in developing and implementing standards, and other 'start-up' tasks, have all contributed to the fact that certifiers have been making profits only in the last year or so (if at all).

- *The Forest Stewardship Council*: The FSC is one of the most significant forestry institutions of recent years. The FSC, and its accredited certifiers (above) offer the only established international system of forest certification. FSC was established precisely for the purpose of certification. It operates a complete package: global Principles and Criteria of forest stewardship, an accreditation programme for certifiers, a trademark to be used in labelling, a communication/ advocacy programme, and a multi-stakeholder governance structure. It was designed by leading foresters, with WWF very much as a godfather, to deal with contemporary forest problems, especially environmental degradation in the tropics. It reflects both the sustainable development paradigm (its governance structure being split into equal 'economic', 'social', and 'environmental' chambers) and current trends in international environmental diplomacy (there being parity between Northern and Southern votes within its membership).

The story – certification as a 'soft' forest convention

Whilst we have listed the FSC above as an actor in certification, in many ways it is also the principal *product* of a 'soft' international policy process. It embodies, in effect, a civil society convention on good forest management, with a market-based approach relying on voluntary implementation. It directly addresses failures of government and of a large part of the private sector, and thus asserts civil society rights to a part in decision-making on forests.

The analogy with a legal convention might also be extended to what appears, at first sight, to be FSC's 'top-down' nature. Its relatively rapid negotiation, its global prescription of forest management standards (with its heavy reliance on the work of a few experts), and its centralised board structure were all legitimate and necessary responses to the worsening global forestry situation, and an appropriate way to get started with much momentum. But in spite of the diverse membership and many provisions for participation, these characteristics do describe a (top-down) council (reflected in FSC's name) rather than, say, a decentralised network or federation. This might be contrasted with the development of IFOAM, the International *Federation* of Organic Agricultural Movements. The latter was the result of a much slower, diffuse process – the development of multiple and very different organic standards in many countries and for different products, over decades, and the eventual need for mutual recognition.

Whilst, FSC has not yet had time to develop so slowly – the policy response had to be different – FSC has been introducing measures to ensure it responds better to local situations, and to emerging situations and knowledge. Whilst its room for manoeuvre is limited by the threat that rapid change might dissatisfy certain members, or chambers (see below), and lead to schisms in membership, a number of provisions have been put in place, e.g.:

- The establishment of National FSC Working Groups, principally to develop national standards based on the global P&C. Once these standards are available, all certification in the country has to be performed against them – which diminishes the problems that have arisen when a certifier both interprets global P&C in a certain way and then judges forest management according to that interpretation. Working groups have indeed formed in many countries, including Papua New Guinea and Costa Rica, although they are often keen not to align themselves wholeheartedly with FSC.

- The establishment of many working groups on the concerns of different producers or products, such as small group certification, and percentage-based labelling for production where it is not possible to have 100 per cent certified product in the mill.

- Its revision of governance structure (from two chambers, one with economic interests having 25 per cent of the vote, and the other with environmental and social interests having 75 per cent of the vote; to three chambers, with the three interest groups having 33 per cent votes each).

- The 1998 memorandum of understanding with the Indonesia Ecolabelling Institute to work *towards* mutual recognition between their two, rather different certification programmes. FSC recognises the need to bring in more bottom-up perspectives (in terms of recognising more land use systems, and perhaps more parallel certification initiatives from different countries).

The last point is significant, for the certification story has broadened from that of FSC as the central plot, to one with multiple plots. Whilst the FSC initiative, and those of certifiers who were eventually to become FSC-accredited, really took off after 1993, some actors saw themselves as not gaining by engaging in the FSC policy process. They started their own equivalents. There was soon an initiative by ISO members. There was also a number of national and regional certification schemes spurred on by particular groups of actors – variously forest owners, government or forest industry – with others starting even now. Each of them have their own protagonists, with different motivations.

The *International Organisation for Standardisation (ISO)*, through its ISO 14000 series, offers a framework for the certification of environmental management systems (EMSs). From 1995, Canada and Australia led an initiative to encourage the application of ISO 14000 to forests – a process which NGOs and actors from smaller countries found difficult to counter because of the costly and somewhat closed nature of ISO working groups and meeting schedules (Mankin, 1998).

The resulting approach – ISO 14000 certification of the Environmental Management Systems of forestry companies – covers similar ground to forest management certification, except that it does not specify forest management performance standards, and does not permit a label to be attached to products. The EMS is certified, rather than the forest. Although not strictly a forest certification programme, the ISO approach offers much potential for indicating the environmental objectives of forest management and how they are being achieved, and a number of companies have applied it (notably plantation-based enterprises). In response to criticism from the 'FSC camp', an ISO Technical Committee Working Group prepared an 'information document' on the various forest performance standards available, including FSC's, to help enterprises incorporate relevant standards into their EMS. The ISO approach has been strongly promoted by some larger companies, especially those involved in plantations, which have been used to ISO quality and environmental management procedures in many other operations and sectors. ISO is also well-recognised by government authorities and by WTO, which considers ISO standards to be acceptable technical barriers to trade.

As Mankin (1998) notes:

> It is not ISO, or an EMS approach per se that concerns environmental NGOs. Indeed, many of the FSC's procedures are based on ISO guidelines, and the FSC's Principles and Criteria contain several management system components. Furthermore, the FSC's performance standards and forest certification requirements can easily be combined with an ISO EMS; this is a particularly important option since the ISO EMS standard lacks any requirement for a specific level of performance. Regrettably, the heated ISO debate has made it very difficult for proponents of either approach to acknowledge and promote this inherent compatibility.

National certification approaches have been developed, or are in preparation, in Canada, Finland, Norway, Malaysia, Indonesia and Ghana, and there is a regional Pan-European approach. These are significant forest countries, in which the private sector has considerable policy influence (see section 5.4). These countries have tended to develop hybrid certification approaches including local performance standards, with some consultation by local stakeholders in defining these standards, and a management system approach that allows an encouragement and recognition of improvement. Governments have been much more active in these processes than they have in FSC's approach (as have larger companies). As such, the national systems often integrate the principles and criteria to which government has committed itself, e.g. those of ITTO in the case of Indonesia, the Montreal Process in the case of Canada, and the Helsinki Process in the case of Finland and the Pan-European system. In this way, the national approaches can be seen as much closer to 'hard' policy than FSC's. For example, the evolving Ghanaian certification system takes the emerging national forest management standards (very much a government-led affair, albeit with some participation) as its basis.

The involvement of government in national certification systems also means that certification tends to fit more closely with wider national policy objectives, as opposed to being an instrument aimed principally at improving global forest management using export markets:

- A principal purpose of the *Ghanaian* system is to provide a competitive means to reduce the annual timber harvest by one-third. There was an intention to sort out a system that is right for Ghana's forests and development, and building on the recent development of a Ghanaian accreditation body, before exposing the country to FSC influence – getting their own house in order and thus avoiding the perceived potential pitfalls of an internationally-defined approach.

- One purpose of the *Indonesian* approach is to cut down on government bureaucracy and inspection – privatising certain public functions in terms of forest planning and monitoring.

- In *Costa Rica*, some forest units have been certified by external certifiers since 1993 but a new Forest Act in 1996 provided for a national system to be developed based on national standards. At the same time, an FSC-inspired civil society working group was set up. A government committee has since been collaborating with the working group and other project-based initiatives. The first standards developed were for large plantations, but since then the core focus has been on standards against which groups of smallholders can be certified. Thus certification is being seen as a tool to enhance the recent policy emphasis in Costa Rica on smallholder forestry (see section 4.1 and 4.5).

- The UK Woodland Assurance Scheme is the result of a very close marrying of governmental audit needs (required to approve and monitor felling (planting permissions and related grants) and FSC requirements. Whilst there are efficiency and market-related reasons for this, the whole process of marrying the requirements has been centrally concerned with building relations between three groups of stakeholders – government, producers, and environmental groups. Indeed, its wider impact may be highly significant. The UK experience demonstrates the possible advantage of undertaking a home-grown policy approach to considering the purposes of – and possible options for – certification, rather than working within the confines of a prescribed FSC national working group alone. In the UK, FSC helped the process, but as one actor, rather than as script-writer, stage-manager and star.

In all of these national certification policy processes, certain individuals within government and certain producers have been key. So also has been the perceived policy 'threat' to nationally-important forest industry from what was believed to be a quick-fix international (and non-governmental) approach to certification (FSC). The engaged producers are not necessarily the largest ones – although in Canada, it was the Canadian Pulp and Paper Association which was a major protagonist for a national approach. In Europe, small producers have been significant – MTK in Finland considered that its smallholder members could not support FSC approaches, largely because of its cost; the Pan-European Forest Certification Scheme was developed to suit small forest owners, and will work through group or 'umbrella' certification.

The development of some of the national approaches illustrates an interesting alliance of producers with government. Where before producers would complain about regulatory requirements, now they were extolling the virtues of them (both in terms of high forest management standards and in terms of their efficiency in achieving forestry objectives). Some, in reality, may have been wary of the perceived higher standards and costs of FSC certification, and of the stringency that a privatised approach to audit would bring. The national certification programmes have drawn heavily on government commitment to various C&I approaches, and are themselves a manifestation of the concerns for sovereignty which led to the C&I approach in the first place. As such, the national certification programmes may be major vehicles for the implementation of C&I.

Table 5.2 A summary comparison of FSC, ISO and national approaches to forest certification

ISSUE	FSC: a performance-based approach	NATIONAL CERTIFICATION PROGRAMMES mixing performance and process	ISO 14001: a process-based approach
Main protagonists	Environmental and some social NGOs; Retailers and organised buyers' groups	Governments, producers and some trade associations of (major forest-exporting) countries	Industry, especially large producers which undertake ISO certification in other fields; Governments; WTO recognises ISO
Inherent values	Industry, especially large producers which undertake ISO certification in other fields; Governments; WTO recognises ISO	Building on sovereignty concerns (governments) and producer group needs (especially small groups). Build on Rio-influenced C&I approach and subsidiarity principle	'Value-neutral': modernist; enterprise-focused; continuous improvement
Purpose	Define good forest stewardship and accredit certifiers; third party certification essential; Labels and chain of custody can be provided to the market	Define good forestry in a national context; demonstrate consistency with intergovernmentally agreed C&I; present an alternative to FSC (varied reasons behind this)	Specifies elements of management system to improve performance; third party certification optional; Certification permits general publicity, but no labels
Standards	Performance standards based on global P&C, encouraging compatible national standards; normative	Some mix performance standards (based on intergovernmental C&I e.g. Canada and/ or national government standards e.g. Ghana) with a management system approach; others focus on performance, such as Brazil, Finland and Malaysia	Management system standard; No forestry performance standards specified – but a working group's Information Document suggests options

Governance	An NGO; NGO/ private members; equal economic, social, environmental chambers, with North/ South balance	Quasi-government body or NGO; may be connected to national standards or accreditation body; usually some consultation/ representation mechanism	An NGO; members are national standards bodies
Accreditation	An international accreditation body itself	Through a recognised national body (Canadian Standards Association, Ghana Standards Board)	National accreditation bodies
Approach to SFM	Stresses high environmental and social performance – challenges the manager; Avoids 'sustainability' – emphasises stewardship as activities which might eventually be proven sustainable	Varies, but based on understanding of best national practice (Ghana) or national commitments to an international C&I set (Canada)	Stresses management capacity and continuous improvement; Enterprise chooses performance standards; social standards difficult to integrate
Credibility with stakeholders	High with NGOs/ buyers; Lower with some governments; Mandate problems; risks of 'monopoly'	Credibility within the country can be high, but not necessarily with local NGOs or with producers who aim at FSC certification; currently weak/ unknown credibility with international trade. Little current credibility with buyers groups.	High with intergovernmental bodies and industry; low with NGOs/ others; Narrow participation; no chain-of custody certification reduces market potential
Trade distortions	Environmental standards may be considered too high by WTO; Social standards may be considered unwarranted	Potential confusion between different country labels; may be possible to minimise this through standards equivalency procedures, or mutual agreements with FSC (Indonesia), or harmonisation	WTO-Technical Barriers to Trade policy recognises ISO; ISO standards are not considered unnecessary trade restrictions

Principal source: Bass, 1998b

Box 5.2 Stumbling blocks for certification in Sweden

The recent attempt to agree an FSC National Standard for Sweden proved highly divisive. Forest industry and environmental NGOs supported the process. The big forestry companies embraced FSC certification and many of their estates were approved for certification soon after the standard was agreed. The Swedish Federation of Forest Owners' Associations, however, withdrew from the working group after 18 months of discussion, citing various reasons:

- proposed standards were perceived to discriminate in favour of the forest industries with significant economies of scale and vertical integration;
- standards would be more onerous in southern Sweden which is more ecologically diverse, and which is dominated by private forest ownership;
- perceived lack of understanding of the manner in which policy is traditionally negotiated in Sweden – the decision to persist with a 'national' standard despite the non-involvement of forest owners representing 60 per cent of production was criticised for breaking the consensus approach to policy;
- perception that the working group was over-representative of those who have a clear interest in FSC certification;
- chain of custody arrangements remained expensive;
- resentment of extending rights to the Sami people. This was characterised by the federation as undue political interference in an issue that democratic government legislation had already addressed.

The forest owner associations are now in the process of developing their own certification systems, possibly based on ISO 14001 or EMAS, and are exploring possibilities to cooperate with equivalent organisations elsewhere in Europe. It appears that the Swedish Federation of Forest Owners' Associations accepts the forest industry companies' decision to participate with the FSC standard as a commercial judgement.

Source: Roberts, 1999

Table 5.3 Certified forests to January 1999 under FSC's P&C

1. Total number of forest management certificates: 154
2. Total area of certified forests: 15 million hectares
3. Country concentration:
    - Sweden, USA, South Africa, and UK count for 58 per cent of the certificates
    - Sweden, Poland, USA and Zambia account for 85 per cent of total certified forest area
    - Sweden alone accounts for 52 per cent of the certified area
4. Regional concentration:
    - developed countries account for 80 per cent of certified area; developing countries 20 per cent
    - developed countries account for 66 per cent of certificates; developing countries 34 per cent
5. Types of certified companies and the percentage of global certified area:
    - industrial companies: 66 per cent
    - state operations: 30 per cent
    - communal/ group companies: 2 per cent
    - non-industrial companies: 0.8 per cent
    - resource manager certificates 0.3 per cent

The policy impacts of certification

In comparison with many intergovernmental initiatives, certification, and the FSC approach in particular, was both based on a highly aspirational set of goals, and was also designed to be implemented from the very beginning. It is fast leading to many changes. Whilst as of January 1999 there were only 15 million hectares of certified forests at present, and 80 per cent were in the North (Table 5.3), many more millions of hectares are in the process of certification. The prospects (or 'threats') of certification are being hotly debated in many countries. And momentum is always being added: the World Bank has recently called for the certification of 200 million hectares, divided between North and South, by 2005 – one result of a potentially powerful alliance with WWF at the very highest level.

Thus the most significant impacts of certification will soon need to be assessed – or anticipated – at local, national and global levels. These may be:

1. At the *forest* level – specifically changes in actual forest management, and in stakeholder capacities, rights, responsibilities, revenues and relationships, and on the relative distribution of costs and benefits. Which groups will really improve their forest management as a result of certification – or will it merely reward a few currently-responsible, well-resourced timber producers or countries, as at present?

2. On the evolution of *sustainable markets* – how far is certification influencing actors further 'up' the supply chain from forest producer to final consumer, to consider social and environmental factors? Where will it affect the market mainstream, or will it merely define a niche?

3. On *evolving policy and institutional* arrangements and frameworks that are generally required for SFM at the national level (Figure 5.5).

There is little evidence as yet about the local and market impacts (although IIED is leading research in the area, building on e.g. Markopoulos, 1998). There are already some indications, however, concerning policy and institutional impacts. There are two basic types of country where policy/ institutional impacts of certification are already evident: those where there has been much certification activity; and those where there has been much debate about certification (which tends to be those with significant dependence on import/ export of wood, and/ or problems of public trust in forestry). In both, we can begin to see evidence of certification's contribution to good policy:

Figure 5.5 Certification in relation to the foundations of SFM

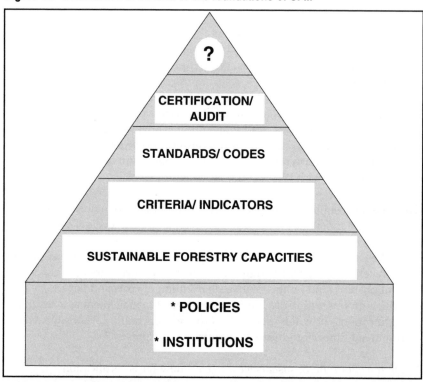

Certification depends upon a solid foundation of good policy, capacity and meaningful standards. But its market basis may also confer it with power to fast-track the development of these fundamentals. (Note how the pyramid is topped by a question-mark – as new instruments are bound to arise with increasing institutional sophistication).

- In terms of *participation* at the international level, the invention of the FSC provided a new kind of forum for forest policy issues that was not encumbered by intergovernmental procedures and norms. At the national level, the various types of certification working group have brought together different actors who had yet to meet. In *Ghana*, for example, this group has potential to address more forest issues than those immediately concerned with certification. In *South Africa*, those companies which have achieved certification take considerable pride in it and have gone on to make internal procedural improvements, and have become more involved in national and international policy processes, as a result. However, whilst discussions on certification may both have brought people together and brought policy debates into sharp focus, the actual development and practice of certification is not necessarily the direct answer to all these debates. Sometimes policy discussion of a specific tool can lead to more progress than discussion of an issue.

- In terms of *privatisation and decentralisation*, the experience of certification has made it clearer how certain functions – notably audit and extension – can be hived off from government.

- In terms of fast-tracking *national standards* for commercial forestry, the global momentum for certification has provided a good impetus for bringing disparate (and often rather dilatory) initiatives together and rationalising them, as in Ghana, Costa Rica and Indonesia. To a considerable extent it has taken decision-making power away from some vested minorities. It may help to encourage the development of its own policy/ institutional prerequisites.

There remains the interesting issue of the *relationship between certification and regulation*, especially in countries where FSC approaches are operative and government has had little involvement.[43] Some governments do not hold great store by civil society approaches, which are not encouraged. In contrast, in other countries civil society approaches are considered legitimate, and concerns relate more to the relative efficiency between certification and regulation in achieving the aims of SFM.

As a market-based instrument (MBI), certification is considered potentially capable of internalising the costs of environmental protection with greater efficiency than conventional legal regulation. In practice, the distinction between certification and a regulatory regime based on performance standards may be a narrow one. Both offer producers a choice between meeting previously-defined targets or facing possible penalties. In the case of certification, penalties are exclusively financial (e.g. the possible loss of competitive advantage); in the case of regulation, they are also administrative and judicial.[44] However, although certification is nominally an MBI, the environmental objectives it encapsulates are not determined within a closed market system, but by public debate (with varying degrees of openness and participation), including by groups which may not participate formally in the market. This is what makes certification so legitimate from a policy perspective. Nonetheless, it also means that there is a trend towards transferring a great deal of the control over policy direction

---

43 Thanks to Matthew Markopoulos, Oxford Forestry Institute, for material on this subject.
44 Regulation forces producers to comply, whereas certification only offers inducements. However, certification could become equally coercive if the market adopts it as a de facto product standard.

previously held by governments to civil society and market constituencies. Although wider participation increases the strength and credibility of the approach, the degree to which government influence can, or should, be reduced is limited. Not only do government interventions play a major role in shaping and influencing timber markets, but also the widespread adoption of certification will be difficult, if not impossible, without a legislative framework that encourages the development of sustainable forest-based capacity and enterprise, and standards which apply to all forests, including those aimed at certified markets.

The fact that certification and conventional regulation are synergistic, rather than antagonistic, perhaps supports the case for their closer integration and a clear role for governments in the development of certification (viz. the legal backing that is given to organic agriculture standards and certificates in many countries). The synergies will take different forms in varied countries. In some cases, certification will not be possible without the 'foundation stones' of good institutions, capacity, etc (Figure 5.5). In others, it may be that the momentum of certification will help to fast-track the establishment of these fundamentals.

If legitimacy and efficiency reasons point to a place for certification alongside regulatory regimes, a final policy concern is *equity*. Here, the picture is less clear-cut. Some of the social and environmental standards integral to certification would seem to support the case that certification can lead to local livelihood improvements[45] – although this tells us little, yet, about intra-community benefit distribution or the patchy nature of impacts amongst different communities. However, there are concerns about equity between enterprises and countries. This is partly because we are already seeing that most certified producers are large companies in rich, northern countries, those in the South being largely plantation-based companies or small, aid-supported operations. There is an impression that certified producers are an élite club, supported by richer (northern) consumers, and with membership restricted because of the high costs associated with certification and management to those standards. Should small producers be held as accountable as multi-nationals?

There are several dimensions to the issue of equity:

1. The nature of *any market-based instrument* means that competition is recognised and encouraged; there is therefore a limit to how inequality can be addressed by those involved in certification.

---

45 There is some evidence of certifiers not sticking purely to inspection of the immediate social impacts of forest management, but rather getting actively involved in issues of enterprise ownership, local rights, and social organisation – which could be counterproductive if the certifier does not understand the context very well.

2. There are structural/ governance-related inequities in many countries which, again, actors in certification can do little about.

3. Participation in the *development* of certification schemes and standards; as we have noted, often there is broad participation, which confers certification with its legitimacy, but this is not always the case.

4. The *standards* themselves and what kinds of groups they favour in relation to 4-6 below; clearly, the choice of social standards, and the applicability of all standards to different forms of forest use and technologies (not just those of rich companies), is important.

5. The ability to apply *capital, skills and other resources* to improve forestry and/ or to meet certification standards, and the ability to bear associated risks; this will favour larger companies, unless there are group schemes or forms of government support or underwriting.

6. Access to *information* about certification and markets for certified products; again, this may require the development of new services and government support to redress inequality in access.

7. The actual *impacts* of certification, or of the process of certification, on the distribution of costs and benefits in relation to livelihoods and capacities; here, close monitoring and the development of mitigating measures may be warranted.

Beyond integration with the national regulatory framework, a final set of questions relates to the integration of different certification initiatives internationally – the links of certification to intergovernmental C&I processes; and issues of assessing equivalency, negotiating mutual recognition, and working towards harmonisation, of standards and processes, in ascending order of ambition.

Intergovernmental institutions seemed initially to treat certification with suspicion, an inevitable result of their mixed government memberships which wanted neither discrimination on a country basis, nor the favouring of particular types of forest operation. ITTO's approach has been to keep a watching brief on certification (resulting in some papers by Baharuddin and Simula, the excellence of which may have partly been due to the 'detached' position they took). Recently, ITTO's C&I have been adopted in national certification schemes e.g. in Indonesia and Malaysia.

Regarding 'harmonisation', it is clear that the various certification initiatives are relatively young, and a period of continued experimentation, adaptation and indeed competition between them could help to improve their overall efficiency. However, ultimately some form of 'harmonisation' may be desirable for legitimacy and equity reasons, as well as to smooth trade flows. Here, the proposal of the WCFSD for a 'forest management council' to offer a forum for learning, impact assessment and debate amongst the many certification and C&I initiatives is helpful.

The point of such a council, however, is surely to recognise that certification has to be able to deal with local differences, and thus to reconcile the legitimacy of specific local standards with the need for 'translation' or mutual recognition, rather than forcing one approach.

5.4 Multi-national private sector influences – policies for investment in SFM, or for asset-stripping?

'Merchant princes' (Ghana), *'robber-barons'* (PNG), *'timber mafia'* (Pakistan)… the worst companies in the timber industry appear to have a deserved reputation for their predatory, gangster tactics in seeking out weak regulatory regimes, exploiting the loopholes they find, and overseeing cut-and-run operations which trash the forest and leave nothing but a few collapsing bridges and divided communities in their wake. But not all of the private sector is like this!

In all the countries studied in this project there are good examples of companies with long established track-records of fair management, decent environmental reputation and positive contributions to local and national development. They are not the norm, and policy processes need to better understand motivations and dynamics within the spectrum of private sector practice in between these extremes. Only then can frameworks of regulation, incentive and partnership stand a chance of effectively penalising bad behaviour, and catalysing good practice. In this section we focus on the multi-national contexts and actions of an increasingly globalised private sector in relation to policy processes.

The context and the actors

Rising demands on forests, globalisation and privatisation:

Global demand for industrial wood is expected to grow by nearly 20 per cent in the next fifteen years (IIED, 1996).[46] The private sector will be an increasingly important actor in meeting (and indeed creating) demands, as it offers many potential advantages, notably: production efficiency due to exposure to competition; technological development and transfer; and the ability to undertake long-term investments (Bass and Hearne, 1997).

As governments adopt more liberal market policies and are under pressure to reduce budget deficits, they are both relaxing their direct control over private sector activity and encouraging firms to provide many forest goods and services traditionally supplied by the public sector. Throughout the world, the private sector is taking a more prominent role in timber production (Table 5.4), and increasingly also in other forest goods and services – biodiversity, watershed and recreational management in a few areas – and the provision of forest management services. Over the last ten years, the pace of reform has reached unprecedented levels, and forestry sectors around the world are being transformed by moves to increase private sector participation. IIED's *Privatising Sustainable Forestry, A Global review of trends and challenges* (Landell-Mills and Ford, 1999) shows increased private sector roles in most countries, and in all the 23 countries which were examined in detail.

Significantly, governments are also restructuring their own roles to suit. The private sector has usually had a strong role to play in this whole process. Whilst this has often been positive, it has also led to losses, due in part to a concentration of political power in some companies. Either way, private sector tactics have usually been a rational business response to prevailing market forces and existing policies. These conditions may encourage short-term forest asset-stripping, or alternatively they may encourage forest management.

Those companies who want access to cheap forest assets thrive on weak policies, or chaos in the policy arena; forestry regulations are immaterial as long as the enforcement capability is weak. In contrast, those companies who want long-term investment require long-term policy stability, and thrive on an enabling approach which allows them to achieve SFM at the

---

[46] FAO's Global Fibre Supply Model of 1998 and Global Forest Outlook Study of 1999 puts further detail on the *supply* side of this picture: industrial forest product *output* by 2010 will be a quarter higher than at present (but only 10% higher than the 1990 peak). With the exception of Africa, supply potential exceeds projected consumption in every region of the world. Of particular interest is the huge increase in non-forest fibre sources (including trees on farms) in many parts of the world; the growing importance of plantations and the expected decline in the importance of natural forests (FAO, 1999).

Table 5.4 Private sector share of commercial timber extraction[47]

Country	Share of total output (percentage)
Africa Cameroon Ghana RSA	>90 per cent; 100 per cent natural forests; government utilises 42,000 hectare plantations <100 per cent; a third of concessions in reserve forests awarded to state companies 88 per cent of total commercial roundwood (1996/7); state enterprises supply over 60 per cent softwood sawlogs, planned privatisation of commercial harvesting
Asia China India Indonesia Malaysia	No data, but the private sector's share is growing Scattered private plantations produce 30-90 per cent for different states <100 per cent; private sector has 90 per cent of concessions in production forests; 10 per cent are held by state companies 100 per cent; 60 per cent from national production forests, 40 per cent state forests
Central America Costa Rica Honduras	100 per cent; all from private forests <100 per cent; private and community groups in state forests and extraction from private and community forests. State harvesting ended in 1992.
Eastern Europe & CIS Latvia Poland Russia Slovenia	100 per cent; all state and private forest harvesting done by the private sector 81 per cent total (6 per cent from private forests, and 80 per cent of the 94 per cent of total extraction from state forests) About 70 per cent (95 per cent of forestry enterprises privatised, but state retains shares to about 30 per cent of companies) 100 per cent; 37 per cent from privatised Forest Management Enterprises and 63 per cent from smallholders
North America Canada Mexico USA	100 per cent; 81 per cent from provincial/ territorial forests and 19 per cent private forests Just under 15 per cent of ejidos and indigenous communities with forest land sell standing timber to private loggers 100 per cent; 94 per cent of total from private forests, and all extraction from state forests
Pacific PNG	100 per cent; 88 per cent of exports by companies in community forests and 12 per cent by land-owning companies
South America Bolivia Brazil Chile	<100 per cent; all from state forests and private forests; uncertain share of output from community forests 100 per cent; 75 per cent from native forests, 25 per cent private plantation 100 per cent private; all from private plantations
Western Europe Finland Ireland UK	91 per cent from private forests and small share of state forest extraction 100 per cent; 5 per cent private plantations, 95 per cent from state forests 46 per cent from private forests; 27 per cent state forests auctioned standing; most of the state Forest Enterprise extraction contracted out.

Source: Landell-Mills and Ford, 1999

47 These figures include private sector extraction from private, state and community forests.

lowest cost. The space provided by forest policy alone is not enough to both stop asset-stripping and encourage good forest management. Hence the need to bring forestry concerns to 'higher' levels, such as macro-economic policy.

'Asset-stripping' companies

Asset-stripping companies are principally interested in underpriced, high-value resources – as in natural forests with good timber stocks. Stock markets, which value listed companies on a daily basis, place a higher premium on companies which can secure such assets at the lowest cost (which means those with lowest social and environmental provisions).

It is rare that long-term investment is made in natural forests in the tropics, principally as growth rates are low, the special qualities provided by some of the timbers can now be substituted by engineered products e.g. from woodchips, and natural forests are increasingly subject to environmental and social demands that companies cannot meet, or risks that they are not prepared to take.

Companies with a longer-term view; and the importance of plantations

Hence those companies which take a longer-term view are moving towards plantations – or highly intensively-managed forests with plantation-like characteristics. Plantations present practical, logistical and tenure advantages. They are low-cost, low-risk, high-yield and with a uniform and predictable product, which can be used for a wide range of finished goods thanks to recent technological developments (Sargent and Bass, 1992). Most of the highest-yielding forests in the world today are now owned by corporations, which have access to the genetic resources, technology and other inputs needed to achieve such yields. However, it is notable that most of these are devoted almost entirely to single/ few species for wood, and no other outputs have as high a priority as wood in management objectives.

Increasing size of forest companies and their need to control policy

Whether 'asset-strippers', investors in SFM, or something in between, large companies are now dominant. Perhaps fifty corporations control over 140 million hectares of the world's forests, through ownership and leases/ licences. Most of the 5.9 million hectares of tropical forest, which were logged annually during the late 1980s, were harvested by the private sector. Nationally, a very few companies can dominate – as one Malaysian company does in PNG, for example.

Whilst many companies are already large, there is a trend towards even greater size and the formation of multinational companies (Table 5.5).

Table 5.5 Reasons for globalisation amongst SE Asian forestry companies

Reason	Result
Log export bans and processing incentives in Malaysia/ Indonesia.	Companies requiring logs or material for expanding processing industries must look to other countries.
More effective law enforcement/ tax regimes in Malaysia/ Indonesia.	Fines, revoking of concessions, and scrutiny of accounts more common in Malaysia and Indonesia.
Weak enforcement/ tax regimes elsewhere.	Companies approach foreign governments.
Need to gain equity.	Securing more concessions to improve credit ratings.
Fear of losing market share.	Attempts to supply home industry with lower-priced logs from abroad.
International trade accords.	Developing countries promoting free trade provide incentives to foreign investors – reducing the net real investment.
Availability of under-priced wood; and low labour costs outside SE Asia.	Strong incentive to globalise operations.

Source: Sizer and Rice, 1995

There are diseconomies of scale, however. Large companies are vulnerable to changes in demand in certain sectors. For example, the European forest industry is highly susceptible to down-turns in the construction industry; and the global pulp industry suffers boom-bust cycles resulting in part from the huge size of every new pulp mill, which substantially increases the quantities of pulp available when it comes on stream, with consequent price reductions (Bass and Hearne, 1997). The development of heavily capitalised production systems requires consistency of throughput, and may not be well adapted to major changes in the external environment, particularly economic shocks. Down-turns in one sector, such as house-building, for example, may drastically reduce a part of an integrated forestry corporation's system – such as sawmilling, which in turn may have effects on other parts, such as reducing the raw material feed to pulping activities. Under these conditions, integrated firms will seek to develop new wood-based products and materials either to provide outlets for surpluses or to make greater efficiencies in times of relative raw material shortage.

Hence there are a number of general objectives which the private sector may seek through an influence over policy. Many of them are market-, trade- or finance-related. They include:

- increasing or maintaining access to markets;
- elimination of commercial competitors by, for example, denying them access to markets or to sources of raw material;
- increasing security over the resource base or capital investments;
- reduction of costs;
- obtaining direct benefits from government, such as subsidies for afforestation or infrastructure development; and
- reducing the role in policy formulation of adversaries, such as environmental groups.

Any instrument of policy which has a bearing upon the private sector's activities is likely to attract their attention. These include: regulatory instruments; standards; trade policy; financial, fiscal and incentive systems; research, training and education programmes; and tenure systems.

The size of companies is not necessarily a guide to the amount of political 'clout' that they yield in exercising such influence, but it is undoubtedly the case that larger corporations have greater opportunity for access to, and influence over, decision-makers; and as companies have become larger, so the requirement to support the necessary infrastructure and start-up costs has become ever greater. As Carrere and Lohmann have noted, given the large size of some of the forestry sector corporations:

> *It is hardly surprising that many of the largest are important political as well as economic actors... Because today's immense mills cannot generate profits without a large-scale re-engineering of their social and physical surroundings, the pulp and paper industry relies heavily nearly everywhere on political campaigns to capture handouts from the state and public.* (Carrere and Lohmann, 1996)

Processes of private sector engagement with policy, and integrating institutions[48]

We have noted that a very broad policy framework circumscribes private sector behaviour – to do with investment, employment, access to land, and taxation, and not merely forest policy. And sometimes there is a lack of policy which suits some companies: as yet, there is a lack of clear competition policy at the international level which would prevent increasing

---

48 Much of the material on private sector policy activities and tactics for this section is taken from a report commissioned by IIED from Simon Counsell.

concentration of industrial production in forestry (or indeed other sectors). Such a broad scope is effectively in the hands of politicians, rather than the forest bureaucracy which has had limited impact on the 'extra-sectoral' influences. Whilst there is little involvement in politics or extra-sectoral policy by other forestry actors, large or well-connected corporations can, in contrast, have a major influence on them.

Private sector engagement outside the formal policy framework

Many companies operate outside the formal forest policy framework and/ or steer clear of involvement, particularly those which adopt asset-stripping approaches (Bass and Hearne, 1997). Their tactics include:

- Exploiting a *policy, regulation and enforcement 'vacuum'* – conducting operations in the absence of effective authority, so that eventually the 'custom and practice' of the private sector may become the de facto forest 'policy'. This 'wild west' approach is still being attempted by (SE Asian) corporations in Surinam, Guyana, and Central Africa.

- Finding and exploiting *loopholes in policies and regulations* e.g. the planing of one edge of rough-sawn lumber to get around a prohibition of exports of non-processed wood from the Philippines; and treating (low) fines and reforestation bonds as a routine cost of forest operations, with no intention of meeting the requirements of legislation or to reforest (as in Indonesia).

- Building and exploiting *high-level patron-client relations*. Dauvergne (1995) points out that systems of authority operate more often through informal channels than through the overt formal mechanisms, especially in SE Asia where such relations are central to the politics of poor forest management. Some forms of corruption have been well illustrated and documented in the case of PNG where the 1989 report of Justice Barnett's inquiry into malpractices in the timber industry found that "*it would be fair to say, of some of the companies, that they are now roaming the countryside with the self-assurance of robber barons, bribing politicians and leaders, creating social disharmony and ignoring the laws in order to gain access to, rip out, and export the last remnants of the province's valuable timber*" (Marshall, 1990). There is a growing body of information detailing such malpractices throughout the tropics (see for example, Friends of the Earth, 1992; Friends of the Earth International, 1997; Environmental Investigation Agency, 1996).

Often the range and combinations of such tactics is staggering in its ambition and reach (Box 5.3).

Box 5.3 Personal power and influence in Indonesia's forests

Ten companies control the $20 billion-a-year Indonesian forestry industry. At least five of these have been owned by Mohamad 'Bob' Hasan. It has been said that not a single log is exported from Indonesia without his say so. Recently he said of himself, "I need 13 million m³ of raw material every year, and I employ 4 million people, directly and indirectly" (BBC1 interview, broadcast September, 1998). (Perhaps this importance is over-stated – the new Consultative Group on Forestry and CIFOR estimate between 500,000 and 2.5 million people are employed directly and indirectly.) Hasan was the golfing partner and long-time buddy of former President Suharto who become Minister of Trade in Suharto's last cabinet. When Suharto resigned in May 1998, Hasan took cover. Since July 1998 the Attorney General, Andi Ghalib begun to demand answers to questions about why US$100 million which was collected by government through a reforestation levy was spent on subsidising Hasan's pulp mills, and why millions of dollars of producers' association (APKINDO) funds were funnelled into Bank Umum Nasional, Hasan's own bank and lent to other companies within his business empire. Despite apologising for breaching intergroup lending legislation and promising to repay government money used to fund his pulp and paper company, PT Kiani Kertas, Hasan has avoided criminal proceedings.

The power of this kind of polarised wealth to cause both environmental and social forest problems was exemplified during 1997 and 1998 in Indonesia. Companies, who had used their political muscle to gain control over forest land and displace the indigenous Dayak communities, then used the occurrence of the super-dry conditions created by *el nino* to set fire to the forest, thus cutting the costs of clearance prior to establishment of agro-industrial plantations (which employ relatively few). Meanwhile, the International Monetary Fund has, in effect, propped up both this extreme inequality and these specific forest problems through its insistence that timber exports keep flowing to ward off financial crisis. However, the IMF has also called for greater transparency in concession allocation, more community involvement and for some concession agreements to be rescinded.

There are also many ways in which the private sector influences broad opinion:

- *Influencing science and academia.* As many of the policy processes involved in the forestry sector have become highly technical, so access to appropriate techniques and scientific research has become very important in policy formulation and implementation. Scientific authorities and inspectorates, university forestry research departments and standards organisations have consequently all become important targets for private sector influence and control. Control of expertise, information-gathering technology and data can be important determinants in the private sector's ability to negotiate terms with forest agencies, particularly where these are lacking in such information themselves.

- *Influencing general public perceptions.* Civil society and the public at large can also be very important targets of influence, especially where there are well-developed and articulated 'counter-views' to that of the private sector. The need to proactively convey a message and image to the public is now widely recognised within the forestry private sector, which has made increased efforts to shape the public's view in a way which accords with its own practices and perceptions. 'Public relations' is increasingly seen by the private sector as providing the catalysing element in the political process; *"(Politics) provides the packaging and the vehicle to achieve the industrial objectives...There are two elements to the political subsystem... the message and the target. The message needs to be short; for example, 'Trees are good. We need more trees not less'. Our objectives should be to create and move inside an ever-increasing friendly circle of public opinion"* (Fernandez Carro and Wilson, 1992). Annex 1 provides further discussion of the use of messages and discourse to achieve influence in the policy process.

Industry claims about the sustainability of forests and forest products have become ubiquitous. These started with simple, and often misleading statements such as: "we plant two trees for every one we fell"; and have progressed to a much more positive, and transparent, approach – that of third-party certification (Section 5.3). To some extent this is due to effective engagement by NGOs and the media; but it is also due to the recognition by leading companies that certification can both reduce social and environmental risk, and define a new 'brand', enlarging market share.[49]

Private sector engagement with national processes
Other companies aim to influence the formal forest policy framework, especially those who are investing in plantations. The larger their investment, the more conducive policy needs to be for them. The results may well be policy changes which are positive for most stakeholders (such as where corporations are intent upon influencing the policy environment to obtain longer-term tenure of forest resources).

A principal arena has been the national political, policy and legislative framework. Clearly, forest ministers and their staff are prime targets of influence. However, advisory bodies, policy consultation fora and legislative committees, at the local as well as national level, are also of great importance, as are ministers of trade. The incentive and subsidy structures within forest policy are of particular importance to the private sector. These may include explicit provision of grants and subsidies for the establishment of infrastructure, processing plants and reforestation, as well as implicit subsidisation through low stumpage fees, and other forms of forest rents and trade tariffs. The maintenance of such benefits has required intimate and consistent engagement in the political process. Some tactics employed by private sector groups to exert such influence include: engaging in legitimate multi-stakeholder forest fora; forming private sector forestry

---

49 A key question for such companies is then the interaction of the certified brand with other brands: does it reduce demand for non-certified products?

alliances to lobby national policy-makers; forming coalitions with interests outside the forestry sector; and party political funding. These are employed by domestic companies, but more significantly also by multinationals and foreign companies.

Private sector engagement with international policy processes

With the globalisation of forest industry, and the proliferation of MEAs covering biodiversity, carbon storage, etc, the private sector has also become involved in formal *international policy processes*. Here, their underlying concern tends to be the maintenance and expansion of free trade. Dudley and others quote a vice president of one large company:

> *"the forest products industry… share(s) a common interest in maintaining open international markets for trade in pulp and paper products. To this end, pulp and paper producers should work together to preserve the integrity of the open global trading system as embodied in the rules based WTO/GATT (sic). Proposals for changes in existing trade rules and related initiatives to promote the use of trade measures as a mechanism to enforce environmental progress pose new threats to the integrity of the open trading system"* (Dudley et al, 1995).

In contrast to this, Palo and Uusivuori (1999) demonstrate the correlation between free trade policies and deforestation in many developing countries.

Thus the private sector has exerted its strongest international policy influence in trade-related policies, notably:

- The development of the ISO approach to certifying environmental management systems of forest companies, through direct submissions and support of national delegations;

- EC ecolabelling, through corporate boycotts of applications for the EC standards for tissue paper – followed by strong lobbying for lower standards for fine paper;

- CITES regulations on trade in certain species, specifically the negotiations against listing of Brazilian mahogany under Appendix II, through influencing UK and US timber trade bodies;

- ITTO's decision-making processes, through involvement on national delegations and by chairing council, committees and working groups, and financial contributions to the organisation's funds from Japanese corporations. Indeed, some suggest that ITTO has been strategically important to the private sector. Thus, the private sector has argued strongly that CITES should defer to ITTO.

Private sector influence has been weaker on the very large intergovernmental policy processes that deal with forests in their broader senses, notably the Intergovernmental Panel on Forests (IPF) and the conventions. Where its influence has been significant, this has been through involvement in national delegations. For example, the Canadian private sector has been one of the strongest advocates in favour of a global agreement on forests, a position which was expressed by the Canadian government at the IPF.[50]

Direct engagement of the private sector in intergovernmental processes has been very limited. This is partly because these processes have few openings for any form of non-governmental involvement. Much the same applies for the private sector as for NGOs in terms of access to the process: there are no real integrating institutions for bringing private sector together with other actors internationally.

The paucity of engagement is also partly because of the lack of coherent organisation of the private sector internationally. Competition between private sector actors is the norm; and most of their influence is through individuals or companies rather than associations, often to more self-interested ends – with some very notable exceptions of leading companies who understand their potential role as sustainable development agents.

Indeed, there is a trend of progressive companies coming together to influence those international processes that concern SFM, although this is still much more tentative than equivalent government or civil society efforts. Most notable among the integrating institutions is the broad-interest, multi-national WBCSD (see Box 5.4), but the International Chamber of Commerce has also played roles in the IFF (with the WBCSD) and there is an International Forest Industries Round Table.

Whilst organised international private sector initiatives have yet to make a big difference on formal international policy, it should be noted that international policy initiatives have had a major impact on the embryonic private sector efforts. The Rio Earth Summit catalysed the formation of the WBCSD. The international C&I initiatives have strongly influenced private sector debate and responses, in terms of codes of practice, and their ability to strike deals with their own governments. And the influence of the FSC has helped to move private sector associations from first and second party

---

50 International agreements have the advantage for the private sector that they necessarily generalise about forest conditions and prescriptions, thus providing scope for interpretation which can be used to justify the present status quo.

Box 5.4 The WBCSD – a debutante in international forest policy

The impetus for forming the World Business Council for Sustainable Development was largely the lack of private sector inclusion in the preparations for the 1992 Rio Earth Summit. Stephan Schmidheiny, the Swiss industrialist, launched the BCSD alongside its book *Changing Course*, a 'global business perspective on development and the environment'. This book was written with the assistance of IIED and similar groups who had seen the potential to 'green' investors and producers (having begun 'greening' governments and civil society). The WBCSD went on to sponsor key work, for example on the *Sustainable Paper Cycle* (IIED, 1996) and on internalising environmental and social costs in markets. Made up of members from (usually large) corporations from both North and South, and operating in many sectors, it has a working group on sustainable forest industry. This group has made inputs to the discussions of the IFF, but thus far usually on the margins, i.e. special sessions. It has been a little more successful in its engagement with the World Commission on Forests and Sustainable Development (which has been perhaps the first significant international initiative to examine the potential positive and negative roles of private sector forestry).

The WBCSD's political and strategic muscle is improving, and it should prove to be a significant force, provided that it can ensure high levels of commitment to sustainable development amongst its members, both individually and collectively. Other stakeholders need to be continually assured that WBCSD represents positive-minded companies i.e. with internal rules and sanctions concerning company practice, and transparency in their relations with other groups.

pronouncements on forest practice, to realising the importance of transparency and third-party verification: the American Forest and Paper Association is a case in point.

The key private sector challenges

Better forestry requires positive interaction amongst the 'sustainable development triad' of government, civil society, and the private sector. Each of the three groups has something to offer in the balanced production of public and private goods that is required for sustainable development. The two clear challenges for government and civil society regarding the private sector are:

- how to smooth the path for companies wishing to invest in SFM?
- how to block the path for companies wishing to asset-strip?

Clearly, the former group of companies needs to be better engaged in policy processes in a practical sense, and not merely those aspects which are concerned with trade and investment. More integrating private sector institutions and partnerships could help them to do so. Groups such as the WBCSD are valuable, and the private sector needs to be encouraged to form them. Companies need to explain more clearly the policy, legal and financial impediments to better forestry practice – as well as those signals which help other companies to get away with bad practice. But just as weak and corrupt governments are not excluded from most international policy processes, ways must be sought for poorer-performing private sector actors to learn (at least) from international policy processes.

5.5 What future roles for international processes?

Getting issues addressed at the right level

The global context is changing rapidly, in economic, environmental and social systems. There are many uncertainties in the outcome. On the other hand, globalisation ensures that there are also increasing commonalities across countries, for good or bad.

But there are also many views on what is 'global'. Some international actors stress global services (biodiversity, etc) as the stuff of international policy. Others wish to deal with global causes of forest problems, the sticky issues of debt, foreign investment, trade, etc. There is confusion between common national problems (what we might call 'worldwide' issues) and truly global issues; thus some actors try to take national problems 'up to' the global level because they appear not to be resolvable locally. In contrast, others stress that what had been treated as a global issue e.g. better control over biodiversity 'hotspots' may actually need to come 'down' again, e.g. to become a decentralised effort where local people's rights, capacities and rewards are improved for maintaining global bioquality.

As with national policy, the power relations of international actors have been critical to the outcome. If the USA has not ratified an international agreement, it will have limited impact. The World Bank has had disproportionate influence on matters of structural adjustment and forest sector financing. WWF and the 'Washington mafia' of environmental NGOs have had better access to international processes than NGOs and CBOs from India. And well-organised international corporations of European or North American domicile have had more influence than small forest associations in developing countries.

But the room for manoeuvre is growing. Joint initiatives between the World Bank, WWF, and CEOs of big companies are now opening up to participation by smaller groups in developing countries. NGO self-selection for participation in international negotiations mean that there is a little more democracy within civil society than before. The World Commission on Forests and Sustainable Development was a truly multi-stakeholder initiative, which made admirable attempts to reach some of the grass roots through its regional hearings.

Getting good forestry embedded in environment and trade agreements

All the PTW case study countries show that international governmental, private sector and civil society processes are increasingly influential on the national scene, with those most involved in forest products trade (PNG and Ghana) or forest services trade (Costa Rica) being more open to such influence.

There is also now a trend for each type of process to start to introduce actors central to the other types. However, there is no continuing global forum which could be said to be truly multi-stakeholder, or truly multi-sector (in the way that recent national forest fora, or sustainable development councils, could claim to be).

With the prospect of international payment transfers for environmental services such as biodiversity conservation and – especially – carbon offsets now somewhat closer, following a period of voluntary markets, mulitlateral environmental agreements (MEAs) may become increasingly important for the future shape of forests (Stuart and Moura-Costa, 1998). This will especially be the case if international lending institutions effectively catalyse international private investment, as the World Bank wishes to do with carbon trading. It is thus even more important that the MEAs understand how good forestry should be carried out, in relation to the security of forest goods and services that are important at local and national levels. In other words, there are dangers if MEAs remain transfixed by global services alone. The recent international policy processes that have generated C&I can provide the principles and conditions that the MEAs need. (Box 5.5).

Trade issues, the only policy area in which private sector forestry actors have had significant influence and NGOs have been relatively insignificant, will become increasingly important. There are potential clashes between the WTO and those MEAs which include trade measures, but also potential solutions, such as a special WTO environmental agreement. These have yet to be really tested or resolved. There is real scope for the private sector itself to bring social and environmental issues to the table, if given adequate incentive. The recent partnerships with NGOs may help them to bring a balanced or joint perspective to trade debates – getting away from the perception that environmental and social factors are a barrier to trade. Indeed, such factors are insignificant compared to subsidies, quotas, tariffs and other conventional barriers.

Box 5.5 The Kyoto Protocol and global forest issues

The Kyoto Protocol, under the Framework Convention on Climate Change (FCCC), has addressed only a part of the carbon cycle: not unnaturally, parties were concerned about burning fossil fuels that have taken geological time to develop. But there are other segments of the carbon cycle to be dealt with, notably carbon sinks and the sequestration services of forests. Perhaps less attention was given to these because intergovernmental forest agreements have been weak – no 'steer' was given on how good local forestry can contribute to carbon storage. There were also uncertainties about the relative strengths of different forest (management) types as sinks – which meant that Kyoto's number-crunching sessions, rushing to develop formulae, avoided the difficult equations surrounding sinks.

There is a real danger that the Kyoto Protocol and the associated Clean Development Mechanism – if they yield to pressures to handle sinks – will be able to deal only with simple forestry models. Afforestation/ reforestation and/ or the set-aside of protected areas provide big, easy-to-measure blocks. Furthermore, the development banks see themselves as catalysts for greatly increased private sector investment in carbon offset forests (possibly through stock/ commodity markets). The question is where this will help the security of forest goods and services sought by local interest groups, and where will it result in corporations capturing even more value and land. Will it lead to pressure for deforestation followed by subsidised afforestation; and will it push out natural forest management, rotational shifting cultivation, and farm forestry?

Certainly, international carbon offset protocols could have enormous implications for:
- the siting of forests
- the composition of forests
- 'permissible' activity in forests
- who gets forest benefits
- who is effectively in charge of forestry

The language of good forest use to secure desired goods and services needs to be introduced into the carbon equation – C&I as the 'words' and bottom-up national forest programmes as the 'syntax'. Trees on farms, agroforestry, shifting cultivation, natural forest management, may be difficult to measure in climate terms, but they sustain livelihoods and other forest benefits. Managed forests such as these could also store/ sequester more carbon than simpler plantation or set-aside systems.

The irony is that it is now well-understood that the development of new global markets for forest goods and services should include social and environmental externalities. The danger is that they will be forgotten in the rush to develop the carbon market.

Finally, the world is beginning to realise the costs, the inefficiencies, and often the irrelevance, of addressing some forest concerns at the global level. Global negotiations are often of the relationship-maintaining type, rather than purpose-led. *Regional* initiatives look like becoming more effective than global – groups of countries with similar concerns, or clear transboundary problems (of pollution, or investment), getting together. The Central American Forest Convention is one such purpose-led initiative.

Key items for the international agenda

We should build on the proven and promising international tools to date – the C&I, the NFPs, certification, the specific MEAs, and purposeful regional agreements. We should get more involved in the 'extra-sectoral' big issues that matter, and notably trade. Eight key 'agenda items', that have become evident to IIED, are suggested below:

1. Agreement on a simple set of C&I which provide an accepted international *lingua franca* for SFM. Such C&I should become integral to MEA protocols, payment schemes and certification programmes regarding the production of global forest services; to trade policies regarding standards equivalency/ mutual recognition and legitimacy of technical barriers to trade; and to *investment*, so that finance authorities can lend, and judge performance, on the basis of environmental and social behaviour and not merely financial performance.

2. The extension of *certification/ verification services to global* environmental services (there are already carbon offset certification schemes, and moves towards certifying forest management for biodiversity conservation), and to other real forest issues (perhaps domestic markets in countries facing severe forest problems).

3. A review is still sorely needed of the *potential and performance of existing intergovernmental instruments and institutions*. There is much scope for applying the Climate Change and Biodiversity Conventions, and some other existing environmental agreements, to certain aspects of forestry. This would have to be preceded by a performance gap analysis of existing initiatives. The same applies to the intergovernmental agencies.

4. Use of the concept of *security of specific forest goods and services* at different levels, from household to global, in international processes.[51] It is what people get out of forests – and the equitable distribution of associated costs and benefits – which matters more than targets concerning e.g. extent of forest area, or deforestation percentages (dimensions which have dominated much international discussion). It should become apparent that a range of instruments is needed to apply to individual goods and services, both intergovernmental and operated by civil society.

---

51 The concept of *food security* has long been key in forging policies and provisions to avoid and address food shortages at different levels.

5. Better integration of the international initiatives on *NFPs and National Strategies for Sustainable Development*. This should improve the way that cross-sectoral links to forests are addressed; through better integration of both into those institutions and policy processes which really determine the allocation and use of resources, e.g. business strategies and local community rules. This demands a recognition of 'what works' at national and local level, and building on this 'second best', as opposed to promoting unachievable perfection.[52]

6. *Involvement of important, but thus far marginalised*, actors in international processes, and notably representatives of local forest-dependent groups and the private sector. The current, rather ad hoc, rules on participation and representation do not provide for efficiency, continuity or adequate representation.

7. Improving *institutions and methodologies for representation and participation*. It would be a useful start to review the lessons from experience at local to global levels, and to share successful approaches. (At present, the higher 'up' the local-to-global hierarchy, the more depauperate the means of representation and participation.)

8. In particular, there is a need to find *better means of representation and involvement of the private sector*. Real efforts are needed to form private sector bodies for constructive engagement with other groups, to explain the impediments to SFM, to improve their own commitment to SFM, and to help all groups work out *practical* ways forward.

Finally, and as we stressed at the beginning of this section, for international processes to stand any chance of success it is essential to *improve national policy processes and capacities*, as the main link between the forces of globalisation and localisation, where 'win-wins' can be identified and trade-offs made where necessary. This capacity is needed to push the global agenda in the light of local needs; to interpret global obligations in the local context; and to ensure that there is adequate local capacity and incentives to undertake global obligations and provide global services. *The Policy That Works* project has given us a good idea of what the national process and capacity needs are. We summarise these in the next and final section.

---

52 There is a need to distinguish issues concerning SFM, from those concerning the role of forests in sustainable development more broadly. SFM is concerned with individual forests, and usually involves environmental and social constraints on their use. Sustainable development, in contrast, allows the specialisation of forest use to achieve overall optimisation of economic, social and environmental benefits. This implies the legitimacy of liquidating some forest capital, and of importing forest goods and services rather than producing them all on domestic forests. Progress here might best be made through the NSSD agenda.

6 Conclusions

"Just like the beautiful flower which has colour, and also has perfume, are the beautiful fruitful words of the man who speaks and does what he says."
> The Buddha's Teachings (Dhammapada, 3rd century BC)

"Everyone understands how laudable it is for a prince to keep his word and live with integrity and not cunning. Nonetheless, experience shows that nowadays those princes who have accomplished great things have had little respect for keeping their word and have known how to confuse men's minds with cunning."
> Niccolo Machiavelli (The Prince, 1513)

"To combine public altruism with private self interest and unwitting hypocrisy is quite a trick, but I think I have it now."
> Ivor Cutler (Maturity, 1986)

These are strange days indeed. In the old days of powerful religious teachings and princely states, policy messages seemed fairly clear-cut and everyone knew their place. Today our leaders emerge from more diverse sources, there is a wider array of policy signals, and many people are confused, cynical, or heading off in ill-advised directions.

For forestry and land use there are many complicated policies available. But they rarely seem to 'fit' well even with the rather limited number of rigidly-structured institutions charged with implementing them. We need to turn this around – we need straight-forward, motivating policies that people believe in and organise themselves to implement. This will enable the emergence of a greater diversity of more flexible, still learning, integrated institutions.

'Policy inflation, capacity collapse' syndromes need replacing by simple, agreed policies with vision, with strong capacities to interpret and implement them. This calls for nothing short of a reinvention of policy in the current context of forestry and land use. We aim to contribute, by making the case for policy as the way to bind national level decision-making to other levels of decision-making and to real practice. This requires engagement with the different actors demanding forest goods and services, and with those in a position to produce them – not just engagement amongst authorities and élites. Policy, at its best, can provide the 'A list': agenda, actors, aspiration, agency, arena, adaptation, and action.[53] In short, policy is where the dis-connected can find the tools to re-connect.

> 'Real world' policy (in contrast to formal policy documents) is the net result of a tangled heap of formal and practical decisions by those with varying powers to act on them. Effective real world policy connects local action to plans and programmes through integrating institutions and top-bottom linkages. These linkages comprise information flows, debate and partnerships.[54]

6.1 The significance of context

"Well, it depends on what you mean by the word 'is'…"
Bill Clinton (to the Starr Inquiry, and on TV around the world, 1998)

Our collaborative studies of six countries and international initiatives convince us that *context* is all-important. Examination of the local context – involving local actors and analysts in the process of analysis – is itself a prerequisite to 'policy that works'. No outsider is adequately qualified to pronounce or intervene in policy until an understanding of context is in place. Indeed, without this, there cannot even be an understanding of what the word 'policy' means locally. Within a country, too, there are differences in individual actors' contexts and perspectives. Sharing information on these is key – all of which points towards the need for a healthy interaction of multi-stakeholder processes within the given context.

Beyond this fundamental principle on context, there are common findings about useful policy *processes* and related *institutions that bring stakeholders together*, and just a few findings on the choice and application of specific *instruments*. But these also take the form of principles or frameworks, rather than detailed specifications for individual policies, instruments, processes or comprehensive menus.

---

53 Currently, the 'B list' is more prevalent: bureaucracy, blueprints, bullying, burdens, blunt instruments, blockages, and boredom!
54 More eloquent calls for the 'unity of theory and practice' are voiced by champions of pragmatism and found in the traditional literature of Marxism.

In other words, there are no magic bullets – no evidence that particular instruments such as auctions or certification should be universally applied in preference to any others; specific participation processes may not always be helpful; and extensive 'menus' such as the TFAP framework may not necessarily fit with the policy context (see Section 6.3).

Our review of experience has shown certain contextual factors to be most significant:

- *History and power structure.* Policy tends to be a reflection or judgement of history and social relations, legitimising or condemning them; uniting or dividing actors; clarifying issues or acting as a smokescreen for other actions. The historical experience of policy fundamentally shapes different stakeholders' expectations of what change is, or is not, desirable or even possible. Many of the new policy processes have been most helpful in identifying and dealing with historical inequities. Clear methodologies are needed to ascertain the historical reasons behind today's policies.

- *Forest asset base.* As well as very different forest endowments, countries and localities vary in the scale and type of demands put on their forest assets, and vary in the other resources which can be accessed to support any forestry endeavours. An understanding of the shaping of the forest asset base and the demands on different forest goods and services is important. Very different streams of policy and institutions are associated with countries emphasising plantation development, as opposed to those still at the stage of liquidating natural forest capital to generate other capital assets. Too many studies look at causes of deforestation, without assessing countervailing forces of forest conservation and afforestation; the balance is far more revealing than one trend alone.

- *Ecological influences.* A country's or locality's particular vulnerability to external ecological influences and global change will determine how forests are viewed. Low-lying islands increasingly value their mangroves in the face of tidal surges and sea level rise; desert countries now see the protective role of forests in relation to the Desertification Convention.

- *Economic and financial conditions.* The degree of internal/external market exposure, and the role which forests are expected to play in livelihood, sector, and national income generation, are highly significant. An understanding of where sources of finance are coming from – foreign or domestic, within the forestry sector or outside – is critical for deliberations on where the best leverage points for SFM might be.

- *Social-cultural influences and conflicts.* There are fundamental cultural differences in approaches to forests, e.g. as lands to which people belong (as in India), or as lands to transform (as in 'pioneer societies' such as much of the Americas). Forests may be owned or settled by particular groups, and forest policy is strongly coloured by how those people are viewed by those who set policy, and by conflicts between groups.

- *Institutional norms and precedents.* The strength of state-civil society relations is very significant for determining what kinds of participation are possible in policy processes. Such relations also influence the policy instruments employed, i.e. enabling or regulatory approaches. A significant issue is whether and how groups learn and undertake change. Institutional norms tend to fall into two types – those that maintain the *status quo*, and those that exhibit more dynamism and focus on purpose (the latter not necessarily being better, certainly not without democratic checks).

- *Scope and scale of changes in all of the above.* One of the most important factors is what room there is to make decisions about how to deal with change, and whether participating institutions stress continuation of what has gone before (or incremental change) or quantum leaps forward. The policy context may be characterised by decrees, by the cumulation of small negotiations/ decisions, or by large periods of (deliberate) indecision/ imprecision. Time can be a significant factor: different groups and nations seem to favour different paces of change, or to consider greater or lesser time horizons past and future in their decisions. There are current tensions between the long time horizons needed for good forestry, and the shorter time horizons needed by poor people who are obliged to follow multiple livelihood strategies in times of great flux.

6.2 Opportunities amongst the problems

Globalisation and localisation are stronger policy imperatives than 'forest crisis' – and all need taming through national policy

The globalisation of economies, and the increased demand for forest goods and services of all kinds, offer both opportunities and constraints to policy, forest and land management. Pressures for international policy and market regimes compete with those for more local control. The respective roles of governments, civil society and private sector continue to evolve in this dynamic environment, with its emerging synergies and tensions. Policies at national level offer the principal points of linkage between global and local imperatives, and are crucial in addressing the tensions between them.

International initiatives help where they are able to accommodate local perspectives, convening multiple stakeholders, and allowing local interpretation and subsidiarity in their implementation. The previous approach – of international initiatives dominated by certain governments – has been considerably modified by greater participation and better information on impacts. But progress is still constrained by weak intergovernmental institutions.

Argument over policy tools is where the positive action is

Such debate appears to be more effective than argument over forest 'crises'. Major changes in practice and the balance of power between stakeholders have occurred through implementation of, and/ or reaction to, policy tools such as log export bans, certification and TFAPs. Whilst it is possible to argue that getting the policy process right first will mean that the right policy contents will follow, in practice it is often argument over particular policy instruments that brings people together in the first place.

'Fortress forestry' is slowly crumbling

The command-and-control, fines-and-fences approach to forestry is no longer working. In recent times, forest institutions have further suffered from an overload of objectives and collapse of capacity. In many countries, this helps the forest asset strippers to have more free reign than ever. However, many of the old building-blocks are still needed, and a process of building more adaptable forestry institutions, with open doors, is slowly under way in many countries – but fora or market places, would be better than castles.

For environmental and social values of forests to compete with economic values on an equal footing, a new language of valuation is still needed

That forest problems, especially forest reduction, are due to inappropriate values is a truism, but while economists fret over imperfect techniques, real decision-making is going on elsewhere. Others doubt whether the social determination regarding what constitutes a forest good or a service can ever lend itself to valuation approaches. Nonetheless, as long as the information base and the institutions which manage it remain focused solely on forest area figures and timber values, big decisions about forests can only be legitimised by reference to them. The information base needs to be extended to spread understanding of the broader range of social and environmental as well as economic values. For this to be possible, new approaches and language which value alternative 'worldviews' need to be installed in policy (see below).

Cross-sectoral policies and market signals can be hazardous, but new approaches are emerging which may help

It is remarkable how problems derived from extra-sectoral policy can dominate forests and forest stakeholders, but solutions are still largely forestry-based, ignoring that factor. Yet there has been a gradual cumulation of processes and tools which have at least helped to ensure widespread understanding of forest values to sectors such as tourism and agriculture, and of what is being lost through bad forestry, and at best have helped to diminish specific forest damage resulting from extra-sectoral causes. Environmental impact assessment of projects and strategic environmental assessment of policies have drawn attention to the values of specific forests. National conservation strategies and environmental action plans have, at the very least, offered massive training courses in the links between forests and other sectors. These approaches, although tending to be given stronger political backing since Rio (indeed, National Sustainable Development Strategies are a fundamental national commitment), remain essentially technocratic and not yet central to either government expenditure (Pakistan is one exception) or the investments of the private sector. More needs to be done in terms of:

- Systems of due process (at least), or due diligence, to be exercised by government bodies in dealing with cross-sectoral links (as a minimum, charters based on such systems could be developed and applied in NSDSs and sectoral policy development processes)

- Information on cross-sectoral values of forests being fed to the political and market actors in various sectors – in other words, ways to influence the all-important resource allocation systems (here, the current development of NGO watchdogs based on models such as MineWatch looks to be promising)

- Development of systems of incentives e.g. in the tourism, water supply, and farming sectors, to sustain forest values.

Markets can help us make the transition to SFM if they have the right information

Stock markets still value companies daily on the basis of their assets and likely profits, with no direct reflection at all of their social and environmental performance. This will continue to encourage companies to seek out underpriced forest assets, and to minimise environmental/ social costs, staff and facilities. However, the finance and insurance sectors are just beginning to realise the short- as well as long-term value of environmentally and socially sound operations. Equally, certain retailers are creating demand for products that derive from SFM. With NGOs involved as watchdogs, and government setting the broad rules, there is potential for some markets to metamorphose from SFM constraints to opportunities. However, more needs to be done to deal with the continuing constraints imposed by WTO's ability to proscribe environmental and social barriers to trade – perhaps an environmental convention attached to WTO.

Informal policy – what institutions actually do – has positive as well as negative aspects

Negative features are the secretive and unequal use of patron-client relationships to pursue inequitable agendas. But positive progress may result from the informal peer networks which cut across institutional boundaries – making practical articulation with, or branching out into, civil society. One of the indirect advantages of many multi-stakeholder processes is the informal groupings that form, and then are used by their members to get things done through activities, both great and small.

Formal policy sometimes catches up with informal policy

If the process is bad, formal policy reinforces the status quo or alternatively over-responds to civil society pressures – with the pendulum swinging too far towards new preoccupations. If the process is good, the gulf between local forest actors and national 'policy-makers' breaks down – and formal policy begins to learn from local projects, practice and innovation.

Uncertainty, complexity and long time frames in forestry mean that effective policy narratives and clear strategies for influencing policy are needed

Forestry, like other forms of environmental management, is a long-term affair. Decisions affecting peoples' behaviour towards trees and forests – including efforts to rectify mistakes – may take many years to reveal their consequences. Politicians and other actors who often control policy, on the other hand, have short pay-back times. Forests may thus fail to register at all on their list of priorities, or to be dealt with as 'others have always dealt with forests'. Many attempts to raise forestry issues up the priority list rely on stories of 'crisis' to try and grab attention and force quick action. Indeed, the 'forest crisis' has been regularly proclaimed over the last twenty years, and is now old news. However, effective policy rarely comes from quick responses to immediate crises. This is because forestry, like other environmental issues, is also a complex and uncertain affair. Solutions are more about building resilience in production and planning systems, in order to cope with change in circumstances of incomplete information.[55]

---

55 Policy analysts – often blamed for getting things wrong – frequently seem on the edge of despair at not having enough information or being unable to predict accurately enough what is going to happen. The 'law of unintended consequences' may apply as much in forestry as in many other fields. This law – the comfort of civil servants everywhere – states that a policy will have consequences which cannot be predicted, but if it does not do exactly the opposite of what was intended, then it can be claimed as a great success.

Given these long time frames, the general slippage of forest issues from political priorities, and the inherent uncertainty of the enterprise, alternatives to the crisis narrative need to be promoted. Indeed there is a wide range of ways in which use of narratives can influence policy more strongly (Annex 1 describes some of these – notably building constituencies which can, one day, bring about wholesale policy change). All of these ways of 'playing' policy effectively rely on telling a story that 'strikes a chord' with people engaged with the policy process, and thus lodges itself in the policy 'discourse'.

6.3 Policy processes that work

Preceding sections have shown that, where historically forest policy was the affair of national forest authorities, now we have to contend with policy from civil society, other sectors, and local and global levels. We now look at what we have learned from the processes behind these new forms of policy.

As we noted in the opening pages, one of the impetuses for our inquiry was the mushrooming of international forestry principles, obligations, programmes and prescriptions of recent years. What works at international level?

International policy processes that work

Section 5.5 discussed a number of international policy initiatives that appear promising. All of them need to evolve, however:

- Some of the *multilateral environmental agreements (MEAs)* which focus on specific global forest services, and include (under-utilised) implementation provisions – but which need informing about good forestry and need to be better recognised in key trade fora

- The *criteria and indicators* (C&I) processes, which encompass broad SFM needs, and allow for local interpretation – but which need application to the key areas of trade, investment and MEAs

- The process of developing and implementing *certification*, which can provide real incentives for SFM – but which needs to continue to improve their 'fit' with local policy, livelihood and land use realities, so as to solve real problems and not only service the needs of particular markets

- Country-led *national forest programmes* (NFPs), which could be a major vehicle for reconciling pressures of globalisation and localisation – but which need to be built on local knowledge and institutions as well as the internationally-agreed elements such as the IPF Proposals for Action

- Focused *regional* agreements, which offer the right political and operational level for integration of local and international needs – but which need to ensure that they are strongly purpose-led, and not become a vehicle for other agendas.

The policy processes attached to these initiatives tend to have three common features:

- They link global with local issues, i.e. whilst having a clarity and unity of vision for the global level, they can be 'owned' locally by accommodating local conditions and capturing local needs. Thus they act on the principle of subsidiarity i.e. a common strategy made operational through decisions taken at the lowest effective level

- They involve more than one of the main groups of government, civil society, and the private sector, with provision for means of participation

- They involve catalytic institutions, which can cross the boundaries of the main countries or other groups in the process. One example is progressive bilateral donors, which are both engaged in international policy on behalf of the (OECD) country in which they are based, and are active in pursuing practical sustainable development at national/ local level in developing countries. Worldwide networks, such as IUCN and some of the working groups of intergovernmental agencies, have also had good 'linking' effects – although often only at the margins.

As we shall see, these sorts of characteristics are also found in successful national and local processes i.e. wide participation, clear goals, change champions, and strong links. But, so far, international (and especially intergovernmental) processes often miss other attributes which are perhaps more characteristic of successful national and local processes:

- Institutions for experimentation and learning (as opposed to the staid intergovernmental institutions and a lack of performance review processes)
- Transparency (as opposed to an acceptance of member countries' voluntary reporting)

- Parity between government, civil society and business (as opposed to one or the other leading, and especially the first two)
- Getting more marginalised forest-dependent groups to the table (as opposed to single-field initiatives and élites)
- A rich range of brokers, institutions, and methodologies for participation (as opposed to the highly limited palette used internationally)
- Better coordination between policy instruments (as opposed to independently-operating conventions and programmes)
- Highly accountable authorities subject to democratic pressure (as opposed to 'fortress UN' institutions)
- Informal institutions and approaches (as opposed to highly formal initiatives only)

To be fair, those international processes which miss many of the possible ingredients of success tend to be intergovernmental, and there are very good reasons why they should not be too 'light on their feet'. Yet it does appear as though we are reaching the limits of what can be achieved by intergovernmental effort in the forest sector alone. By the same token, the really big extra-sectoral problems – WTO, debt, foreign investment, technology access, etc – can only really be dealt with intergovernmentally. They are too big for the forest sector alone to handle effectively. International forestry initiatives clearly need to make specific inputs on these 'big issues' to the CSD, OECD, WTO/GATT, etc, and be fully coordinated with them.[56]

National policy processes that work:
International processes tend to be relatively easy to understand: they have involved more or less clear, time-bound, written policies with well-documented participation and decisions. In contrast, national policy processes are a more opaque mix of decisions, overt and covert, often with murky pasts and uncertain intents. Nonetheless, some of the characteristics of policy processes that work have become increasingly clear from our country and thematic studies (see Box 6.1).

The characteristics framed in Box 6.1 will read like plain common sense to some people and will seem like a Utopian wish-list to others, whilst others again may find sources of aspiration – useful ingredients – to use in cooking up better policy. Certainly, if you are a current holder of policy, or are someone trying to tell holders of policy what to do, we would urge you to consult this list of conclusions. However, for many readers, the more

---

[56] This will require clearer analyses of the genesis of the issues and their links with forests, e.g. links between forest management and debt, poverty, or health.

Box 6.1 Characteristics of policy processes for better forest management

We conclude that good policy will:

- Highlight and reinforce forest interest groups' objectives
- Clarify how to integrate or choose between different objectives
- Help determine how costs and benefits should be shared between groups, levels (local to global) and generations
- Provide signals to all those involved on how they will be held accountable
- Define how to deal with change and risk when information is incomplete and resources are limited
- Provide shared vision, but avoid over-complexity
- Increase the capacity to practise effective policy
- Produce forests that people want, and are prepared to manage and pay for

Processes which can help to achieve good policy include:

1. *A forum and participation process*: to understand multiple perspectives, negotiate and cut 'deals' between the needs of wider society and local actors, and form partnerships. Government may organise the forum, but it needs broad involvement of stakeholders, and strong links both 'vertically' (local-national-global) and 'horizontally (between sectors and disciplines). The forum is regular, as cyclical, continuously-improving policy is a useful goal in itself. The resulting policies are 'owned' by stakeholders broadly, not just the forest authorities. They are 'alive processes', not 'dead papers'.

2. *Agreement on national definition of, and goals for, sustainable forest management*: focusing on the forest goods and services needed by stakeholders, and which can be met from forest and non-forest sources; and thus on sustainable development objectives, rather than forestry *per se*. Criteria and indicators can provide helpful categories under which to organise policy analyses and consultations, as well as being useful products.

3. *Agreed ways to set priorities* in terms of e.g. equity, efficiency and sustainability; as well as timeliness, practicality, public 'visibility', multiplier effect (or other criteria). It is never possible to do all things which people may want to see; without agreed approaches to setting priorities, an overly-comprehensive win-win policy may arise but be ineffective.

4. *Engagement with extra-sectoral influences on forests and people*: using strategic planning approaches, impact assessment and valuation, but also emphasising information and advocacy in political and market processes. An effective ingredient is cross-sectoral pilot projects that feed into the policy process itself to improve learning and demonstration of the ultimate policy's intent, and other partnerships that work across institutional boundaries.

5. *Better monitoring and strategic information on forest assets, demand and use*: adaptive research which can deal with change, and feed information on policy impact back into policy and management.

Management information systems on forests are needed at both forest management unit and national levels, and should be linked. Information systems may be better dealt with as the 'hidden wiring' in a policy process, rather than as overt objectives in themselves.

6. *Devolution of decision-making power to where potential contribution for sustainability is greatest*: Decisions are best made and implemented at the level where the trade-offs are well understood, and there is capacity to act and monitor. Thus large forest allocation decisions need to be made at national level; SFM targets and management agreements at local levels.. Experimental collaborative approaches can be effective first steps towards this.

7. *Democracy of knowledge and access to resource-conserving technology*: Openness to information from all sources is important, as is communication of both information used in policy-making, the policy itself, and information on policy impacts. This increases awareness amongst all stakeholders for empowering effective forest stewardship.

pertinent question is likely to be: how do we get there from where we are now? We outline four critical steps to make the transition to the kinds of policy process described in Box 6.1:

• Recognise multiple valid perspectives and the political nature of the game
• Get people to the negotiating table
• Make space to disagree and experiment
• Learn from experience, get organised and fire up policy communities

6.3.1 How to improve policy, in practice

Step one: Recognise multiple valid perspectives and the political game

"A tree's a tree. How many more do you need to look at?"

Ronald Reagan

Policies are based on assumptions. For a policy to have a chance of mobilising people and resources for action, it needs to be based on a clear narrative, or story. Such a narrative depends on assumptions; and the simpler and clearer a policy narrative is, the more likely it is to be resting on some fairly major assumptions. In forest policy, typically, these assumptions have remained untested and unchallenged until quite recently.

The challenge is to promote recognition of different conceptions of what the problems and priorities are. People's priorities for forests should be judged not on whether they are 'true' or 'rational', but on the level and degree of social commitment which underlies them – who 'subscribes' to them, and what impacts that has. In many forest situations with multiple perspectives and considerable uncertainty, blueprints for policy are unlikely to work, and policy designers may have little idea whether or not their policies will meet their objectives in practice, or have unintended consequences.

Differences in people's priorities and world views are not just the result of inadequacy or ignorance, so they cannot be easily resolved by passing on more information, training and 'awareness raising'. More information will not necessarily lead to a situation where decisions become obvious – where general agreement will be reached. Generating a greater mass of information may simply lead to bewilderment or paralysis. Often it will be more important to develop mechanisms by which diverse perspectives can be brought to bear on existing information to enlighten understanding of an issue.

Step two: Get people to the negotiating table

> *"Power is participation in the making of decisions."*
>
> Lasswell and Kaplan (1950)

Examples of practical application, in forest policy processes, of promoting and accommodating multiple perspectives are rare as yet. The work of this project in PNG makes a start (see Section 4.8). However, there are signs that the approach is spreading, with notions of 'pluralism' beginning to emerge in forestry discourse[57] as well as in other natural resource management fields (Lee, 1993; Röling and Jiggins, 1998).

With many world views of the same 'reality', two possible negative results can be anticipated: firstly, no policy at all emerges[58]; secondly, it could play into the hands of the already powerful – a few heavy-weights dominate policy, either directly or by exploiting any involvement of others so as to lend credence to their own views. However, at the moment it is common for the dominant policy players to have only one set of strong opponents, e.g. those who condemn the locals as forest destroyers versus those who champion them as holders of forest-saving indigenous knowledge; or the donors versus the loggers in PNG. Such polar extremes commonly frustrate attempts to make progressive change. In a more complex arena, each group of actors needs to present their priorities in ways which they can 'sell' to others,

---

57 Pluralism has begun to be adopted in writing on forestry (Daniels and Walker, 1997; Sulieman, 1997; Wiersum, 1997; Anderson, 1998). The FAO's latest *State of the World's Forests* report adopts a definition of pluralism: "the existence within any society of a variety of groups with different, autonomous and sometimes mutually conflicting interests, values and perspectives. Further, these differing views cannot be reduced to a common perspective by the reference to an absolute standard" (FAO, 1999). An edition of FAO's forestry journal *Unasylva* (No.194, 1998) is devoted to the subject of institutional arrangements for pluralism in forestry.

58 Policy paralysis is occurring in the US – where conflicting interests are straining traditional decision-making mechanisms. Even after public participation efforts have been implemented, large numbers of forest management plans have been legally protested against by a range of groups with different views, and are stalled in the courts.

framing their arguments in narratives (and counter-narratives) to try to convince others. *Policies that work are based on narratives that work.*

This should not be read as a recommendation to 'sit back and watch the social drama of policy unfold' nor as a simplistic exhortation to 'get everyone participating and democratise the policy process'. Rather it is a call for recognition that current inequities, forest asset stripping or stakeholder stalemate may persist because of misunderstandings or lack of knowledge amongst stakeholders of each others' perspectives, powers and tactics, and the potential for change in these (PNG PTW; Malla, 1998). Stakeholders who prioritise better understanding of other relevant stakeholders' views, approaches and powers, are more likely to be able to harness policy processes to bring about change in those views and powers over time.

When shared vision is possible... In some cases, work to analyse stakeholder claims and narratives will reveal common threads, shared values and beliefs – indeed, a 'public interest'. Processes which help identify and build shared vision or consensus on key goals can be effective. Cross-institutional forestry working groups in Ghana and Zimbabwe, the Sarhad Provincial Conservation Strategy in Pakistan and the Joint Forest Management institutional support network in India, have all made notable progress on this, although all are based only on a sub-set of forest stakeholders. In Canada, Round Tables on forestry and the paper industry offer some lessons (Duinker, 1998). These Round Tables have developed many ideas, but were constructed as neutral bodies and therefore have no powers, relying on their members' powers through other institutions (Bass *et al*, 1995). The national working groups set up to define national forestry standards, often at the instigation of the FSC, have resulted in some important 'soft policy' such as certification standards and procedures, and have been useful in-country focal points for developing shared understanding of what is good forest management, who should be accountable, and how to monitor them.

Whilst in some circumstances efforts to build consensus can be highly effective, in others people resist recognition of alternative views, and efforts to expand the range of policy actors face major hurdles. Policy may involve people with completely different levels of power and resources, often with a history of disagreement or hostile political interaction which prevents easy agreement. It therefore behoves the committed policy worker to reveal the hidden social and cultural assumptions underlying different world views and policies, allowing people to start relating to them openly.[59]

Multi-stakeholder processes in forestry, some of which have put considerable emphasis on intersectoral and public participation, have yet to make a big

---

59 The effective policy worker is often a combination of analyst and non-partisan activist (see Annex 1).

difference to policy in practice. They assume that societal consensus is possible but generally have grossly under-estimated the time and resources (of goodwill and money) needed to generate or refine such a shared vision. Some suffer from intersectoral cooperation from other ministries and key parliamentarians which in practice is wafer-thin, with the result that decisions which strongly affect forests are made without cognisance of the forestry debate. This phenomenon is evident in decisions about land allocation in Zimbabwe, and mining in forest reserves in Ghana. Others suffer from the domination, however subtle, of one group's vision over others. Such processes can merely ossify the existing power balance by dissipating other groups' energies on issues which do not challenge it, and glossing over divisions. *True multi-stakeholder processes require transfers of power.* The process of developing and pursuing the National Forestry Action Programme in South Africa appears particularly promising in addressing this key axiom (Foy *et al*, 1998).

Widening the ownership of policy requires that stakeholder fora recognise that people have different power and potential contributions to make to better forest management. If an agreed solution is then sought, some people need to be empowered to make positive contributions, and others need to be restrained from making destructive contributions. Forms of 'weighting' of groups according to their power and potential need to be hammered out (Colfer, 1995; Mayers and Bass, 1998).

Step three: Make space to disagree and experiment

"The truth is that we understand fully what we do not want. But as to what we do want and how to get it our ideas are necessarily vague. They are born out of practice, corrected by practice….it is from setbacks that we will learn."
Samora Machel (speech at Mozambique independence, quoted by Anstey, 1999)

When there is no agreement… Consensus can also be illusory, disabling or merely a sham. In some contexts, 'consensus' ends up as synonymous with 'conventional wisdom' – remaining stuck with the patchwork of untested assumptions discussed above. Emphasis on consensus gives the impression that nothing is wrong, and can lead to cynicism and disengagement from policy as people feel unable to change things. Consensus may thus impede creativity and innovation, inviting mediocrity and providing a disincentive to productive effort.[60] Many people in Western Europe, for example, consider government policy to be cooked up under the influence of corporations, the political manoeuvring of 'spin doctors' and a media which constantly feeds them on a diet of trash public 'consensus'.

---

60 Emergence of consensus is frequently taken as a measure of success of participatory methods – but it may sometimes be a result of coercion, capitulation and a highly uneven playing field.

Where people are at odds with each other (but not actually at war) on the methods or content of forestry or policy, it can result in greater richness of debate and of needed checks and balances. Situations of 'bounded conflict' (Lee, 1993) can allow the interplay of differing groups with differing objectives to flag errors and provide corrections.

Non-consensus based approaches are often needed, which can accept dissenting views. Such approaches may temporarily manage conflicts, but they seldom permanently resolve them. Collaborative management approaches in forestry are in some cases – such as in Ghana, Zimbabwe and parts of India – being treated as *collaborative learning* processes. *Adaptive management* is another key principle in this context. This assumes that, "because human understanding of nature is imperfect, human interactions with nature should be experimental" (Lee, 1993). Forestry actions and policies should thus be treated as experiments from which we must learn. The learning element is critical: policy experiments cannot be whims, but require deliberate monitoring by stakeholders with different views, and an orderly process to consider adaptation and review – the 'cyclical' notion of policy processes which was picked up by most PTW country studies (Section 3).

Stakeholders may not agree with each other, but through involvement in a policy community they can learn about other stakeholders' perspectives, power and tactics, can recognise why they disagree, who is currently 'winning' and 'losing', and why. Fall-back positions can be shored up. Through this experience, a few opinions will be swayed, ideas will emerge and the sort of information, 'sharpened-up' stories and organisation required for the losers to fight their corner more effectively next time, can be identified.

Unfortunately, few institutions are yet able to nurture these policy communities.[61] Creation of such institutional frameworks is a fundamental challenge.

Step four: Learn from experience, get organised and fire up policy communities

"One thorn of experience is worth a whole forest of instructions"
Mikhael Gorbachev

Using information and generating learning. Contexts where multiple interests prevail require better communication of information if negotiations between parties are going to make progress. But this means more than transferring facts and knowledge from those who have them to those who do not. Using information better means moving away from the 'banking approach' – where

---

61 Schanz (1997) notes: "no institutional arrangements have been set up thus far that are able to deal with conflicting views of reality, which might be the reason why crises are still perceived as threats rather than opportunities, and why reacting still prevails over acting in forest policy today."

knowledge is deposited in the learner's head – to approaches that help learners' pose and solve problems and evaluate information for themselves (Anderson, 1998). Generating and utilising information to establish processes of 'experiential' or 'reflexive' learning is key.

We have concluded at various points in this report that good policy allows *local experimentation and initiative* to thrive and aggregate at national and international levels. Experiments with different forestry pilot projects and trials of policy tools are vital for stakeholders to explore each others' claims, make mistakes, learn, and make changes for themselves. Power differences may even change a little in the process. Debate and experience of specific policy tools can lead to progress on specific issues, and can improve the policy process itself. We have noted how national certification processes have brought stakeholders to the table to define 'good forestry' and who should be accountable for it, a discussion which has extended beyond the boundaries of forestry aimed at environmentally-discriminating export markets.

Initiatives to develop criteria and indicators of sustainable forest management appear in general to be positive steps in building policy communities. Multiple worldviews are often being accommodated in these efforts – with no one possible solution dictating the scope and pace. These initiatives show a lot of promise in developing standards where multiple interests are at play, and just might go the next step and foster institutions capable of generating policy commitment around them. Key factors for their success would appear to be:

• the creation of neutral fora;
• the acceptance of local differences;
• negotiation with a wide range of actors;
• explicit integration of social, economic and environmental factors; and
• acceptance of the need for constant testing and revision.

Policy debates and projects can help to move the policy process out of the exclusive hands of foresters and consultants, spread information, and allow mutual recognition amongst stakeholders of power, claims and potential. Flows of this sort of information lend themselves to improved negotiation which in turn improves learning, capability and organisation.

Clear tactics are therefore needed to change policy. In some cases policy work will mean working directly with the current 'policy makers' to improve policy where opportunities arise. Well focused, often highly detailed, analysis may be needed to get the mix of policy instruments and options right. In other situations, effective policy work requires pointing to new information, challenging deeply-held assumptions and contributing to a new vision of what policy should be aiming for. It is becoming increasingly apparent in many forestry contexts that this requires collaborating on analysis and organisation with those who are currently marginalised from the policy process, so that they can muscle in on policy in the future.

The call for stronger participation should not be seen as a swing away from good analysis towards 'policy by brainstorm'. It is not just a matter of inviting everyone to the party and hoping for the best. Clear tactics for analysing and influencing policy are needed (some of these are discussed in Annex 1).

To sum up, the four 'steps' describe a learning, adaptive process brought about by a regular forcing open of the policy debate by stakeholders and their ideas, and a continuous sharpening of priority problems and proven solutions. A premium is placed not on one-shot planners' dreams but on step-wise approaches that notch up shared experience – making visible progress and building momentum for broader change.

6.4 Policy instruments that work

Processes are the crux of this report. The *Policy That Works* project was not a comparative analysis of policy instruments – irrespective of process. Whilst likely pros and cons can be pointed out for a range of instruments[62], this tells us little about whether they will work if applied. In short, policy instruments are even more context-specific than policy processes.

However, it is possible to make some conclusions on instruments from the work of the project reviewed here. The first task is to note the types of policy instruments that are now known to be effective in at least some contexts, and to refer the reader on to other sources of information on them (Box 6.2). But here we wish to draw special attention to those policy instruments which are not mere implementation tools, but also play roles in the policy process itself – making it iterative and cyclical rather than static and linear.

- *Research and extension brokering*. In *Sweden*, where a strong public interest in forest prevails, government has put high priority on access to good information in the policy process. The forest authority's major role is disseminating guidance and information about policy, while another body – Skog Forsk – was set up specifically to act as a brokering agency between forest owners and users, large and small, and researchers. Skog Forsk's membership covers most of Sweden's forests enabling it to channel its

---

62 Progress in this area has been made by Bass and Hearne (1997), and Merlo and Paveri (1998).

Box 6.2 The policy tool kit: sources of information about their use in different contexts

Five main types of instrument for implementing policy can be distinguished:

1. **Regulatory/ juridical.** Constitutional guarantees; laws, by-laws and regulations on forest practices, rights, tenure, trade; legally-binding international conventions
- *(Mayers et al, 1996; Bass and Hearne,1997; Merlo and Paveri, 1998; Tarasofsky, 1995; PTW Ghana, Costa Rica and Zimbabwe)*

2. **Economic/ market.** Taxes and revenue systems, subsidies, fees/ rebates, tradable permits, compensation, prices, auctions, certification
- *(Repetto and Gillis, 1988; Hyde et al, 1991; Bass and Hearne, 1997; Karsenty, 1998; Grieg-Gran et al, 1998; Panoyotou, 1998; Landell-Mills and Ford, 1999; PTW PNG and Costa Rica)*

3. **Information.** Research and information systems (technical, marketing, socio-economic), training, consumer information, land-use planning
- *(IIED/WCMC 1996; PTW Pakistan, Dalal-Clayton and Dent, 1999, forthcoming)*

4. **Institutional.** Mechanisms for dialogue, partnership, out-sourcing, institutional reorientation, intersectoral integration, common property regimes and local management, extension
- *(Mayers and Peutalo, 1995; Bass et al, 1995; Bass et al, 1998c; PTW PNG, Ghana and India)*

5. **Contracts/ agreements.** Access/ management agreements, direct works, non-legally-binding international agreements
- *(Landell Mills and Ford, 1999, PTW Costa Rica)*

The full range of contextual factors summarised in Section 6.1 will determine: the 'job' for which tools are needed; the availability of tools for the job; the degree to which choice between tools is possible; and, who will be able to use them! Where choice is possible, criteria for choosing the appropriate mix may include (Mayers and Bass, 1998):
- timeliness
- relevance
- credibility to stakeholders
- equity amongst stakeholders
- ease of understanding
- level of uncertainty and controversy
- information needed to operate the instrument
- cost effectiveness – economic efficiency
- time requirements
- reliability
- reproducibility

It is not easy to find instruments that will meet all these criteria well for any situation. Mixes may be needed. For example, forest certification is getting attention as it is timely, has credibility to many groups, is an efficient way of moving certain (larger) producers towards sustainability, and emphasises reproducibility. But there are worries over equity and efficiency for other producers (Section 5) which suggest that other instruments may be better for some circumstances. For example, fair trade certification may be more equitable for small groups, and informational instruments may be more efficient in building capacity for better forest management.

Within the parameters of available choice, getting the right balance of **coercion, persuasion and incentive** – implied by the above tools, to different degrees – requires a strategy for developing and maintaining the tool kit, or **policy tools mix** (Merlo and Paveri, 1998; PTW Ghana).

members' needs to researchers and, in turn, make research information useful. These tools for democratising information have created a high degree of engagement of forest owners and users in influencing and implementing policy, and seem to underlie the widespread consensus which typifies policy processes in Sweden.

- *Collaborative management experiments feeding back into policy change.* In *Ghana*, a forestry departmental unit was set up with a specific mandate to develop understanding of local capabilities for forest management, and to undertake experiments which modified foresters' roles in relation to those of other local stakeholders. The innovations in the experiments undertaken and the communication skills of the unit staff were very effective in attracting the interest and support of senior ministerial and departmental staff. These policy makers were keen to associate themselves with the experiments and this association catalysed considerable learning amongst other 'high-level' staff. The results are now being seen in a broader process of institutional and policy change in favour of local forest management capabilities.

- *Trials with by-laws and devolved authority.* Under recent legislation in *Zimbabwe*, some local government agencies have begun to link village-level realities and national policies by involving villagers in formulating local government by-laws covering village-level land-use plans, orders controlling tree cutting and protecting sites, and appropriate sanctions. Other legislation seeks to place the proprietorship of natural resources on communally-held lands with local communities by granting 'appropriate authority' to local government structures. Both of these legal instruments suffer from the resistance of cash-strapped local government agencies to devolve any revenue generating possibilities below district level.

However use of these instruments shows much promise since they have focused attention on rights, responsibilities and capabilities of different players at local level.

- *Legal, finance and information mechanisms for increasing local negotiating capacity.* 'Public interest' objectives for forests need to be balanced against conflicting private interests through location-specific negotiation. Similarly, only through negotiation can potentially good forest managers at local level, currently marginalised from the policy process, hope to achieve the capacity to protect their interests in the long term. In such contexts, experience in *Papua New Guinea* suggests that state agencies should take the lead to: scrutinise the plans of developers; publish model contract provisions; legislate for court review of manifestly unfair contracts; and create finance arrangements, where landowners can borrow against future income to pay for professional advice.

- *Progressive land taxation: has particular potential in some places.* Although still with more potential than actual impact, experience in Costa Rica and Zimbabwe suggests that a progressive land tax may be an effective tool for redistributing agricultural land to those who really need it, and taking pressure off remaining forests. If effectively levied and managed by local government it may also provide institutional linkages to, and locally-controlled resources for, better land use.

- *Property rights changes: difficult, but not impossible with practice. Papua New Guinea* serves as a salutary lesson that local security of resource tenure, by itself, is not sufficient to ensure long-term sustainable forest management. When customary tenure is not backed up by sufficient local institutional strength – both to be able to deal with outsiders (whether they be offering 'development' or 'conservation' projects in the case of PNG), and to maintain the local side of the bargain in any deals made – the long term management of any piece of forest land cannot be guaranteed. But it can be done! New legislation, in places as diverse as *Ghana, China* and *Scotland,* is tipping the balance in favour of more control of trees and forests by local farmers and communities. However, trees on farms in many places are either retained and managed by farmers, or removed, depending primarily on their effects in the farming system. Thus, improved formal tenure is only part of the story here too. The considerable technical problems of integrating timber and forest trees with agriculture also needs to be addressed – hence the close linkage of tenure change with research and experiment, and with information, extension and support systems.

All of the above policy instruments are, effectively, 'power tools'. They both implement policy and increase its information base and reliability, by providing feedback. In so doing they are instruments of change, helping to unblock situations of entrenched excessive power and stifled creativity.

6.5 Summing up – get into policy work!

Linking the corridors of power to local reality

Whilst some forests and people are doing well in the context of globalisation, conventions based on trust and a sense of fairness are being undermined in other situations. The results are cronyism, gangster methods and the predatory business practices of timber kings; poor worker conditions and exploitation; one-sided forest revenue shares; and loss of 'location' through forest evictions or nomadism in forest employment. For those who can afford it, insurance and armed guards in protected enclaves are available. Increasing numbers of those who cannot seek ways of opting out of a global economy which is overwhelming them; losing commitment to non-violent norms of behaviour, and increasing demands for local autonomy.

Meanwhile the shapers of the global economy – the US Treasury and IMF, deregulators and merchant banks – are unable to respect individual national circumstances. They have compelled countries to embrace the laisser-faire model as fast as possible, and have badly under-rated the importance of local political cultures and histories.[63]

How can we reassert a social morality and political philosophy which has gone out of fashion? A few widely-accepted public values or virtues – such as loyalty, trust, accountability, security, equality and freedom – need to be (re-) established as the foundation for policy debate. Institutions with strong cultures are sorely needed. Success here would be signalled by people solving shared problems and satisfying economic, spiritual, recreational and other needs, at levels which change for the better over time.

All of this may sound way beyond the realms of forestry, but we believe forestry can at least contribute. By legitimising a broader range of narratives about what forests are for, forest policy can internalise long-term goals and get forests managed by more people who can agree with each other – through contracts and leases with agreed management systems. The notion of stewardship – according to agreed criteria rather than laisser-faire – is being pioneered in forestry and has become a significant galvanising force,

---

63 Dominance of the laisser-faire economy, 'creates a world order without order'. This is an economic problem in itself as "global markets are panicky, easily influenced mechanisms (viz the recent stock market falls and collapse of currencies) which respond to certainty and decision ...but there is nobody in charge" (Will Hutton, Observer, 1998).

becoming accepted as a way to achieve compromise between different needs.

The key to making policy real is to find ways to link the corridors of power to local reality. This requires understanding of the complicated area between policy pronouncements and practice, and to explain the difference between what people agree to do and what people actually do.

To improve policy we need to unite decision-making with its consequences, such that policies, plans and strategies are not separated from practice, but are linked to it. This means that they benefit or suffer from it; that they learn from it; and that they improve it. Both policy processes and instruments are needed to make such links. Good policy becomes defined, and refined, through experience of those who have the potential to deliver good forest management and work for equitable livelihoods – often the very people who are marginalised by current policy processes.

The challenge for all those who can get their teeth into policy for forests is to find the right 'power tools' for the right people. They will make their own policy space (see below).

Even foresters can act politically

There is a common perception amongst foresters that the fate of forests is determined by forces beyond their control. Whilst this, to varying extents, may be true, the fatalism which it engenders needs to be overcome.[64] In the face of these extra-sectoral influences, foresters are inclined to retreat into their shells and encourage the illusion of stability: as the determining forces are beyond control, it is appropriate to ignore them. Extra-sectoral influences are treated as temporary aberrations from which the forest must be protected. Romm (1985) noted that this results in a recurring tendency to focus on the 'best' way to treat the forest – ecologically, technically and economically – and to generate policies which are merely 'normative statements of purpose that seem inevitably thwarted by society's possibly malicious refusal to stand still'.

But foresters can make progress which engages and tackles some extra-sectoral influences. *Policy that works* showed that much progress has, in fact, been made by policy processes learning from local solutions to forest problems, both indigenous and project-driven. It has also been made by local user groups and farmers coming together to tackle local forest problems, and by 'policy-makers' giving them the chance to experiment. This has widened the ownership of policy and formed larger policy communities.

---

64 The 'founding father' of US forestry, Gifford Pinchot, used to urge foresters to be technical and stay out of political debate; he was eventually sacked for political reasons! Another forester, Jack Westoby, on the other hand, made calls from the early 1980s onwards for foresters to speak out on issues of access to land, participation and social justice. Westoby was highly influential internationally – so the call for foresters to engage with power is not new.

The type of work now needed is collaboration on analysis and institutional change *with* those who are currently marginalised from the policy process, so that they can present their views and experience, and make their claims, more effectively.

In a sense, this means turning the conventional approach on its head, i.e. we need more policy *process* challenges for the powerful, and policy *content* analysis for the marginalised.

We need to find ways in which policy analysis can be harnessed and developed by potential good forest managers who are currently marginalised from policy – the 'concealed forest stewards'. To be effective, such work needs to respond to demand from marginalised groups, and to understand the networks within which such 'field-level decision-makers' carry out their activities. It requires identification of the individual and organisational choices that are the hub of a problem, and tracing the rules, structures, policies and other signals which affect them. In this way, people on the ground can either become 'policy-makers' themselves, or (at least) be supported by the instruments and resources to help them make effective inputs to current policy-makers. It also implies that work needs to be better targeted such that policy-makers can learn, and be subject to checks, balances and incentives from below e.g. due process/ diligence.

Almost every aspect of forestry is a political activity, and the type of forestry which dominates in a country presupposes a political direction to development. The recent stress on (or at least rhetorical adoption of) participatory forms of forest management may signal a direction of social change towards equitable livelihoods. However, forestry as a social exercise has generally been reactive rather than leading the field – forestry tends to piggy-back on change in other sectors or parts of society, or at least incorporates the political buzzwords generated by them.

Those strategic alliances, which turn out to be vital developments for forestry, may have formed for reasons which have nothing to do with forestry. Or they might form around just one catalytic forestry issue which can bring useful incremental change without having to take on the whole agenda at once. Either way, foresters should engage with them as early as possible. In some countries, the lack of political profile of forestry (or a decline in its profile – as in Papua New Guinea with the current Southeast Asian economic crisis) can also be an opportunity for far-sighted foresters to make changes. But to declaim a 'lack of political will' is generally to duck the responsibility which foresters' considerable powers give them. Forestry can and should be an activity which changes the political environment for the better.

Doing policy work

A1 Introduction

A1.1 What policy work is, and why it is worth doing

The aim of this Annex is to offer some guidance on approaches and methods for engaging with policy. Here, 'engaging' implies the messy business of 'locking horns', rather than the anticipation of a happy marriage.

The main text of this report urges a wide range of people concerned with forestry and land use to get involved in shaping and implementing better policy (see Sections 2.4, 3.2, 6.3, 6.5). This work cannot be left solely to professional 'policy wonks'[1] partly because there are rather few of them in forestry and land use and they never seem to be around when you need them, and partly because they may have got it wrong anyway!

We have also noted that 'policy' can be described as 'what organisations do', and that further definition of the 'policy content and process that matters' is specific to the context and actors involved, and should be the first task of those who wish to engage with policy. Getting involved in policy – 'doing policy work' – has several possible elements. We offer some working definitions of these elements in Box A1.1.

The main text is full of examples of different sorts of policy work that have been effective in improving policy for forests and people, e.g.:

* Ghana – work with the collaborative forest management unit reinstalled the notion that local benefit is what forestry should be for
* Costa Rica – JUNAFORCA's engagement with policy provided compelling evidence that smallholder forestry works, and laid the paths for improved conditions
* Scotland – policy work has stopped excessive upland afforestation and reversed the Forestry Commission's policy on land disposals
* Globally – several landmark policy studies (by individuals, institutions or commissions) had important influence at international level (Section A1.2 below)

---

1 A term coined in the US for policy obsessives.

Box A1.1 Defining 'policy work'

Work on policy may involve various different activities. The policy literature is replete with varying definitions of these. For our purposes we use the following definitions:

- *Policy research*: investigation aimed at increasing knowledge useful for policy.[2]

- *Policy analysis*: a type of policy research aimed at examining or tracing the component parts of policy context, content, process and impacts.[3]

- *Policy advocacy*: making a particular argument about how policies should be made or implemented, and/ or about policy content. This may or may not utilise policy research.

- *Policy influence*: having a hand in changing or maintaining policy content and/ or the policy process. This may or may not utilise policy research.

To the above, we add a final over-arching definition:

- *Policy work*: the range of actions which have an explicit link to understanding or influencing policy.

This final definition reflects our recognition that the policy process is broader and more muddled than the focus of much public policy research and analysis would suggest. More actors are involved in policy than just those in government, and some actors are involved in attempting to both research and influence policy. These activities may be carried out with the current *policy-makers* and *policy-implementors*, but they may also be carried out with those who are neither. The term '*policy holders*' may thus be used to recognise that power over policy may not lie in policy-making but also in policy-implementing, and that the holders of this power may change over time.

A1.2 Introducing the policy research literature

Table A1.1 is an attempt to categorise and characterise the literature on policy research concerned with forests.[4] It is also a guide to a few key reference sources on policy research in forestry and on the wider fields of natural resources management policy and development.

Some approaches to policy research are far more prevalent in the literature than others. Although Table A1.1 gives some key references for each of the approaches, it should be noted that 'rational choice' and 'content-focused' analyses dominate in forestry literature, whilst 'pragmatic pluralist' approaches are also gaining ground. The other approaches are, relatively, in short supply, although intersectoral policy analysis is increasing (with accompanying trends for resource valuation). A greater depth and diversity of research using these other approaches is needed.

2 This is a broader definition than many used in the fields of policy studies. For example, in a book giving guidance on how to do policy research in the social sciences, Majchrzak offers the following definition: the process of conducting research on, or analysis of, a fundamental social problem in order to provide policy-makers with pragmatic, action-oriented recommendations for alleviating the problem (Majchrzak, 1984). This assumes the existence of clear problems and well-defined policy- makers, but this may not be the case. Policy research often aims primarily to enlighten – to help identify and spread understanding – of problems rather than solutions, and may aim to do this among actors who are not policy-makers (but who may influence policy-makers or one day become policy-makers themselves).
3 Policy studies – mostly concerned with *public* policy analysis – contain a wide range of definitions. 'Classic texts' on public policy analysis include: Dror, 1986; Dye, 1976; Hill, 1997; Hogwood and Gunn, 1984; Jenkins, 1978; Lindblom and Woodhouse, 1993; Wildavsky, 1987. Majchrzak defines policy analysis as: *the study of the policymaking process – the process by which policies are adopted and the effects of those policies once adopted... typically performed by political scientists* (Majchrzak, 1984).
4 'Literature' these days is not confined to printed matter, but is also to be found on web-pages and in email, etc.

Table A1.1 Policy research approaches – characteristics and key references

Policy research approach	Characteristics	Key references	
		Forest policy analysts	Natural resource/ development analysts
Rational (public) choice	Promotes development of sectoral policy statements and recommends policy contents and processes	Gane 1987; Westoby 1989; FAO 1987; World Bank 1991; Gluck 1995; Tikkanen and Solberg 1995	WCED 1987, UNCED 1992, World Bank 1992
Instruments/ content-focused	Analyses and recommends only content of policy – its instruments and mechanisms	Repetto and Gillis 1988; Grut *et al* 1991; Hyde *et al* 1991; Grayson 1993; Wibe and Jones 1992; Bass and Hearne 1997; Merlo and Paveri 1998	Bresser and Klok 1988; Pearce 1994; Panayotou 1992, 1998
Pragmatic pluralist (process/actor/ networks focused)	Recognises political dimensions, assumes social groupings influence policy, and proposes solutions emphasising participation	Cubbage *et al* 1993; Sizer and Rice 1995; Barber *et al* 1994; Cortner *et al* 1995; Ellefson 1992; Anderson 1998	Grindle and Thomas 1993; Rees 1990; Ascher and Healy 1990; Long and Long 1992; Lee 1993; Röling and Jiggins,1998
Inter-sectoral	Examines influence on the focal sector of policy in other sectors	Gregerson *et al* 1993; Kaimowitz and Angelsen 1998, 1999; Contreras 1999	Munasinghe and Cruz 1994; DFID 1998; Dalal-Clayton and Dent 1999
Political economy – structuralist	Emphasises existence of strong political-economic forces determining policy	Dauvergne 1997; Utting 1993; Peluso 1992; Barraclough and Ghimire 1995	Swift 1996; Blaikie 1985; Gadgil and Guha 1995
Anthropology of policy and power	Examines policy discourses, narratives and power of actors	Fairhead and Leach 1996; Filer with Sekhran,1998	Roe 1994; Hoben 1996; Shore and Wright 1997
Historical	Traces forces and events causing policy change over time	Dargavel *et al* 1988; Buttoud 1997; Perlin 1989	Grove 1995; Schama 1995

Literature designed *solely to advocate or influence* policy is not generally covered in Table A1.1 This is largely because such literature tends either to be very context-specific, or to include insufficient analysis. Examples of literature produced by groups advocating content or process in forest policy might include:

- campaign literature from a host of international and national NGOs;
- private sector groups' lobbying or campaigning material;
- government documents of various kinds;
- multilateral agency documents, notably in forestry those spelling out grant and loan conditions required by the World Bank and the IMF; and
- bilateral donor strategy documents.

Literature on *how* to advocate and influence policy for forests is not so common, but is increasing. Examples in forestry include: Juniper 1998; Institute for Development Research 1997; websites and literature of the Rainforest Action Network and the Rainforest Foundation.

In 1998, 162 individuals responded to a survey, administered by CIFOR, which asked them to list their top five publications that have influenced debates on policies affecting forests over the last twenty years. The individuals were participants in an e-mail forum of forest 'policy experts' – *Polex*. Most participants work in developed country organisations and international agencies; rather fewer are developing country policy-makers or researchers. The three publications that were mentioned most frequently were:

- Repetto and Gillis (1988). *Public policy and the misuse of forests.* Cambridge University Press – on forest concession policies and trade restrictions that promote unsustainable logging
- Poore *et al* (1989) *No timber without trees: sustainability in the tropical forests.* Earthscan Publications – on whether tropical forests were sustainably managed for timber, and how they could be
- Peters *et al* (1989) *Valuation of an Amazonian Rainforest.* Nature – on the value of non-timber forest products

The *Polex* responses suggested that the documents which respondents considered to be influential did *not* affect policies directly. In most instances, they seem to have influenced the general 'conventional wisdom' in international, policy, academic, and funding circles on different topics and this eventually filtered down to policy makers in specific countries. However, documents with major direct influence in particular countries – government commission reports, action plans and official policy documents – were also noted by some. Other documents in the 'top ten' included three key products of the 1992 UNCED Earth Summit – Agenda 21, the Forest Principles, and the Biodiversity Convention – and the Brundtland Commission report 'Our Common Future' that preceded UNCED.

In his summary of the survey, the *Polex* mailing list coordinator noted; "the responses give the impression that conventional wisdom tends to associate each major forest-related issue with a handful of publications that have crystallised public interest in a topic, given it greater legitimacy, or synthesised previous research on it" (Kaimowitz, 1998). The discussion of approaches to understanding and increasing policy influence – through use of documents and other means – is taken up further in Section A5 of this Annex.

A2 Recognise the political game: theory, value, language and power

Section 2 of the main text introduced the forestry 'players' and the SFM 'plot', and Section 4 related individual stories from several countries. From this, it should be clear that real forest policy abides in the realm of politics rather than deep in forestry text – books. In this section, we take a little more time to explore some of the theory which may help us to understand policy.

A2.1 Policy is a slippery concept

"Policy. A course of action adopted and pursued by a government, party, ruler, statesman, etc.; any course of action adopted as advantageous or expedient".

Oxford English Dictionary

'Policy' – the word developed from both Greek and Latin roots – first came to mean both the art, method or tactics of government and regulating internal order. This second meaning split off with the formation of Robert Peel's 'new police' in Britain in 1829 and the administration becoming the domain of 'policing'. We can note in passing that forestry seems to have retained the linkage between policy and policing since most national forest policies have traditionally put a strong emphasis on maintaining a national forest estate by 'policing' against exploiters and encroachers.

The Oxford English Dictionary also reveals uses of 'policy' which are now obsolete: "a device, contrivance...stratagem, trick". The former meaning of policy – as the art of government – has also gone through changes, from its former pejorative meaning as cunning, deceit, trickery to become more respectable. For Shakespeare, policy encompassed the arts of political illusion and duplicity. Show, outward appearance and illusions were the stuff of which power was made. He employed the terms of Machiavellian philosophy...Power cannot be sustained purely with force. It needs 'policy'. Indeed whilst the English, Dutch and German languages insist on two different words, 'policy' and 'politics'; the French, Italian and Spanish do not feel the need for this distinction.

A2.2 Problems with the 'rational systems' model

A model of the policy process as a series of stages – e.g. information-decision-implementation-evaluation – is a useful way of chopping up a complex and elaborate process for the purposes of analysis. The 'stages' idea can of course be presented in other versions which may be useful for explaining things, e.g: analysis; anger/provocation; persuasion; consensus; action; analysis..."

Figure A2.1 The notional 'policy cycle'

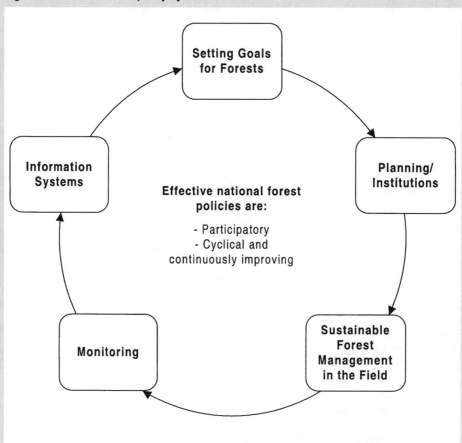

But there are problems with such a 'rational' model, whether it is described as a linear form or as an improving 'virtuous circle' form (Figure A2.1)

- There are few blank slates – current policy processes are usually products of long histories; a realistic starting point for analysis may thus be far back in time, or difficult to discern at all.
- Stages may not be sequential; they may occur simultaneously or in apparently random order. Policy initiation may start anywhere in the system.
- Stages are not insulated from each other and there may be various overlaps and interactions between them.

It is quite legitimate to employ a rational model as an analytical device – in order to 'map' what elements exist at present, and where the 'entry points' for policy work might be. However, many practitioners and policy wonks have tended to treat it as a prescriptive framework – how policy *should* be made, implemented and assessed. The rational systems notion has been conferred with a status as a normative model, "a dignified myth which is often shared by the policy-makers themselves" (Gordon, Lewis and Young, 1993). But, to talk about 'rationality' (as so much policy analysis literature does – often taking its cue from economics), without reference to ends or to the issues about who has the power to determine these ends is at least beside the point, and at worst dangerous.

A2.3 Irrational alternatives?

The literature provides a range of other models, each attempting to show the limited practical usefulness of the search for 'rationality' implicit in many attempts to devise 'objective' tools for policy analysis. Three distinct types of model are described below:

- incremental 'muddling along'
- bureaucratic process; and
- political bargaining.

The *incremental 'muddling along'* model has it that it is rare to be able to identify a clear-cut group of decision-makers, or an event which can be pinpointed as the moment when the decision was made. Therefore, policy is a continuous bustle of activity (or even a period of inactivity), and it is only in retrospect that people become aware that policy was made. 'Decision-makers' – bounded by their skills, knowledge and habitual modes of thought – thus muddle through. Weiss (1986) breaks the incremental model down into eight types of behaviour [our interpretation]:

- reliance on custom – do what has always been done
- improvisation – do something off-the-cuff which seems to fit the bill
- mutual adjustment – make small changes in response to others
- accretion – wait until things build up, then do the obvious thing
- negotiation – get together with others and bargain a solution
- move and counter-move – do something tactical in provocation or response
- implementing pet remedies prior to identification of a problem – never mind the context, just do what you fancy
- indirection – leave it to others

The *bureaucratic process* model suggests that policy decisions emerging from institutions are fundamentally affected by the way in which these institutions work. This is due to three main reasons. First, once individuals become part of a bureaucracy, they acquire goals or interests distinct from those of their professional independent selves and quite separate from those of their political masters or the general public. Second, bureaucracies endeavour to retain a monopoly over information and then utilise this to ensure that their own interests are protected. Third, organisations are coalitions of interest groups. Internal horse-trading and compromise rather than the rational evaluation of evidence will characterise final decisions. Resulting policies are seen as, e.g., maintaining internal or external relations, rather than purposefully addressing a problem.

The *political bargaining* model proposes that policy and practice are not the product of individual and organisational choice processes, but the outcomes of a political struggle between interest groups within society. *Pluralists* argue that no one group achieves a dominant position in the longer term. *Elitists* argue that establishment groups can bias the whole policy formulation and implementation process towards their own vested interests. More radical *structuralists* argue that only one élite holds power – the capital-owning class. Public service and regulatory agencies are seen to operate in support of private capital, by reducing social conflict, providing essential services, etc. They operate to maintain a social system which is conducive to capital formation and in which the economic development interests of capital dominate (Poulantzas, 1973; Habermas, 1976). When viewed from the political perspective, the key question changes from how policy decisions are made, to who has the power and influence to make the effective policy choices.

We argue that each of these models has some explanatory power for particular aspects of policy or for the policy process in particular contexts. But none of the models is sufficient on its own to fully explain the complexity and messiness of what is often going on. Some further realities about policy processes include the following considerations:

- Both formal and informal rules and procedure determine: who participates, paths available for action, rules of the game.
- There is rarely one person or group of people who is sitting on a set of policies which could be changed.
- Policy decisions are often the cumulative result of interactions, conflict and cooperation. The 'policy-makers' – those who currently 'hold' policy – may be a clear 'policy élite' or they may be a combination of less obvious groups.
- Decision-makers see different faces of an issue – depending on concerns which are ideological, professional, personal, concern for clients or relationship with others. In other words, they themselves have several identities, and this may be reflected in how they exercise their role as 'policy-maker'.
- Interactions reflect power relations and include: overt exercise of power by some groups over others; decisions to do nothing; non-decisions (keeping conflict from flaring up); refraining from an overt statement of policy in order to maintain flexibility; keeping certain subjects out of the policy arena in order to maintain a personal power base on that issue; and the shaping of perceptions such that conflict is prevented from arising.

We pursue some of these issues further below.

A2.4 Policies as 'myths', 'dominant symbols' and 'demons'

Anthropologists have likened the notion of policy to that of 'myth', in the sense of the term as a *guide to behaviour*. Malinowski's study in the 1920s of Trobriand society in Papua New Guinea used the notion of myth in this way. Unpicking this idea further, policy has been described as a '*charter*' for action, as a *commentary* that either justifies or condemns action, and as a focus for *allegiance*. A *political myth* has been described as the pattern of the basic political symbols current in a society (Lasswell and Kaplan, 1950). This concept is close to others e.g. Marx's 'ideology' (Marx and Engels, 1998) or Mannheim's 'Utopia' (Mannheim, 1936). Given this range of functions, policies may "encapsulate the entire history and culture of the society that generated them" (Shore and Wright, 1997). Indeed, key policies can reveal the nature and structure of cultural systems. The Truman Doctrine – 'containment' of communism – in the USA of the 1950s, has been described as one such 'dominant symbol'.

At a more day-to-day level, the words or 'labels' applied to issues and problems by policy-makers or development administrators can embed a particular angle or version of reality which may bear little relation to that held by others, and can be very difficult to dislodge. This labelling is generally done in an innocuous manner, apparently for simple convenience, yet it is often highly political – since it may fundamentally influence the creation of agendas or access to resources. In discussing 'common sense' or 'rational' models of agricultural policy, Clay and Schaffer (1984) also note that such models are full of "apparently innocuous but ultimately pernicious concepts such as 'target groups'".

In forestry, symbols suggested by words such as 'monoculture' or 'native forests' – which have come to have strong connotations – can also be the enemy of consensus. Maughan (1994) describes an advantage of organised conflict resolution (in US watershed management) in terms of removing the 'symbols and demons' which get in the way of constructive debate.

A common example of a label applied in policy contexts, is the so called 'gap' between 'policy' (statements) and 'practice'. Such a gap is rarely a void, but a space already crowded with perspectives and biases, and thus 'full' already of preconceptions and misconceptions. Rather than trying to understand these crowded spaces and better connect statement and theory to practice – the real challenge – the notion that they are knowledge gaps serves to bring forth hasty new policy pronouncements and prescriptions about what is needed.

A2.5 Policy as language and discourse

The language of policy (and of policy research) functions as a type of power. This power is exercised through styles of expression – "power comes as much from the barrel of a phrase or sentence as a gun" (Apthorpe, 1997).

When the primary aim of policy language is to persuade rather than inform, '*goal language*' is used. This inspires, uplifts, gains support, defines parameters, or offers a 'badge' to wear.

Policy statements are unlikely to depend on a weighing of positions and evidence, but to rely on presentation of a position that is held to be exemplary in some way, and in a style chosen mainly to attract, please and persuade. Style can be as powerful as substance.

As policy protagonists use language with symbols and labels to convey their ideas, 'discourses' can be discerned. Discourses have been defined as "configurations of ideas which provide the threads from which ideologies are woven" (Shore and Wright, 1997). 'Dominant' discourses work by setting up the terms of reference and by disallowing or marginalising alternatives. Some policies can be seen in this way, as they set a political agenda and give institutional authority to one or more discourses.

Apthorpe (1997) discusses an example of two competing discourses about rural livelihoods under green revolution technologies – 'ideal ruralism' and 'radical realism' – which failed to see eye to eye at all. Ideal ruralism is preoccupied not with any actual pattern of rural livelihood but with deducing only an ideal type – to ensure it avoids falling into the 'local bias' of which realist case studies are accused. Radical realism pursues local detail, proposes solutions based on it, and charges its idealist rival with being too selective in its perspectives and relying on only negative characterisation – the rural poor are described as landless, stockless, feckless, etc. What one side took in formal fashion as objectives, was taken by the other not as objectives but perspectives.

While the ideal ruralists' prescriptions were being converted into policy documents, the radical realists were still preoccupied with substantive details, little recognising that they were being as selective in their perspectives as the ideal ruralists were. In Apthorpe's example, the realists' solutions were Utopian, e.g. concluding that planners are to blame for poverty, not local people, and therefore that planners' offices should be restaffed or abolished. This did not go down well with the officers concerned. Neither could policy-makers deal with the particular and specific nature of the realists' conclusions, which ill-suited policy's characteristic concerns with transferrability and replicability (Apthorpe, 1997).

A2.6 Policy as 'political technology', oversimplifying and stereotyping people

Foucault coined the term 'political technology' as the means by which power conceals its own operation. Others have noted that policies can be seen in this way – as instruments of power for shaping individuals' sense of self. "The political nature of policies is disguised by the objective, neutral, legal-rational idioms in which they are portrayed. In this guise, policies appear to be mere instruments for promoting efficiency and effectiveness." (Shore and Wright, 1997). The objectified person "is seen but he does not see; he is the object of information, never a subject in communication" (Foucault, 1977).

The metaphors of the individual and society which are used in policy shape the way individuals construct themselves – as 'citizen', 'professional', 'stakeholder', 'criminal', etc – and influence the way people behave. 'Governance' – the processes by which policies not only impose conditions, but influence people's norms of conduct so that they themselves contribute (not necessarily consciously) to a government's mode of social order. Although "imposed on individuals, once internalised, [these norms] influence them to think, feel and act in certain ways" (Lukes, 1984). Basic categories of political thought are reconfigured to create new kinds of behaviour through notions like 'popular capitalism' and 'active citizenship'.[5]

Policies thus provide a means by which consent is 'manufactured' – conditions are engineered so that, seemingly, consent of the public comes 'naturally'. In this way policies also have a legitimising function – serving to buttress the authority of rulers – one cannot successfully argue against 'the proper order of things'. It is also evident that policies themselves can function as a vehicle for distancing policy authors from the intended objects of policy and for disguising the identity of decision-makers.

These days, the manufacturing of consent is not solely the preserve of governments. For example, as North (1995) points out, many influential groups (government and the private sector, and to some extent the general public), are beginning to distrust the green movement's definition of issues. The 'prophecies of catastrophe' which is the *modus operandi* of so many groups have not, in fact, been followed by the prophesied problems.

Hecht and Cockburn (1989) suggest, with reference to the various solutions that have been proposed for the *Amazon*, that 'knowledge systems' are systems of domination – the question of *who* defines a situation is critical. In recent years, NGOs, particularly the green movement, have been adept at defining situations in ways which make the influential listen. Yet there are "a number of pitfalls that lie in the line of march staked out by the 'green' movements in the First World. By de-emphasising 'old-fashioned' concerns with political economy, property relations and distribution, they extol the [NTFP extractive] reserves as environmentally sound solutions where the good rural life can continue. But all reserves are far more precarious than their current popularity would suggest."

The intentions of big industry have moved from simply manufacturing goods, to manufacturing markets (through advertising) for those goods, and now to manufacturing consent in favour of the ethical and policy conditions under which they would prefer their markets to evolve. Monsanto, for example, heavily made the case for genetically modified organisms (GMOs) as contributing to the elimination of hunger in developing countries, and as environmentally desirable through e.g. the reduction of chemical usage they would bring. In this, however, there was inadequate recognition of the public's fear of science and a failure to realise that consumers become suspicious and vulnerable when they are starved of choice. The 'overselling' by Monsanto has now blown up in the company's face.

Agendas can generally be more easily controlled if policies can be used to over-simplify issues. Simplistic problem definitions often lead to the domination of policy by a single group or institution that has the required muscle provided by money, relevant mandate or technical expertise. If other concerns are introduced into the simple story, this group is likely to perceive

---

5 Rose (1992 – quoted in Shore and Wright, 1997) has even argued that the idea of 'freedom' acts as an instrument of government control in the construction of 'free market' and 'free society' which requires: "a variety of interventions by accountants, management consultants, lawyers, industrial relations specialists and marketing experts....[to] make economic actors think, reckon and behave as competitive, profit-seeking agents, to turn workers into motivated employees who strive to give of their best in the workplace, and to transform people into consumers who can choose between products."

them as a threat. Ascher and Healy (1990) note how central leaders may get large political rewards from the symbolism of simple large initiatives with an impressive single-performance measure. Even when such grand schemes and policies begin to manifest problems, the political symbolism of the big project often leaves little room for pulling back; the bigger the venture, the more the central leaders' reputations are on the line.

In such contexts, policy tends to be based on highly aggregated and centralised analysis (if it is based on analysis at all) which is likely to be blind to local variation and to mask distributional issues. That is not to say that such contexts have no use for information from the field. As Polly Hill notes, information collected directly from the field "is not some kind of pure substance with inherent validity, [but rather] matter which has commonly been extracted from unwilling informants by resorting to many convolutions, blandishments and deceits [and then] fudged, cooked and manipulated by officials at higher levels, the main purpose being to ensure that the trends will be found satisfactory and convincing by those with still greater authority, as well as to compensate for presumed biases" (Hill, 1986, quoted in Lohmann, 1998).

Most insidious are the policies which create unattractive stereotypes of people whom holders of policy would like to keep marginal – often the very people who are most dependent on forests, or might be able to manage forests best. Hecht and Cockburn (1989) describe how views on forest-dependent people, held by governing élites in Brazil, are fundamental in determining policy towards forests. In the Amazon context, "the portrayal of native peoples as *Rousseauian* creatures has... permitted a view of them as children, incapable of wise decisions or the exercise of adult responsibilities. Until recently the official Brazilian view is that they are wards of the state, unable to participate in political life." This is consistent with other policies which have been exercised towards Amazon forests and their people – the massive 'flooding the Amazon with civilisation' through major government programmes for the region, and the settlement and clearance of forests for other uses when pressures in e.g. cities and industries began to build up.

Colchester and Lohmann (1993) note similar examples in Thailand and Vietnam, where a policy belief that ethnic groups were inferior helped to colour their interpretation of the shifting cultivation that these groups were invariably practising. Coupled with the increasing observation of environmental problems (soil erosion, etc) in highland regions, the result has tended to be a policy assumption that shifting cultivation and its practitioners are destroyers of the forest. Do Dinh Sam (1994) for Vietnam, Rerkasem and Rerkasem (1994) for Thailand, and Bass and Morrison (1994) analysing the regional consequences, have outlined these policies. They tend to aim at settling shifting cultivators, without understanding either the fact that shifting cultivation is sustainable in circumstances of low population density, or that the transition to more settled forms of agricultural or forest management requires much time, and support on many fronts.

A2.7 So, 'policy is a power thing'... but there may be room for manoeuvre

In summary, to understand policy that matters is to understand power and influence. The exercise of power may be obvious and crass, or it may be subtle: *"is it not the supreme and most insidious exercise of power to prevent people, to whatever degree, from having grievances by shaping their perceptions, cognitions and preferences in such a way that they accept their role in the existing order of things..."* (Lukes, 1984)

The balance of power between interests may be highly entrenched, but it is rare for a system to be devoid of room for manoeuvre – for some people at least. Moments of change or indeed crises occur in the cultural, political, economic or natural environment; these cause reactions and create windows of opportunity to put issues on the agenda. Small, well-focused actions in these moments can produce significant, enduring improvements, if they are in the right place. The policy analysis literature refers to: 'access points' and 'critical junctions'; 'high-leverage changes', 'quantum leaps' and 'punctuated equilibria'. Often these actions may be counter-intuitive and non-obvious to many people in the system.

This room for manoeuvre can be identified with the benefit of hindsight, but our interest here is in whether it is recognised in advance. Hill (1997) describes the 'rubbish bin' model of policy change, which assumes that problems, solutions, decision-makers and choice opportunities are independent. Solutions are linked to problems primarily by arriving in the bin at similar times. Changes occur with unique juxtapositions of events and the unique responses of individual actors. With such a model we can do little but sit in the rubbish bin and watch what happens.

In the following sections we pursue an alternative approach in the belief that we can better understand the forces at play in these processes, predict what might happen and get ready to influence it. Here, we offer various methodologies, many of which were tested in the *Policy that works* country case studies.

A3 Develop a strategy: objectives, framework, key steps

"There can no more be only one approved mode of policy research than there can be only one way of learning"

(Wildavsky, 1987)

"One should not be too straightforward. Go and see the forest. The straight trees are cut down, the crooked ones are left standing"

(Kautilya, Indian philosopher, third century B.C.)

Having, we hope, installed a notion that policy processes are essentially political, and dispensed with naive optimism about pluralism and the rationality of decision-makers, we can get on with identifying practical approaches to tackling policy! Utilising a range of approaches and methods is likely to prove productive. Our aim is to help fill a toolbox, but we must stress that not all tools will be needed in any one context. It is important to be selective, recognising the work on the policy edifice that has been done before, by others with their own tools. A basic framework is first needed – to stay on track.

Important conditions, required before undertaking policy work, tend to be:

1. Reason – a clarity on the need and purpose needs to be defined – which means identifying the real issues
2. Timeliness – key people must already feel some need for change
3. Locus – an independent but influential institutional location for coordinating policy work can be helpful
4. High-level support and expectation – that the work will lead to significant changes in important matters such as governance, policy and investment
5. Commitment of key participants
6. Reasonable idea of the tactics required for influencing those who need to agree changes

A3.1 Identify the issues – the problems and the opportunities

It is unlikely that a pool of policy researchers will be sitting around waiting for a problem to arise or a success to analyse. It is more likely that policy researchers will be asked to address a problem or opportunity, or that people recognise the existence of a policy failure or success and want to know how to tackle it. In either case, a preliminary definition of the issue is needed.

An initial assessment is also needed on whether the issue is researchable, and/ or whether there might be room for manoeuvre with it – i.e. whether doing policy work is worthwhile. For example, a problem may be too big, intractable, complex, expensive or dangerous to be worth tackling. Or it may simply be the wrong moment to broach the issue, or there may be others in a better position to work on it. Information might best be gathered from key informants, perhaps in an informal way, rather than throwing the whole thing open to deep consultation at this stage. Once an informed 'gut feeling' that a problem is do-able is recognised, it is useful to capture it in a basic model. A hypothetical example of defining a policy problem follows:

- *Problem:* the forest is being cleared by cocoa farmers

- *Possible cause*: currency devaluation

- *Possible chains of causation*:
  - ○ currency devaluation ➔ cocoa exports more profitable ➔ cocoa farmers better off ➔ more forest cleared by more cocoa farmers to make profits
  - ○ currency devaluation ➔ food and agricultural input prices rise ➔ cocoa farmers worse off ➔ more forest cleared by cocoa farmers to make ends meet
  - ○ currency devaluation ➔ forest officers' real incomes drop ➔ forest officers are less able/ interested in preventing forest clearance by cocoa farmers ➔ more forest cleared by more cocoa farmers

- *Stakeholders involved*. At this stage, it is useful to note the main stakeholders who may be involved in problem causation, suffering its consequences, or solutions.

- *Values and assumptions*: each of the presumed chains of causation is based on a range of values and assumptions which need to be identified, since these imply very different lines of investigation and possible solution. Some may be associated with particular stakeholders.

Further work reviewing existing information and the range of opinions about the issue will allow the related policy factors to be identified. These factors may be policy influences on the problem, or policy influenced by the problem. An initial understanding of these factors is needed to allow development of this basic model and to allow specific objectives and research questions to be identified (see below).

A3.2 Develop initial understanding of five groups of factors: context, actors, process, contents and impacts

Understanding why and how an area of policy is 'shaped', and how it changes (or stays the same), requires consideration of many factors. These factors can be divided into five main groups: context, actors, processes and integrating institutions, policy contents and policy impacts. It is usually important to make explicit investigation of each of these groups and the linkages and interplay between them. The idea is to identify promising policies, actors, initiatives, obligations, and integrating systems which could be employed for further progress – building on what works.

These factors, which are commonly important to an understanding of policy influencing forests are summarised in Figure A3.1 and outlined below.

Understanding policy context. Policy is conditioned and shaped by a wide range of contextual factors relating to the physical, cultural, political and economic environment and to decisions made in the past. These factors include:

- Pressure from forest stakeholders and society at large
- History of forest use and policy
- Institutional capacity
- Tenure system and pattern of ownership
- Economic conditions and changes
- Forest resource conditions

Understanding policy actors. In any one context, various institutions and stakeholder groups will have a bearing on policy. These policy actors and the apparent power structure involved in decision-making need to be identified (this may need more detailed work later – see Section A4.7). It is useful to identify who is pushing for what, and who cannot be 'heard'? Who are the 'integraters' and who are the 'dividers'? The range of influences on policy actors can then begin to be unpacked. These influences include:

- Institutional/ organisational factors
  - ○ mandates, rules, norms, functions, strengths and weaknesses
  - ○ dynamics, interactions, institutional culture
- Individual motivation factors
  - ○ ideological predispositions, pursuit of political objectives
  - ○ position and control of resources
  - ○ professional expertise and experience: adhering to professional standards; promoting own careers
  - ○ institutional loyalties; enhancing the standing of own agencies
  - ○ personal attributes and goals, such as rent seeking

Understanding policy process. Here we are interested in identifying the way in which agendas translate into implementation. Section A2.2 in this Annex describes the problems and advantages of various conceptions of the policy process. Developing a conception of the

Figure A3.1 A framework for analysing policy change

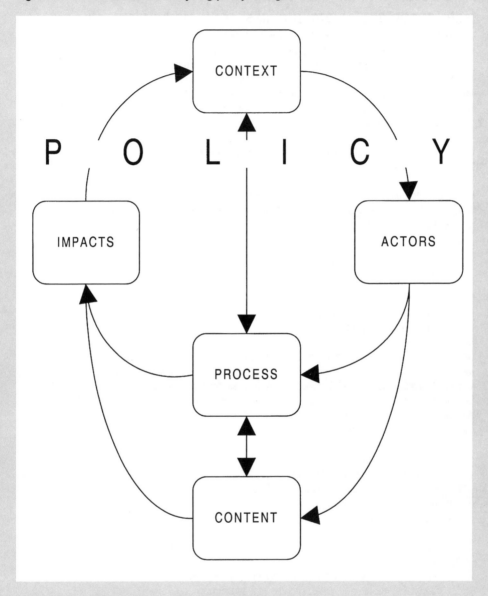

current policy process which makes sense in a particular context is an important step in research. Any of the elements, mapped in the notional policy cycle, Figure A2.1, which exist in practice need to be identified.

A key feature of the process is the way in which the policy agenda is formed. The need for change in policy for forests stems from different perceptions of the agenda amongst key policy actors. These perceptions will be shaped by combinations of institutional position, experience, motivation, ideology and the use of language. We can therefore think of agenda formation as being firmly linked to the origin and maintenance of particular narratives and discourses (discussed in Sections A2.4-6 of this Annex).

A particularly common type of discourse framing what is on the policy agenda is *crisis*. Many policy processes are catalysed by the perception and language of crisis amongst policy actors. 'Deforestation crisis'...'woodfuel crisis'...'forest sector crisis' have been recurrent phrases with much impact on policy around the world over the last decade or more. Other policy processes are subject to more day-to-day – '*politics-as-usual*' – language, whilst others are catalysed by a *breakthrough* into the policy arena of new ideas (such as new taxes or market instruments) and new actors not previously involved in policy. The following are common types of agenda perceived by policy actors in forestry over the last few years. The agenda types are arranged in order of those most commonly emerging from perceptions of crisis, through those which grow from politics-as-usual, to agendas formed by the breakthrough of new policy actors or the perception that their innovations should be mainstreamed.

- *New controls* – major changes in institutional structures, laws and regulations
- *Privatisation* – deregulation and market reforms
- *Decentralisation* – divesting responsibilities, or devolving power
- *Cross sectoral cooperation* – harmonising sectoral policies
- *Civil society initiative* – non-governmental and private sector actors cooking up policy
- *Local innovation* – those previously marginalised muscling in on policy with innovative solutions

Each of the above types of agenda leads, characteristically, to a different process by which policy is negotiated and developed. We need to know how the above policy actors get involved, how priorities are set, what communication channels and key decision points or gateways are involved, and how influence is exercised. Elements of process which are likely to warrant particular investigation include:

- *Policy arena*. For example, this might be primarily the macropolitical arena in the case of making bold new laws following crisis, or it might be fora designed to bring national and local actors together in the case of decentralisation.

- *Institutional procedures*. It is especially useful to identify opportunities and constraints to cross-sectoral and top-bottom linkages, in terms of information flows, consultation, and decision-making.

Table A3.1 Characteristics of some of the main policy processes prevailing in forestry

Discourse	'Crisis'		'Politics-as-usual'		'Breakthrough'	
Main Agenda	'New controls'	'Privatisation'	'Decentral-isation'	'Cross-sectoral cooperation'	'Civil society initiative'	'Local innovation'
Arena of conflict/ negotiation	Macro-politics	Macro-economic stringency, private sector	Bureaucracy-local linkages	Cross-sectoral fora	Private sector, NGO fora	Local politics, national policy élite
Institutional procedures	Central, high-profile	Central, competitive	Incremental, administrive	Periodic, consultative	Tactical, collaborative	Devolved, experimental
Determinants of imple-mentation	Legitimacy and stability of regime	Fiscal efficiency, degree of élite consensus	Efficiency and strength of national support	Catalysts for convergence of interests	Strength and credibility of private/ NGO institutions	Viability of local forestry options
	Single or group of policy changes	Structure and interest of private sector	Viability of local institutions	Contingency of budgets/ incentives on cooperation	Extent of 'gap' left by government	Strength and equity of incentives, leaders and organisation
	Degree of élite consensus/ control	Degree of realignment/ horse-trading in private sector	Degree and equity of devolution of power	Level of cross-sectoral consensus	Viability of proposed forestry options	Degree of support from enlightened national élite
Examples	*Papua New Guinea* – new forest law and revenue system	*South Africa* – restructuring of government forests	*India* – handing over forest responsibilities to panchayats and local committees	*Pakistan* – National Conservation Strategy	*International* – progress with forest certification	*Costa Rica* – locally-developed smallholder forestry spread by organisation

• *Determinants of implementation.* Factors involved here might include the strength of central government support, the degree of devolution of power, the viability of institutions, etc.

Table A3.1 illustrates some of these elements of process under the different types of discourse and agenda types described above.

Understanding policy contents. The contents of policy are generally the central focus in the above processes, and are often the 'meat' of any policy research. Policy contents are highly specific to particular cases. Typically there may be tools, instruments and mechanisms involved which are of one or more of the following types: regulatory, economic/ market, informational, institutional, contracts/ agreements. It is useful to ascertain whether there is general agreement over the contents and, even if there is agreement, what is the level of 'policy inflation' in relation to actual capacity to implement real policy, i.e. to think, debate and act strategically.

Understanding policy impacts. Policy processes and contents may have dramatic or inconsequential impacts on forests and people. In very general terms, there are three types of impact which need to be borne in mind:

- Environmental
- Social
- Economic

Each of these three impact areas might be assessed in terms of (provisional) criteria and indicators for good forestry, and/ or for sustainable development, perhaps as expressed in overarching commitments such as an NSSD.[6]

Policy impacts may be the expected ones, or they may be quite unexpected. They may be seen quickly or only be revealed in the long term – hence the importance of reviewing policy and impact regularly and building up a time series. Often the link between policy and impact is very hard to ascertain. The work of tracing causes from effects, and effects from causes is a key part of policy research. These impacts are likely to shape, or become part of, the context for any future change in policy.

Thus, there is ultimately a fourth type of impact we are seeking – impact on institutional change and on the evolving policy process itself. Each impact study should, therefore, look beyond the immediate confines of the policy in question.

A3.3 Develop a framework: piecing together the key policy elements of the problem/ opportunity

Recognising that policy change is the interplay of context, actors, policy characteristics (process and content) and impacts, the above information can then be assessed and integrated in a framework which best describes the real links between the factors. Figure A3.2 summarises some of the factors to bear in mind in developing such a framework.

---

6 Preliminary criteria and indicators (C&I) could prove useful as a framework against which to classify policies, to conduct analyses, to assess impact, to focus debate, to build consensus on the dimensions that really matter, and to classify information built up during the policy review. Final sets of C&I could then be tailored to show whether critical values are getting better or worse, by assembling a time series.

Figure A3.2 Analysing policy affecting forests and people – the interplay of context, actors, process, content and impact

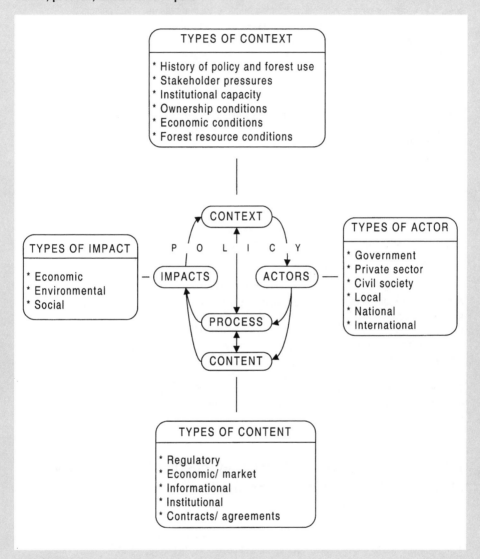

A3.4 Identify type of influence desired – and plan a strategy for achieving it

If any good is going to come from policy work, a clear focus on the type of influence desired is needed from the start. There is a wide range of possible objectives here, ranging from just hoping that someone will listen, to working with a policy-maker for a particular policy decision, to trying to build long-term consensus among groups that might one day influence policy. Policy work may help to think about issues and define problems, rather than to seize on solutions.

One possible typology of tactics for influencing policy follows:

- Dump information near policy-makers
- Draw in policy-makers during analysis
- Service the policy machine
- Stay connected – seize opportunities.
- Convene better policy processes
- Offer do-it-yourself policy review kits
- Build constituencies
- Create vision

Where policy research is involved in any of these approaches, it is important to consider how the findings may be used. Three main ways in which research findings may be used by policy-makers are:

- *Data* – most likely when policy-makers already agree on values, goals and problems
- *Ideas* – most likely when current policy is in disarray or there is much uncertainty
- *Argument* – most likely when there is much conflict – where policy-makers are manoeuvring, justifying positions, delaying decisions, enhancing credibility and personal agendas, etc.

Each of the above tactics is investigated in Section A5.3. Clear identification of the scope and possible tactics for using the planned policy research, i.e. some form of 'dissemination and influence strategy', will ensure that the resources, expertise and specific objectives of the research are well-focused.

A3.5 Match scope of work to available time and resources

'Small and quick' approaches, and 'large and long' programmes may be equally valid, but work best for different issues and for different types of influence.

The advantages of small and quick studies are: timeliness in relation to key events, good political and stakeholder momentum, and the ability to exploit a state of urgency. But they can be too quick for some stakeholders to be involved, they may produce results that are insufficiently well-informed, and they are unlikely to be well-coordinated with other initiatives.

Large and long studies give time to explore issues, time to bring in the right stakeholders and for reactionary stakeholders to see the need for change. They can command resources to do the work well, and use time wisely to produce results that are 'mainstreamed' into all the necessary related processes. But they run into trouble if the money dries up, protagonists change, policy issues are no longer pertinent, and policy-makers can't digest the results. And they can take so long that key actions are delayed 'to wait for the plan' and people lose interest.

Perhaps the best compromise is a permanent forum and process to keep an eye on policy, and the ability to call in short studies as and when needed.

A3.6 Select team members, investigators, advisors

The 'team' is obviously rather dependent on the issues and scope of the process. But, in general and assuming a fairly comprehensive forest policy review process, five types of groups may be needed:

1. *A convenor* of the policy process (high-level and – especially if there are many extra-sectoral issues – a 'neutral' and/ or widely-credible office, e.g. the prime minister's office, or a development planning authority)

2. *A steering group* (multi-agency with government, market and civil society representation at high levels; reasonably catholic, to survive party political changes). This would comprise a mix of policy-connected and policy-affected people, who would review the work (thereby sharing perspectives), possibly in stages to ensure that it is focused

3. *A 'technical' working group* (again, multi-agency/ discipline). They would conduct the analysis and develop technical solutions – but the 'field work' will take place as much in corridors of power as in the forest

4. *Secretariat* (this could be one-off, or in a fairly 'neutral' body). It might comprise:
   ○ a 'neutral' manager who is credible to stakeholders
   ○ economics expertise
   ○ participation/ facilitation expertise
   ○ communications expertise

5. *Key informants* who will need to be kept informed and consulted, individually and in special meetings. These are the 'policy-affected' people, and those with diverse and useful perspectives, such as writers and the media. Stakeholder analysis will help to identify these. Key informants may be involved through e.g.
   ○ local surveys (questionnaires, local meetings)
   ○ interviews
   ○ participatory appraisals
   ○ small working groups

Our proposals on policy processes require – and promote – an acceptance of the principle of participation in policy. Whether or not this is accepted, there are several problems which can occur. One is to do with the *different ideologies* behind participation, e.g.:

- instrumentalist approaches – participation confers higher value, better information, and reduced cost by engaging other actors
- post-modernist – all views are valid and need to be heard
- neo-liberal – reduction of state interference is a good thing in itself
- rights activists – more local/ stakeholder autonomy can be achieved through participation, which leads to claims-making

All of this means that there will also be *different expectations* about the outcomes of participation:

- Apparently 'win-win' solutions will mask or undermine ideological differences
- Bitter experience means many people do not expect any real change
- Participation can be associated (negatively) with party politics

Participation mechanisms need to be selected so as to minimise these problems. It is important to *clarify 'how far' participants can expect participation to go*, i.e. certainly:

- providing information (consultation)
- helping to define priority issues
- confirming findings
- developing options
- contributing to consensus

but it is unlikely that wide participation can be expected in making decisions on priorities, investments and precise policy/ institutional changes.

A3.7 Formulate specific objectives, research questions and methodology

The foregoing preliminary work – on identifying issues, understanding the context, and building a big picture of how the issues fit in this context, needs to be discussed amongst involved stakeholders with a view to:

1. verifying the scope of the issues
2. agreeing the boundaries of the forthcoming policy exercise
3. focusing on objectives and questions that will need to be addressed
4. designing the policy process

These four points form useful agenda items for the early meetings of a steering group or technical group, for example. Without this, the issues addressed in the policy process are likely to expand beyond any ability to handle them, and the process may become discredited.

The idea is to channel attention and thought into what matters, not provide a forum for unending debate. It is especially important that multi-stakeholder groups design the process themselves, for ownership of the process and its results.[7] Key actions include:

- *Select priority aspects of the problem/ issue/ opportunity.* Priorities might be assessed by reference to criteria for human and ecosystem well-being and practicality, e.g.:
  - ○ central to poor people's livelihoods or key economic sectors
  - ○ possibility to act without extra finance
  - ○ key environmental hazards
  - ○ key developmental needs
  - ○ presents major learning opportunity
  - ○ visible to the public/ multiplier effect
  - ○ international obligation
  - ○ high priority amongst key actors
  - ○ timeliness in relation to a pending decision
  - ○ linked to current work – topicality – and skills – comparative advantage

- *Formulate objectives and questions.* Things cannot be left as 'issues', as this does not help to provide direction to analysis or developing solutions. For example, 'watershed degradation' is less useful a formulation than "what incentives have encouraged watershed conservation? And how can we remove perverse incentives to deforest key watersheds?" Questions should:
  - ○ address an important aspect
  - ○ provide a synthesis
  - ○ exhibit policy responsiveness

- *Agree the outputs and who will get them* – it is important that this should not be a surprise once it has been produced, and so stakeholder expectations and political/ legal procedures and implications need to be discussed beforehand. For example, will the output be 'evidence', 'proposals' for policy, a draft policy itself, or a policy and accompanying action plan?

- *Select and sequence methods.* This is primarily a technical task, but the implications of the types of analysis and consultation will be important to the convenor and any policy-level steering group. For example, they may need to prepare the way by encouraging officers to be critical and not 'toe the party line'.

A3.8 Conduct analysis, develop findings, analyse their potential for impact, and revise

See Section A4 for a selection of methods for analysis which we have found to be useful. A possible sequence of tasks in the analysis follows:

---

7 The 1997-9 forest policy formulation process in Grenada was agreed by all stakeholders (interviewed by an independent IIED mission) to have high 'ownership' within the country. Said one: "It started off on the right foot. That make-a-policy process was designed by MANY OF US, not by the [forest] department or any outsiders, and then it talked to EVERYBODY to get a policy which is the NATION'S interest and not just the department's!"

1. *Analyse* particular issues defined earlier – what resources are being degraded and how; who is affected and how; policy, institutional, market, behavioural, international causes of problems
2. *Assess likely futures and vulnerability/ resilience* – accounting for trends (globalisation, technology, etc)
3. *Local consultations* to find out complexities (what effect, who affected, what positive/ negative trends in terms of sector/ livelihoods concerned – basis for indicators)
4. *Build information system* to collate results ('hidden wiring') based around the 'priority' criteria (human and ecosystem well-being)
5. *Synthesise* all evidence and recommendations – common and differing approaches (done by independent working group/ secretariat)
6. *Weigh carefully against priority criteria* – need to do homework to avoid a totally win-win 'additive' recommendation, and instead to achieve a more practical, tactical approach
7. *Assessments of trade-offs between levels* (local to global – buck-passing and 'importing sustainability')
8. *Produce findings, conclusions, recommendations, and implications of recommendations* – in that order of 'ambition'

The above process is not quite as 'linear' as it has been portrayed. When tentative findings have been produced and synthesised, stakeholder positions and institutional factors may need to be revisited and reanalysed to predict the consequences and probability of the findings having impact – such as uptake and implementation. Often it may be necessary to revise the tentative findings in the light of this reanalysis. For example, if the impact desired is a particular policy decision, the following steps can be envisaged:

- Assess power of actors in relation to the targeted decision:
  ○ prune the list of actors compiled during research
  ○ divide the list into those responsible for the targeted policy decision, and those who will try to influence it
  ○ ascertain power of actors to access and mobilise resources and decision-makers
  ○ assess their opinions
  ○ visualise the power structure related to the targeted decision

- Assess institutional factors needed to implement targeted decision:
  ○ organisational structure
  ○ amount of resources needed
  ○ supporting policy mechanisms – existing and required

- Predict potential consequences of findings

- Estimate the probability of implementation. If probability is not high, options include:
  ○ accept the low probability
  ○ change the scope or depth of the recommendations, e.g. from fundamental to incremental change or vice versa; from a desire to change to a desire to obtain agreement on future
  ○ modify the recommendations, e.g. repackage using more appealing terms; modify and

work with actors to create ownership and support; redirect to provoke controversy, deepen public concern and build strong support for meaningful actions

• Prepare final recommendations and check against the agreed priority-setting criteria

A3.9 Prepare findings in optimal output form, communicate and influence policy!

Findings and recommendations need to be driven by the right 'vehicle' to stand a chance. Packaging and presentation are all-important, and are discussed further in Section A5.2.

If communication throughout the study with different potential study 'users' has been good, the ground will be well prepared. But it is important that recommendations are seen to be 'owned' by the broad policy community, not just the author of any analysis. Briefing, debate, and decisions need to take place in the highest relevant forum (which may be Cabinet for a comprehensive overall policy review, or one which addresses very significant issues). Informal briefings with such ultimate arbiters throughout the process can be helpful. But 'bouncing' analysis and ideas in stages with their advisors is crucial (hence the steering group suggested in Section A3.6). At these high levels, oral communication is generally the most effective – any written policy briefs will have to be very short.

A4 Analyse policy – some methods

A4.1 Early and regular consultation with current holders of policy

Development of the strategy for policy work (Section A3) will clarify the balance required between pure analysis and advocacy/ influence. Where influence on, or with, the current holders of policy is needed – and mostly it is – regular consultation with them will be crucial. Policy research cannot generally afford to proceed like a typical detached research project; it needs to engage in a dialogue with key stakeholders so as to create a constituency for the findings. Approaches for generating this early and regular contact include:

- *Interviews* with people from various institutions to gauge diverse opinion amongst different sectors and social actors.

- *Inception workshop* to help define the research agenda – key issues and objectives.

- *Regular face-to-face contact* between the researchers and a range of stakeholders throughout the course of the work, to maintain a two-way flow of information.

- *Advisory committee* – comprising representatives from different sectors, identified in the early stages of work as key actors – to enable regular follow-up on a wide range of opinions and experience, and to build a support-base among key players. Committee members can be consulted individually and in meetings through the course of the study (Section A3.6).

- *Quick write-up and circulation of interim and preliminary findings* – among advisory committee and other peer reviewers – to stimulate debate and garner feedback and further support constituency and consensus.

Repeated consultation and discussion with a range of active 'opinion-formers' and members of the formal policy-making community can generate 'political space' for key issues, and policy opportunities may arise in the course of the work. For example, in the case of the Ghana *Policy that works* study, the Ministry of Lands and Forestry was particularly keen that the opportunity provided by the study be used to explore the potential for forest certification. Thus, an early focus of the study was to contribute to the emerging debate on the appropriateness and potential of forest certification and labelling in Ghana; the background to the issues; and the directions and challenges ahead. This work helped to bring about the emergence of a substantive process which has enabled options and approaches for certification to be developed and debated.

A4.2 Analysis of policy statements and laws

Analysis of policy *documents* is an important part of policy analysis. It cannot give a complete picture of policy – which as discussed must also include dimensions of context, process, intentions and outcomes. The language, style and length of policy documents can tell us much about context and process, although it is only recently that they have tended to give direct information about how they were formulated (such as the Grenada forest policy of 1999, which was formulated through a highly participatory mechanism, and some policies produced through newer NFP processes). However, by keeping these dimensions in mind whilst reviewing documents, we can identify implications, notably implementation issues and potential policy instruments. A desk review of key policy documents might include:

- Gathering policy documents which have a bearing on forests and people
- Cataloguing the contents in relation to the purpose of the analysis, e.g. by criteria and indicators of SFM
- Highlighting inconsistencies, links and overlaps between the documents
- Identifying particular innovations and lessons in the documents
- Comparing the positions in these documents with those of key stakeholder groups
- Noting any conflicts or gaps with respect to international obligations and opportunities
- Identifying issues related to implementation, notably on capacity implications
- Identifying mechanisms for dialogue between stakeholders, for reconciliation of potentially competing objectives and inter-sectoral coordination

Example: Policy documents as a basis for 'sustainable forest management' in Sri Lanka

The two tables which follow below were developed as a way of giving a quick 'interested outsider's' assessment of the extent to which the Sri Lanka forest policy and draft legislation documents appear to provide a good basis for stakeholders to pursue sustainable forest management at forest level (Dubois and Mayers, 1998).

From interpretation of forestry experience in a wide range of contexts, IIED has summarised what it has identified as the *functional needs* of SFM. Table A4.1 relates the two Sri Lankan policy documents to these functional needs of SFM and makes a 'back of an envelope' assessment of the degree to which these documents appear to enable stakeholders to support each need.

Table A4.1 Sri Lankan policy documents in relation to the functional needs of SFM

Functional needs of SFM	Policy	Act
• Clarifying stakeholder roles and procedures	**	*
• Securing property rights	***	**
• Building staff capacities within institutions	**	*
• Integrating multiple objectives	***	**
• Making choices between objectives	**	*
• Building and sharing forest knowledge	*	*
• Dealing with uncertainties	*	*
• Ensuring communication and participation	***	*
• Covering the costs	*	*

Explanation of the columns in the table:

Policy. The degree to which the National Forest Policy of 1995 appears to provide a good basis for stakeholders to pursue SFM.

Act. The degree to which the draft Forest Conservation Act of June 1997 appears to support the National Forest Policy and further contribute to the basis for stakeholders to pursue SFM.

```
***  = High
**   = Medium
*    = Low
```

Such a 'functional' assessment can be taken a step further, to determine how far policies might match up with 'best practice'. For example, IIED has also analysed a wide range of international, regional and national initiatives to define SFM – the various criteria and indicators and certification programmes – and found that they all had the following in common:

- Framework conditions on policy and commitment
- Sustained and optimal production of forest products
- Protecting the environment
- Ensuring the well-being of people

These core elements can be broken down into a number of common sub-elements. These are listed in Table A4.2 for a second 'back of an envelope' assessment. The Table also notes some of the features of the documents which are particularly innovative, and some challenges remaining.

Table A4.2. Sri Lankan policy documents in relation to common elements of international and national SFM standards

Common element of SFM standards	Policy	Act	Innovations	Challenges
Framework conditions				
• Compliance with legislation and regulation	**	**	Poplicy likely to motivate many stakeholders to comply, if well disseminated	Questionable legitimacy of state control of all forests and trees on non-state land. Little provision for international commitments or opportunities
• Securing tenure and use rights	**	**	Multi-tenure approach to permanent forest estate through leases	Lack of provision for conversion areas or improved tenure security in Act
• Commitment to sustainable forest management	**	*	Policy has inspirational strength, if well disseminated	Priorities amongst objectives unclear. Institutional roles unclear Little promotion of incentives cf. regulation
Sustained and optimal production of forest products				
• Sustained yield of forest products	**	**	Multi-user forestry approach on state lands	Lack of provision for transfer of ownership of state plantations in Act
• Management planning	***	**	Forest agreements and joint management potential	Little involvement of stakeholders in planning stages
• Monitoring the effects of management	*	*		Many rules but little emphasis on information systems and flow
• Protection of the forest from illegal activities	**	***	Detailed provisions	
• Optimising benefits from the forest	**	*	Emphasis on promotion/ extension activities on non-state lands	Provisions for Class III, IV and V forests are unclear. Very strict controls on felling and transport of timber on private land

Common element of SFM standards·	Policy	Act	Innovations	Challenges
Protection of the environment				
• Environmental impact assessment	*	*		An objective without provisions
• Conservation of biodiversity	**	**	Strong forest-level protection measures	Lack of provision for stand-level conservation in either document
• Ecological sustainability	**	*	Clearly implicit throughout Policy	Land use planning perceived as purely regulatory cf. incentive- and information-based
• Waste and chemicals management	*	*		No provisions
Well-being of people				
• Consultation and participation processes	**	*	Fairly strong theme in Policy	Proposals for participation but weakened by excessive regulation and state powers
• Social impact assessment	*	*		No provisions
• Recognition of rights and culture	**	**	Fairly strong theme in Policy	Traditional/ existing rights not spelled out/ reinforced
• Relations with employees	*	*		Little on staff development
• Contribution to development	***	**	One of three core objectives	Much still to do to establish vision of forestry in national land use and development

Explanation of the columns in the table:
Innovations. Features of the recent policy and legal documents which strike us as being particularly innovative, and likely to be of interest to others in the forestry world beyond Sri Lanka.
Challenges. Features of the documents which, from our reading, appear to be challenges remaining – potential gaps or issues in need of further policy or legislative attention.

Source: Dubois and Mayers, 1998

Example: Review of legal documentation in Himachal Pradesh

In *Himachal Pradesh*, a major forest sector review is under way, with the aim to produce proposals for new policy. It is known that many of the relevant legal instruments are out of date, and not rationalised – each one being an incremental response to a new situation. In such circumstances, it was felt that the review of the legal documents should be a two-part job: the first to assess the legal instruments available, and the second to look at the legal possibilities and changes associated with any policy proposals.

Step 1: Assessment of current legal situation. This is informed by the main problems raised by an initial scoping exercise. It covers:

- Assessing relevant legal documents, noting their provisions within an eight-part SFM Criteria framework
- Highlighting inconsistencies, links and overlaps between them
- Noting gaps and opportunities with respect to international obligations on forests, environment, human rights
- Identifying particular innovations and lessons in the recent development of legislation on forest-related issues

Step 2: Assessment of legal requirements associated with proposed policy options

- Comparing the provisions of current legislation with the emerging policy options
- Identifying the need for enabling legislation to permit new arrangements, such as partnerships for SFM
- Identifying the need for further regulation on issues related to implementation, notably on capacity to enforce in relation to critical forest assets
- Noting cross-sectoral issues, which may require the forest sector review to engage with authorities in other sectors
- Rationalising legislation, perhaps within the eight-part SFM Criteria framework[8]

A4.3 Policy instrument analysis

We have described in the main text of this report (see Section 6.4) the types of policy instruments and the ways in which they seek to work – by compulsion, persuasion or incentive. Much policy instrument analysis aims to evaluate the impacts of existing policy instruments, or to predict the likely consequences of the use of proposed instruments. However, there are various other reasons why such analysis may be needed, and the approach taken needs to be tailored to the circumstances. There is a wide range of tools for analysing policy instruments, from cost-benefit analysis to environmental assessment and various modelling approaches. The advantages and disadvantages of some of these tools are outlined in Table A4.3.

---

8 In the UK, legislation, grant guidelines, and other incentives were catalogued according to the Helsinki Criteria and Indicators for SFM. The rationalised result was the UK Forest Standards, which subsequently formed a useful basis for converging with FSC certification requirements, to result in the UK Woodland Audit Scheme.

Table A4.3 Comparison of some tools for analysing forestry and land-use policy instrument options

Tool/ Approach	Description/ Purpose	Advantages	Disadvantages
Land Suitability Classification	Distinguish and map areas in terms of characteristics which determine suitability for different uses.	Distils a mass of physical, biological and (sometimes) economic information into a single index of relative suitability for various land uses.	Economic comparisons are rarely made explicit and the relative importance of different factors in calculating the final index may be arbitrary.
Environmental Appraisal or Environmental Impact Assessment (EIA)	Detailed documentation of environmental impacts, adverse effects and mitigation alternatives.	Explicitly requires consideration of environmental effects; ability to monetise does not preempt enumeration of all benefits and costs of an action.	Difficult to integrate descriptive analyses of intangible effects with monetary benefits and costs; not designed to assess trade-offs among alternatives.
Cost-Benefit Analysis (CBA)	Evaluates projects, land-use options and policies based on monetisation of net benefits (benefits minus costs).	Considers the value (in terms of willingness to pay) and costs of actions; translates outcomes into commensurate terms; consistent with judging by efficiency implications.	No direct consideration of distribution of benefits and costs; significant informational requirements; tends to omit outputs whose effects cannot be quantified; tends to reinforce status quo; contingent on existing distribution of income and wealth.
Cost-Effectiveness Analysis (CEA)	Selects land-use option that will minimise costs of realising a defined non-monetary objective.	No need to value benefits; focus on cost information often more readily available; provides implicit values of objectives (e.g. marginal cost of increasing by one unit).	No consideration given to relative importance of outputs; degree to which all costs are considered will be important to judgements as to 'best' approach.
Multi-Criteria Analysis (MCA)	Uses mathematical programming techniques to select option based on objective functions including weighted goals of decision-maker, with explicit consideration of constraints and costs.	Offers consistent basis for making decisions; fully reflects all goals and constraints incorporated in model; allows for quantification of the implicit costs of constraints; permits prioritising of projects.	Results only as good as inputs to model; unrealistic characterisation of decision process; must supply the weight to be assigned to goals; large information needs for quantification.
Risk-Benefit Analysis (RBA)	Evaluates benefits associated with land-use option in comparison with risks.	Framework is left vague for flexibility; intended to permit consideration of all risks, benefits and costs; not an automatic decision rule.	Too vague; factors considered to be commensurate often are not.
Decision Analysis (DA)	Step-by-step analysis of the consequences of choices under uncertainty.	Allows various objectives to be used; makes choices explicit; explicit recognition of uncertainty.	Objectives not always clear; no clear mechanism for assigning weights.
Macroeconomic and Behavioural Models	Econometric programming models used to simulate intersectoral linkages and producer behaviour.	Dynamic and price-endogenous models allow explicit simulation of feedback effects and price movements; best for large-scale projects and land-use allocation.	Tend to be data and analysis intensive; expensive to build and run and often difficult to interpret.

Source: Adapted from Pearce and Markandya (1989), Land Use Classification and Macroeconomic and Behavioural Models added (IIED, 1994).

Rarely are policy instruments used alone, so the *mix* of instruments is often of crucial concern. It is important to work out the boundaries and interrelationships, synergies and conflicts amongst instruments. For example, in *Chile* the 1974 Act (Decreto Ley 701) which established afforestation incentives has been hailed as a success because it also guaranteed the security of forest land ownership – whenever forest investments were carried out – and the free trade of all forest products including roundwood. Thus, the objective of building a successful forest-timber industry, seems to have been met through the right mix of incentives, property rights and trade regulations.

Figure A4.1 is an example from *Ghana* of mapping the impacts of two main policy instruments – forest fees and log export bans. It shows the interrelationships between these instruments and key institutional and market factors, and the resulting impacts on the forest.

Where many approaches to policy instrument analysis founder, or at least fail to break through to having much effect, is in their lack of attention to differences between actors. Policy instruments affect some actors more than others, and often in unforseen ways. Perceptive approaches to analysing the potential impacts of policy instruments focus on how the various actors involved will react to them. The objectives, knowledge and power of each actor needs to be mapped out, and consideration given to the fact that people have very different degrees of access to, and perspective on, information about available policy alternatives and their pros and cons. Once the situations of the different actors are better understood, they can be compared and some judgement made about the likely outcome of the proposed instrument or intervention.

In summary, the use of analysis of policy instruments – to make decisions and trade-offs between possible instruments – is likely to depend on the degree to which it deals with actors' different costs and benefits, agendas (hidden and overt) and powers. The following sections describe approaches which address these issues and can thus make sure that policy instrument analysis is done by the right people, focused on the right issues, and likely to have some effect.

A4.4 Surveys of attitudes and perceptions

Use of the vast array of techniques available for surveying attitudes and perceptions amongst stakeholders has not been conspicuous in the analysis or development of forest policy. However, provided that a clear focus is kept on objectives, the likely biases in responses and the ability to deal with the results (rather than getting carried away with asking questions), such surveys can generate important findings.

In *Papua New Guinea*, a country with a high per capita count of anthropologists, the PTW study team sent a questionnaire to many of them. The questions aimed to draw on the anthropologists' knowledge of local attitudes towards forests and policy. The response rate was high. The questionnaire was also sent to university environmental studies students from all over the country. The two types of respondent proved complementary and allowed a much wider spread of views from local level, than a small number of field exercises would have permitted. Although the interpretations of anthropologists and students are not necessarily

Figure A4.1 Example of policy instrument impact mapping – impacts of forest fiscal system and log export bans on Ghana's timber resource in 1993

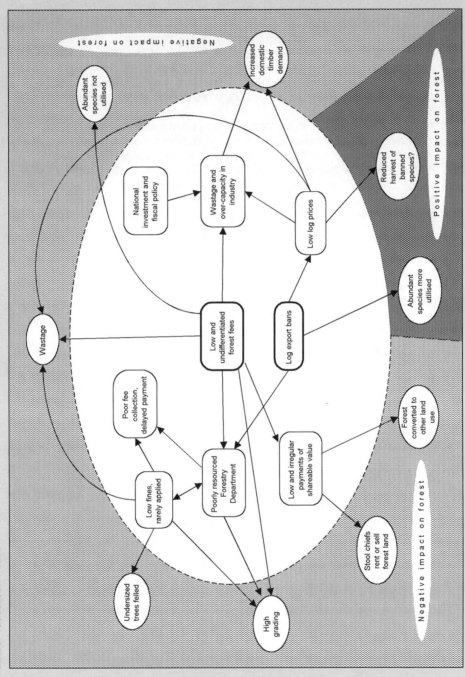

Source Mayers *et al*, 1996

representative of all local views, the results provided the team with a useful basis on which to judge the local linkage and relevance of the 'world' of national-level policy.

In *Grenada*, a small island nation with a high literacy rate (95 per cent), a committee, comprising 18 key forest-sector stakeholders, was established to design and implement a 12-month forest policy review process (September '97 to September '98). The committee decided that the policy should be for the whole nation and that the development process should be highly consultative and participatory, to optimise both content and ownership.

In order to achieve this the Forestry Department was asked to facilitate this process through a combination of community meetings, radio phone-in programmes and consultative sub-sector studies and questionnaires. The main questionnaire (Box A4.1) designed to provide a chance for all citizens to give their views on forest-related issues, was distributed through committee members and printed in all national newspapers. To encourage completion and return of the questionnaire 70 small forest-related prizes were given on a lottery basis. This may have contributed to the high response, 430 or 0.5 per cent of the population. The questionnaire was also used to identify individuals who could be invited to help develop the policy.

The response from this consultation process was extremely useful in demonstrating that the public and forest officials shared similar ideas on forest values. However, the scope of the policy and Forestry Department activities needed to be broadened both technically and geographically. The information collected during the year was fed into a 'Consensus-building' workshop from which a new forest policy was developed. The new policy has given strong impetus for change within the Forestry Department

In *Himachal Pradesh*, however, where literacy is much lower, participatory appraisals and focus group discussions in 24 villages (stratified across the State according to livelihood and forest differences) are being used to obtain the same sorts of information – people's perspectives of forest values, and of the forest authorities and other service-providers. This will emphasise different stakeholder groups rather than treat 'the public' as a whole, as in Grenada.

A4.5 Participatory appraisal – to identify stakeholder vision and priorities

Whilst it can be useful to start policy analysis by discussing with *key informants* (for qualitative, basic information on issues) it is always necessary to move on to (stratified) *sampling of wider groups* for more detailed information on quantities and weighting of issues. The methodology needs to suit the group in question. In A4.4 we noted the value of questionnaires. Telephone surveys and household surveys may also be useful (as used in British approaches to revising forest strategy). For many groups, however, especially at the local level, village/ user group meetings and participatory appraisals are the best way forward, especially where there are problems of representation of the group (Box A4.2).

Communication with local groups, from the early stages of policy work, is important to enable local views to shape the direction and substance of the work. For example, the Zimbabwe

Policy That Works team worked with several communities in important resource and tenure contexts for two main reasons: firstly, to ascertain whether the current collection of national policy statements and laws made any sense in relation to local perspectives and priorities; and secondly, to help the team develop its own 'vision' for forests and people. These local findings and the team's vision were then debated in several local and national workshop exercises.

Box A4.1 Grenada's forest policy questionaire

If Grenada's environment is important to you please take time to complete this:
FOREST POLICY QUESTIONNAIRE

Complete and return to enter a FREE PRIZE DRAW (details at bottom of page)

The Forestry Department (FD) is currently managing a wide-ranging and participatory Forest Policy review process, with assistance from the British Government, and we would like your ideas and opinions. This 'policy' is being developed for use by all Grenadian individuals and institutions, not only the FD, who have an interest in the goods and services that the country's forests and trees provide. The FD is one of the institutions that looks forward to using the new policy to develop and implement a new and responsive strategy to manage forested State areas and assist private land owners, as requested, in forest management issues. The new Forest Policy will also generate new laws and will, hopefully, make a positive impact on everyone who lives here.

The policy development process is being managed by a Committee made up of a wide variety of both Government and non-Government representatives covering areas such as: farming, fisheries, education, hunting, land-use, Carriacou and Petit Martinique, development, extension, water, tourism and others. We, the Committee, invite you to tell us what you think about any issues that concern forests and forest use. Your comments will be highly valued. The questionnaire below is designed to cover many of the issues but please write and tell us what you think about any other forestry matters. This is the only time that such a questionnaire will be published. Please write clearly.

In helping us develop Forest Policy you are directly helping manage and protect our natural forest heritage so that our children's children can enjoy the benefits of a healthy environment that our grandparents passed on to us.

Score the questions below between 1 (unimportant) and 5 (important)   Please circle
1) Should the FD be managing State forest in the hilly lands for the following:

		Unimportant			Important	
a.	Wildlife conservation....................	1	2	3	4	5
b.	Soil and water conservation...........	1	2	3	4	5
c.	Biodiversity (protection)...............	1	2	3	4	5
d.	Eco-tourism / recreation................	1	2	3	4	5
e.	Timber production........................	1	2	3	4	5
f.	Non-timber products....................	1	2	3	4	5
2)	Should the FD be concerned with safe-guarding mangroves?	1	2	3	4	5
3)	Should the FD expand its provision of tree seedlings to farmers or others?	1	2	3	4	5
4)	Should the FD be working with farmers to help reduce soil erosion?	1	2	3	4	5

5) Should the FD be working with hunters
to jointly manage wildlife populations? 1   2   3   4   5
6) Should the FD improve / create
hiking trails and recreational
opportunities in forest reserves?   1   2   3   4   5
7) Should the FD be more involved in
environmental education in schools?   1   2   3   4   5
8) Have you bought timber / fence posts / fencing from the FD in recent years?
YES / NO   (please circle)
9) If 'YES' then: Was the quality: Good * Adequate * Poor *
Did the price seem: High * Reasonable * Low *
10) If you have not bought such items from the FD, why not? Please 'tick'
Do not buy timber / posts etc. *   Did not know that FD sold timber / posts etc. *
Erratic quality * Timber not dried * Limited variety of species * Limited variety of sizes *
Too expensive * Other:

Questions 11 - 14 all ask for the answer YES/NO/please give details
11) Are there other products or services that you would like to see the FD provide?   YES / NO
12) Should forest products that are sold by Government be subsidised ?   YES / NO
13) Do you have particular problems in your area that you would like FD to address?   YES / NO
14) Do you depend on the forest for your livelihood or for some of your income?   YES / NO
15) Do you visit forest areas for recreation?   YES / NO
If 'YES' what activity: Walking / hiking * Picnicking * Hunting *
Bird watching   *   Other activities:
16) If you do use forest areas for recreation how often do you do these activities?
At least once a week   *   2-3 times a month *   4-10 times a year *
1-3 times each year   *   Comments:
17) Do you see much garbage or litter in forest areas?   YES / NO
If 'YES' should anything be done about it and if so what ?
18) Is soil erosion a problem in Grenada?   YES / NO
If 'YES' please tick what the major causes are: Poor agricultural practices *
Clearance of vegetation for construction *   Lack of awareness of problem *
Lack of Government control in upland areas * Other:
19) Does soil erosion affect you in any way?   YES / NO
What should be done about it?
20) Please add your thoughts or comments about any forestry or forest-related
issues, below (or on an attached sheet):
21) How important is it for the public to be invited to contribute to the development
of Grenada's various national policies?   Unimportant   Important
1   2   3   4   5

Your name and address: (optional but required for entry in the Prize Draw:
Any information you can provide about yourself would be useful:
Occupation:_____ Nationality: Grenadian * Other:_____
Sex: M / F (please circle)
Age group: Under 20 * 20 - 29 * 30 - 39 * 40 - 49 * Above 50 *

Box A4.2 Some participatory tools for working with stakeholders

Participatory methodologies comprise various means of obtaining information from local stakeholders, without introducing the bias of the researcher or planner on the one hand, or the leaders or narrow segments of stakeholder groups on the other hand. There are hundreds of such methodologies worldwide – mostly developed in the last 15 years to foster people-first, sustainable development objectives. They have been tested in many participatory forestry projects, especially in developing countries, with the aim of helping stakeholder groups to identify their forestry resources, problems and objectives of both the majority and minorities. A challenge for the next decade is to get these methodologies integrated into forest policy processes. The following is just a summary of the methodologies:

Village/ community meetings. Attend existing village or community groups if they are broadly representative; or call special meetings to give out information and to get feedback. Communicate intentions of a forest organisation at such meetings – especially in the early stages of identifying stakeholder groups and possible impacts. Such meetings are essential when community-wide issues or conflicts emerge.

Focus groups. Convene special groups to discuss a particular topic. For example, farmers wanting land within the forest or hunters and their practices.

Participatory mapping. Provide opportunities for stakeholders to prepare maps of resources/ problems/ conflicts. This can be done on paper or blackboards, or can use local materials such as sticks, leaves, stones, grass, coloured sand, cigarette packets, etc., on the ground. Allow one map to lead to others, as more and more people get involved. Encourage interruption of map preparation to enable more focused discussions to take place. A range of maps can be produced, such as:

- *Resource maps* – depicting villages, forests, farms, hunting grounds and so on.
- *Tenure and rights maps* – indicating who owns, and has rights to, which areas or resources
- *Impact and action maps* – recording where particular impacts occur or actions that are needed.
- *Mobility maps* – showing people's movements to other towns and cities from their community. These can reveal valuable information about seasonal movements, markets used, transportation difficulties and so on.

Time lines. Work with groups to prepare a history of major recollected events in a community with approximate dates, and discussion of which changes have occurred and why (cause and effect).

Matrix scoring. Use matrices to agree ordering and structuring of information, and then for planning. Agree ranking criteria (matrix rows) and relevant issues (matrix columns). Ask stakeholders, usually in a group, to fill in the boxes for each row.

Group contracts. A formal written contract in which a group's members set out their roles and responsibilities, and what they see as appropriate behaviour and attitudes towards one another and towards other groups. Ensure the contract is seen as a working agreement between all group members. This might be appropriate for outgrower schemes and the forest organisation's own liaison committee/ group.

Useful references on participatory appraisal. There is a huge literature in this methodological field. A useful ongoing source of information is *PLA Notes: notes on participatory learning and action*, produced by IIED. Carter *et al* (1996) provides an extensive discussion of the application of participatory approaches to forest resource assessment. Pretty *et al* (1995) offers an extensive and practical trainers' guide. Abbot and Guijt (1998) provide practical guidance on participatory approaches to monitoring the environment.

Source: Higman *et al*, 1999

A4.6 Mapping policy influences

Policies send signals to different actors, and encourage certain types of reaction. Some signals are strong and compelling, while others are weak and almost subliminal. In a way, they can be viewed as concentric circles of influence. Indeed, this is often a good way to visualise them – and such visualisation can help in discussing policy as a mixed group of stakeholders. Some examples are given here.

Figure A4.2 is a generic 'policy influences map', which will often be found to apply. In itself, it can be a useful tool to open up discussion beyond the obvious influence of forest policy alone. Figure A4.3 was drawn by the PNG PTW team. Here, physical metaphors were chosen – overriding policies of structural adjustment and governance 'raining' down on the nation as a whole, land reform policy providing the essential soil on which forest policy must develop – and minerals and agricultural policy either threatening to knock forestry out of the equation, or finding a place alongside forest policy.

Policy influences can extend from one nation to another. For example, the notion of the *'ecological footprint'* can be a useful way of visualising the impact of one country's international relations – through trade, aid, foreign investment, foreign and military policy – on other nations.

Figure A4.2 Generic policy influences map

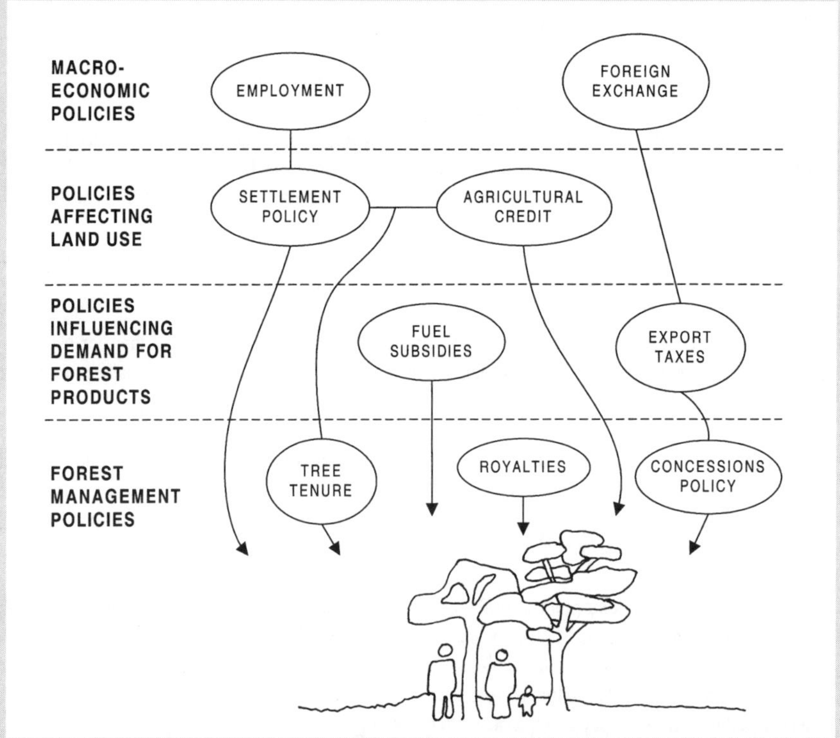

Figure A4.3 Forest policy in the context of other policy domains in Papua New Guinea

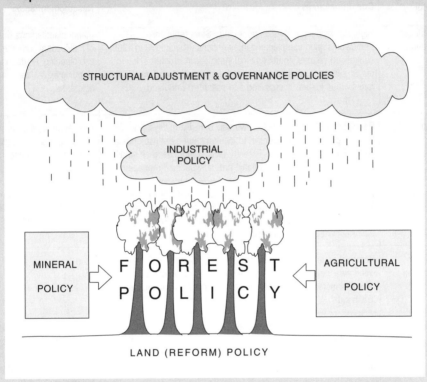

This could again be depicted as concentric circles of influence or, as IUCN Netherlands has done, a series of maps showing the degree of 'heaviness' or 'lightness' of footprints on other countries – clearance of land because of imports of livestock food, intensity of pressure on forests because of timber imports, etc (IUCN Netherlands, 1994).[9]

All of these approaches of 'mapping' are useful first steps in analysis, and can help to focus on key issues, but they invariably always end up begging more questions that require detailed analysis.

An area which often requires particularly detailed analysis is the *impacts of individual non-forest sectors on forests and forest stakeholders* – partly because the effects are often so large, and partly because such analysis has been rare and special efforts are now needed. Table A4.4 highlights these extra-sectoral impacts. The details given in the table are a summary of the work of the Zimbabwe PTW cross-sectoral focus groups which conducted the work. Such a matrix does not look at the links between different extra-sectoral policies, but it does point to the need for action in specific sectors.

---

9 See IIED 1995 for further discussion of ecological footprints.

Table A4.4 Impacts of key policies on woodlands and woodland-based livelihoods in the main land

Key Policy	Communal areas	Resettlement areas
Land allocation, use, tenure	High population densities on poor land. Strict regulatory frameworks: land-use planning interventions (about 1930 to 1960) centralised villages, confused local institutional structures, led to major clearance of woodland for arable production and transformed forest areas to heavily coppiced and pollarded woodland	Planned settlement, but poor level of institutional commitment to land-use planning, leads to forest asset stripping by residents and neighbours
Forestry	Regulation of forest products: 'own use' only. Permit system benefiting RDCs, a disincentive to local management (Communal Lands Forest Produce Act). Forestry extension focused on small eucalyptus woodlots and (recently) natural woodland management	As in communal areas. FC has formal control over cutting but has no capacity to monitor
Decentralisation	Regulatory: licences for any extraction. Land-use plans erode local autonomy. RDCs: many responsibilities but lack of resources and capacity for woodland management. Potential for devolved management eg. through by-laws	Land-use plans not done through participation of locals, absence of viable local institutions
Agriculture (pricing and extension)	Removal of subsidised inputs by structural adjustment. State withdrawal not matched by private sector, lack of information on markets. Increased woodland clearing to maintain agricultural livelihoods	As in communal areas
Livestock	Grazing schemes – isolated success stories, need for designs which take into account multiple functions of cattle in production system. Close linkage between livestock and woodlands unrecognised in policy	Tsetse eradication – opened up large areas in Zambezi valley for settlers, but poor land management, top-down planning initiatives, worsening status of livelihoods
Wildlife	Regulation and granting of 'appropriate authority' to RDCs for sustainable use of wildlife. CAMPFIRE a success, making significant contribution to livelihoods in areas where game abundant	Little impact
Tourism	Expansion of eco-tourism ventures in CAMPFIRE areas where game numbers low for safari hunting. Increase in wood-craft production	Little impact
Rural infrastructure development	Pace slow, huge demand for investment, limits productive potential	As in communal areas
Economic structural adjustment and trade	Decline in livelihoods with: loss of services; increased input prices; making room for those laid off from the formal employment market. Increased reliance on woodlands by poorest. New state programmes to strategically address poverty?	As in communal areas

tenure categories in Zimbabwe

Large and small-scale commercial farm areas	Indigenous forests (state reserves)	Industrial plantations (state and private)
Low population densities on good land, extensive holdings in large-scale sector. Private, relatively secure tenure (although recent compulsory acquisition by state). Remaining areas of woodland often managed under Integrated Conservation Areas. Voluntary regulation	About 1 million hectares set aside as forest reserves, mostly in Matabeleland (in addition to 4.9 million hectares of national parks). Isolated conflicts with communal area neighbours over access to resources	Limited to Eastern Highlands. Isolated conflicts with land-hungry
Guidance and voluntary regulation. Weakly applied FC restrictions over cutting. Ban on export of mukwa and modification of timber concession guidelines in 1988 and 1994	Forestry Commission has full powers to manage, but increasing conflicts with other users hence attempts to co-manage with neighbours	FC in process of relinquishing role of regulator on private lands
Grants, loans and taxes favouring conservation through Intensive Conservation Areas. (See communal areas for RDCs)	Potential for selective co-management with neighbouring communities being explored	Owned and run by companies, companies thinking of promoting outgrower schemes in communal areas
Government continues to subsidise agricultural extension. Shift to horticultural products and non-traditional agricultural exports as beef prices low	Declining agricultural livelihoods in neighbouring communal areas leads to increased pressure on woodland resources in reserves	Declining agricultural livelihoods in communal areas leads to interest in outgrower schemes
Cattle numbers falling due to low beef prices. Isolated examples of sharing pastures with communal areas residents	Grazing for communal area residents one of elements contained in co-management largely to reduce fuel-load	As in indigenous forests
Establishment of private game ranches and conservancies as beef prices low. Favours woodlands	Wildlife management objectives incorporated by the FC. Some safari hunting concessions	Little impact
Photographic safaris and game ranching. Favours woodlands	FC looking to tourism revenue from reserves as timber stocks no longer sufficient for significant revenue	Little impact
Government continues to subsidise. Potential for more demand if land further sub-divided	Reserves mostly quite remote from much infrastructure	Likely to encourage outgrower schemes
Liberalisation – removal of restrictions on foreign currency, import licences and import duties – leading to increased competition and investment. Gains for those who can reorient production strategies quickly. Non-restrictive investment climate may reduce environmental accountability	Reduced funding for forestry management	Relative boom in wood industry – growth in roundwood production. Impetus for development of standards for sustainable forest management

Figure A4.4 Dutch imports of timber from tropical, boreal and temperate zones

From boreal and temperate zones
(mln m³ Round Wood Equivalents)

- >3
- 1-3
- <1

From tropical zones
(mln m³ Round Wood Equivalents)

- >0.3
- 0.05-0.1
- <0.05

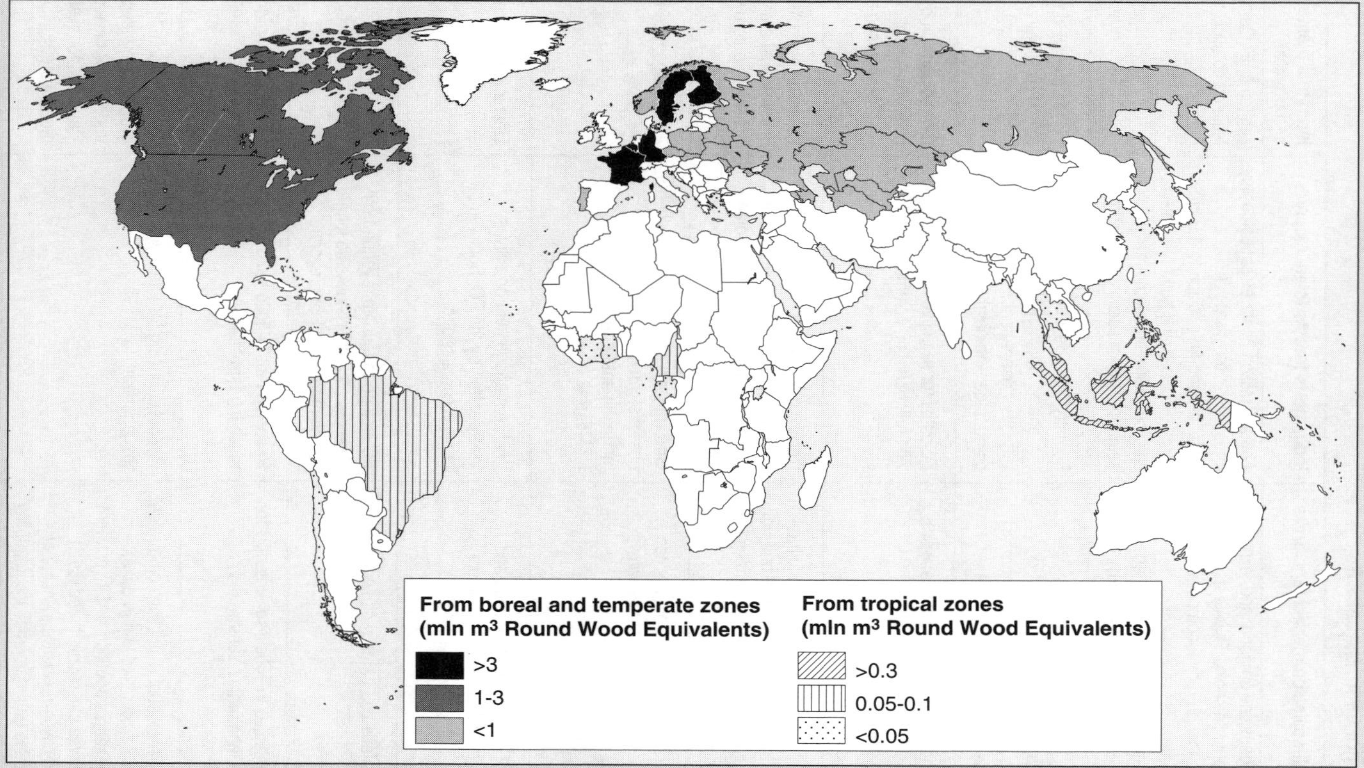

A4.7 Stakeholder analysis

Identifying forest and policy stakeholders
Stakeholders are those who have rights or interests in a system, and related knowledge and skills. For our purposes it is useful to think of forest stakeholders and policy stakeholders. Forest stakeholders – who could be further defined as individuals and groups with objectives and legitimate interests in the goods or services of a specific forest environment or forest resource – might include: people who live in or near forest; people who live further away, who use forest; settlers from elsewhere in the country, or abroad; forest workers; small-scale entrepreneurs; forestry officials; timber company managers; environmentalists; politicians; public servants; national citizens; global citizens; and consumers. All of these people, if their interests in forests are indeed legitimate, should in some way be involved in the making and implementing of policy which affects forests.

However, in practice, policy stakeholders are often only a sub-set of forest stakeholders, or are those who barely have a stake in forests at all – but who nevertheless manage to cook up policy which profoundly influences forests. As the Zimbabwe PTW team put it...

> *"Thus, one of the challenges may be to better recognise both forest and policy stakeholders, and to close the gap between the two. Stakeholder analysis is particularly useful to bring the focus onto distributional issues – understanding winners and losers in policy – and ways to address structural problems and improve effectiveness and social impacts of policy."*[10]

Contexts for stakeholder analysis
The most useful type of stakeholder analysis will depend upon:

*   The institutional level: a national policy analysis will need to engage different stakeholders compared to a regional forest management policy, or local projects – it will involve challenges of 'vertical' representation up and down the hierarchy .

*   The purposes: an appraisal of possible policy would be different from an evaluation/ analysis of existing policy – the former needing to include considerable extra-sectoral representation, and the latter needing to emphasise 'forest stakeholders' perhaps more intensively than 'policy stakeholders'.

Steps in stakeholder analysis
This can take a step-wise approach. An understanding of the total system and stakeholders' overall perspectives is needed first: i.e. – what are the key dependencies on forest goods and services, and the key problems identified by the main groups? Who is closest to forestry issues – this might be mapped as concentric circles of 'primary' and 'secondary' stakeholders, as in the Ghanaian example (Figure A4.5). This provides focus for subsequent analysis – detail can be added in time.

---

10 Stakeholder analysis is nothing new to policy research in the social sciences. Texts in the 1970s, which sought to establish 'policy research' as a focused discipline, put much emphasis on identifying stakeholders and ascertaining the power they have (e.g. Etzioni, 1971; Dye, 1976; Wildavsky, 1987).

Figure A4.5 Levels of stakeholders in Ghana's forests

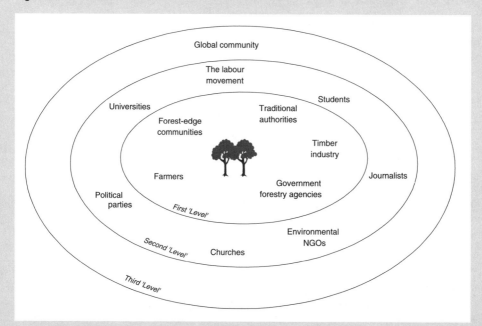

There are several methodologies for identifying stakeholders (Higman *et al*, 1999; Borrini-Feyerabend, 1997):

- *Identification by forest authority staff.* Those who have worked in forestry for some time can identify groups and individuals whom they know to have interests in forest issues and to be well-informed about them.

- *Identification by other knowledgeable individuals.* Land and agricultural agencies may be able to recommend relevant farmers and settlers; local government, religious and traditional authorities, forest agencies and forest enterprises may all be able to identify key representatives of different forest interests.

- *Identification through written records, and population data.* Forestry operations often have useful records on employment, conflicting land claims, complaints of various kinds, people who have attended meetings, financial transactions, etc. Forestry officials may have important historical information on forest users, records of permit-holders, etc. Census and population data may provide useful information about numbers and locations of people by age, gender, religion, etc. Contacts with NGOs and academics may reveal relevant surveys and reports and knowledgeable or well-connected people.

- *Stakeholder self-selection.* Announcements in meetings and/ or in newspapers, local radio or other local means of spreading information, can elicit stakeholders coming forward. The approach works best for groups who already have good contacts and see it in their interests to communicate. Those who are in more remote areas, or are poor and less well educated, and those who may be hostile to other stakeholders, may not come forward in this way. There is a risk that local élites, or others with inequitable objectives, will put themselves forward.

- *Identification and verification by other stakeholders.* Early discussions with those stakeholders who are identified first can reveal their views on the other key stakeholders who matter to them. This will help to better understand stakeholder interests and relations.

It is important that the individuals dealt with actually represent their constituencies. The dimensions of *representation* are:

- *Identity*: does the representative share the views of the group/ constituency in relation to forests? Or will the representative bring other/ multiple identities to the process e.g. tribal/ class or political affinities? Where can such other identities help, and where might they hinder representation and forest management?

- *Accountability*: Was the representative chosen by a particular group/ constituency? And/ or does s/he consult with that group regularly? What kind of specificity and sanction has the group attached to the representative's accountability?[11]

Once stakeholders or their representatives have been identified, it is important to assess:

- Their 'stakes' or interests. Dubois (1998) introduces the '4Rs' approach for assessing stakeholders' Rights, Responsibilities, Rewards (or revenues or returns) and Relationships with other groups. Useful methodologies include:
  ○ semi-structured interviews: cross-checking; identification of common ground; identification of trade-offs; identification of decision-making frames
  ○ oral case histories
  ○ indirect investigation
  ○ use of quantitative data

- The patterns and contexts of stakeholder interaction:
  ○ investigate competing/ complementary interests
  ○ investigate other factors in conflict/ cooperation, e.g: authority relationships; ethnic, religious or cultural divisions; historical contexts; legal institutions

To be useful, stakeholder analyses need to be summarised in a form where everyone's interests and issues can be seen together. Table A4.5 provides an example from Ghana of a summary stakeholder analysis, examining the current stakeholders, interests, means to pursue these interests, and impacts on forests and other stakeholders. (The Pakistan PTW team took a similar approach, and added a column on the constraints and pressures which stakeholders feel they face – as these realities in part determine future directions of policy.) Figure A4.6 provides an historical overview of how the relative influence of Ghanaian stakeholders has changed over time, providing useful context to the stakeholder analysis.

---

11 The use of existing stakeholder associations – professional, commercial, user group, community/ traditional – can help to handle problems of representation. However, there are limits to associations, especially in the private sector (there is a need to find private sector 'leaders' rather than the 'lowest common denominator' which tends to characterise associations generally) and in communities (élites may dominate such associations).

Table A4.5 Example – forest stakeholders in Ghana

Stakeholder Group	Main Interests in Forests	Means to Secure Interests	Main Impacts on Forests and People
Ministry of Lands and Forestry	"Conservation and sustainable development of forest resources for maintenance of environmental quality and perpetual flow of optimum benefits to all segments of society" (MLF, 1994)	Inter-ministerial (intra-governmental) negotiation Policy statements Concession allocation Market mechanisms Laws and regulations Consultation with other stakeholders Provision or control of information Monitoring Much pressure from private forest sector	Dominate policy processes Strong policy control over Forest Department and other forest sector bodies Over-ridden or influenced by some other sectoral policies and impacts
Ministry of Agriculture (e.g. of other central government stakeholder groups)	Source of land for conversion to agriculture	Statutes clashing with some forestry laws/ policies Agricultural extension advice Subsidised pricing of agricultural inputs Fixed crop prices	Conversion of forest land to agriculture – particularly cocoa. Some shade trees favoured. Encroachment on forest reserves
Environment and development NGOs/ lobby groups	Sustainable use Watershed protection Source of biodiversity and endangered species Climate regulation	Influential members lobby government. Access to donor support and international recognition	Some policy influence. Donor support for forest planning and control by Forest Department. Scattered environmental and community projects by NGOs
Forest Department	"Sustained supply of timber and non-timber products in perpetuity and environmental protection" (Kese, 1990)	Forest reserves as power base Allocation of yields Supervision of harvesting Policing role over people around forest reserves Poorly resourced, but significant donor support	Increasingly effective control of logging and farming encroachment in forest reserves. Weak control outside reserves Poor coordination with downstream control structures
Local government (District Assemblies)	Source of revenue through royalty shares	District by-laws Involvement in roadside checks Chainsaw controls	Some increase in law enforcement and protection Increased revenue demands
Traditional authorities	Land – power base. Source of revenue through royalty shares	Tenurial control of land Allocation of land Passive recipients of low and irregular payments of shareable revenue	Stool chiefs sell or rent land in reserves for conversion, allocate lands outside reserves for farming Excess sawmilling

Stakeholder Group	Main Interests in Forests	Means to Secure Interests	Main Impacts on Forests and People
Wood processing industry	Source of logs at low prices to convert to high value processed timber	Strong influence at policy level based on economic muscle Keep forest fees low and poorly collected	Excess sawmilling capacity and wastage in industry Low log prices promoting this
Logging concession holders (without processing capacity)	Source of logs for sale at high prices Only marketable species valued, others may be damaged	*De facto* control over large areas of forest	High grading Undersized trees felled
Chainsaw operators and bush millers	Source of logs for on-site conversion to lumber	Preferred over loggers by farmers More able to avoid royalty payments than loggers Some organised in trade associations	Active throughout high forest zone High grading
Commercial NTFP traders	Source of particular NTFPs for commercial use, e.g: canes, wrapping leaves, chewing sticks, bushmeat	NTFP permit and check system ineffective No local level rights to control access	Locally-based traders may conserve resources Non-local traders likely to over-ride local customary controls and over-harvest resources
Farmers and village-level institutions	Source of agricultural land and creation of fertility Contribution to farming system: shade, mulch, disease control, grazing Forest products for domestic use, sale and exchange (NTFPs may be household economic mainstays)	Marginalised in policy Do not own timber trees on farmed land Farm-level land-use decisions Gain low levels of compensation for farm damage from timber extraction by concession-holders. In practice – are decision-makers about produce taken from forest reserves	Destroy timber trees on farms Variety of tree and forest management practices on farms Encroachment on forest reserves in particular circumstances

Source: Mayers and Kotey, 1996

Figure A4.6 Changing 'shape' of policy, Ghana

Colonial government began reserving forest, through consultation with chiefs, to maintain environmental conditions primarily for cocoa farming.

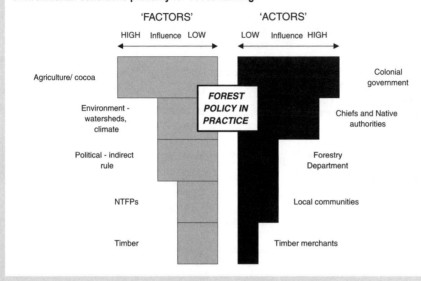

Timber out-turn was prioritised by government in time of World War.

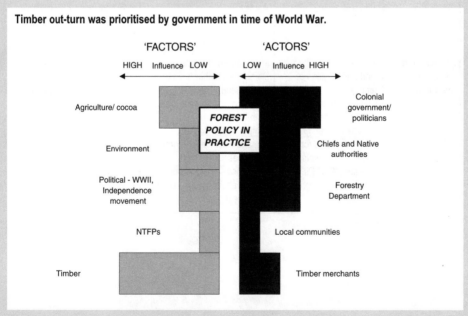

Notes:
- 'Factors' are the major concerns and pressures from within and outside the forest sector which influence forest policy in practice
- 'Actors' are the major stakeholders and organisations who influence forest policy in practice
- There is no horizontal correlation between factors and actors

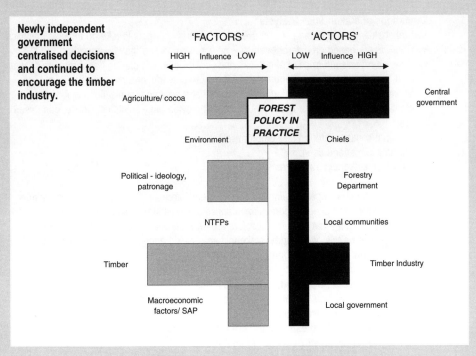

Newly independent government centralised decisions and continued to encourage the timber industry.

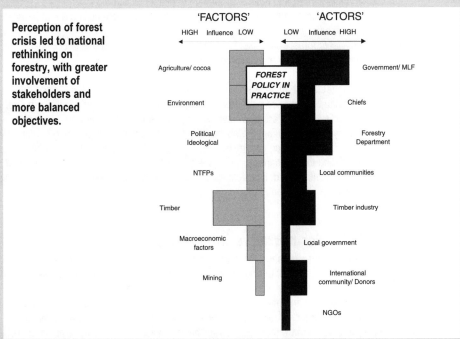

Perception of forest crisis led to national rethinking on forestry, with greater involvement of stakeholders and more balanced objectives.

Limitations of stakeholder analysis
- Stakeholder groups overlap – and even within one group, people take on multiple identities
- Conflicts are based on different values – no common ground may be apparent
- Where it reveals information about less powerful groups, this can be dangerous as it might lead to inequitable actions on the part of the more powerful groups in the process
- Stakeholder analysis is an information tool, rather than a communications tool. It can identify the heart of the problem – but it cannot provide easy solutions. Challenges raised are:
  - what common ground for compromise?
  - how to manage conflicts?
  - which stakeholders' interests to prioritise?

In relation to the challenge of weighting stakeholders' interests, Colfer (1995) has developed an approach for attempting to redress imbalances amongst stakeholders in access to forestry decisions by ensuring that *local forest actors* are fully identified and 'weighted' against certain criteria. Building on this, we suggest stakeholders should be identified, and weight should be accorded to them, depending upon:

- *proximity* to forests, woodlands or trees on farms
- *dependence* on forests for their livelihoods (i.e. where there are few or no alternatives to forests for meeting basic needs)
- *cultural linkages* with forests and uses of forest resources
- *knowledge* related to stewardship of forest assets
- *pre-existing rights* to land and resources, under customary or common law
- *organisational capacity* for effective rules and accountable decision-making about forest goods and services
- *economically-viable forest enterprise* that is based on environmental and social cost internalisation, bringing equitable local benefits

Colfer strongly suggests that an 'inverse' criterion also be used i.e. if a local group has a *power deficit* it should be weighted more heavily (to make up for such a deficit). We can add, conversely, that some stakeholders may have considerable levels of power and influence and interests which may adversely affect the abilities of other stakeholders to pursue good forestry, or even prevent it entirely. In such circumstances, an approach is needed which weights stakeholders according to the degree to which their actions should be *mitigated* or *prevented*. This is, of course, difficult ground. Practical approaches to analysing power are needed. These are investigated further in Section A4.10.

A4.8 Stakeholder narrative interviews
Once the stakeholder analysis has been performed, it will become apparent who are key informants. One approach to get the best out of key informants is that of narrative interviews. This approach allows stakeholders to put forward information in their own way. It can be structured to be able to glean their insights into the key dimensions of context, actors, policy content and impacts – or it can be looser, based on 'telling the story', which allows these dimensions to be brought out without necessarily having to ask overt questions about them.

The interview approach has to be modified for each personality. Policy issues are generally controversial – stakeholders will be wary of how the information they provide will be used. At one end of the spectrum is eliciting anecdotes informally over a beer, or golf; this can be a useful approach with politicians, who need to be engaged in policy review but whose formal engagement can cause problems for others. At the other end is formal, taped interviews with transcripts reviewed for accuracy; this can be suitable for gaining the experience of established professionals – such as senior or retired foresters who have 'seen it all'. A range of techniques can be used:

- presenting different perspectives/ views on a problem and getting interviewees to react to each
- allowing interviewees to leave their own values and definitions unstated (recognising that commitment to a particular perspective may be politically difficult for them)
- using 'if....then' scenarios to determine interviewees' judgements of the feasibility of possible developments or recommendations (people may be more comfortable reacting to hypothetical situations)
- assessing whether further contact/ useful information and commitment to the work can be provided by the stakeholder – some may be flattered or see it as in their interests to provide further advice (one way to build an 'advisory group')

A4.9 Institutional analysis

If policy is taken to mean 'what institutions actually do', it is important to analyse institutional factors in policy work. Three main aspects of institutions generally need to be understood:

- institutional roles – functional mandates of some organisations in relation to others
- internal dynamics and characteristics
- factors shaping institutional change

A4.9.1 Institutional roles and relations

It is common practice for individual organisations to conduct analyses of their strengths, weaknesses, opportunities and threats (SWOT) in relation to their mission. In policy analysis, however, analogous efforts are needed for the forest sector as a whole, or for key institutional arrangements within it, e.g. all those involved in timber production. A 'mapping' approach can help reveal the functional strengths, weaknesses and relationships among formal and informal institutions. Figure A4.7 'institutional analysis for forests' illustrates one possible process to identify and investigate different types of institutions.

Figure A4.8 is an example of a summary of such analysis, demonstrating the linkages between various companies in the forest sector in Papua New Guinea in 1993. The lines, which connect the companies involved, indicate key factors in the relationships between these institutions: shared ownership, management and facilities.

DFID (1999) describe an approach to 'institutional profiling'. Prepared by particular groups or by all groups with an interest in an issue, these profiles can provide a quick visualisation of

Figure A4.7 'Institutional analysis for forests'

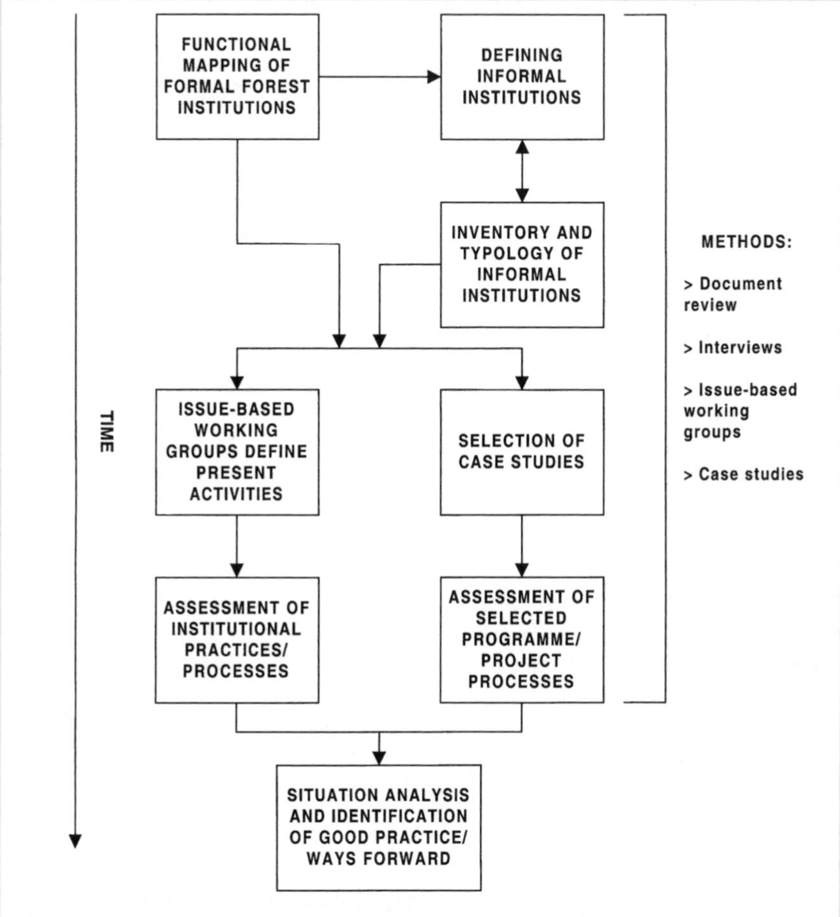

Source: Filer with Sekhran, 1998

the current situation which can be understood easily by all and provoke much discussion. The group first selects a centre point – a policy, a forest, a project, a particular group, or perhaps a particular role or function. Each institution related to the centre point is then represented by a shape – usually a circle – the size of which shows the importance of that body. Arrows are then used to indicate relationships between groups, with the thickness and direction of the line illustrating the strength and direction of influence (DFID, 1999).

Such profiles can help identify blockages, gaps and weaknesses. The process of developing these profiles can also reveal many issues which need further investigation. Profiling can also be used for monitoring, if the exercise is repeated at relevant stages with a view to tracking changes.

Figure A4.8. The putative Sino-Malaysian logging cartel, 1993

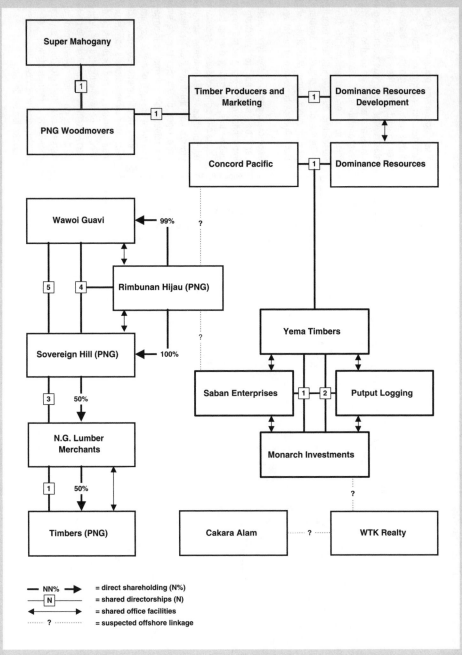

Figure A4.9 Institutional profile – foresters, Karnataka Forest Department

This diagram was prepared by a group of foresters in Karnataka State, India. It shows the wide group of stakeholders recognised by the foresters. Each stakeholder group has been assigned a circle, and arrows are used to indicate the relationship with the foresters – the thickness and direction of the line illustrating the strength and direction of influence of one group on another. The plus or minus signs also show whether that influence is positive or negative.

In other diagrams of this type, circle size can be adjusted to illustrate the importance of the group or the number of people in it. The gap or overlap between circles can signify closeness or distance or relationships. Lines between stakeholder groups can illustrate their relationships, while types of line can indicate whether the relationship is formal or informal, etc. The great strength of these diagrams lies in their development. Cut out circles of paper and other materials can be used in their preparation and participants do not need to be well-educated to participate. Many thorny issues are likely to be raised in discussion over development of the diagram and the outcome is a clear and immediate representation of key relationships which can be very useful to provoke further debate.

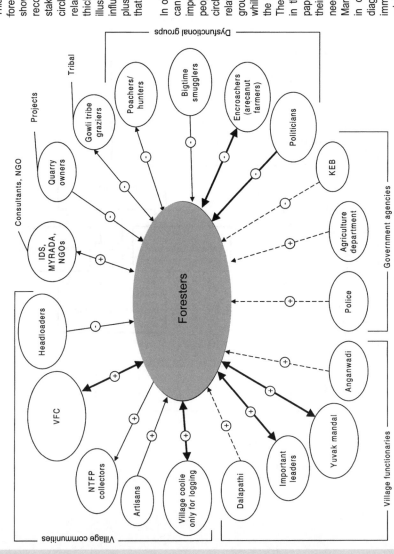

Source: DFID, 1999

A4.9.2 Institutional histories, internal dynamics and actors

Individuals within institutions interact in ways which relate to the institution's formal mandate, e.g. legislation, human resource development and budgets. These interactions are 'visible' and can be planned and managed. But there are also hugely complex, less visible interactions which collectively define the 'institutional culture'. Institutional 'actors' may have diverse motivations (Section A2.2). The combination of such motivations, when linked to both the 'visible' and 'invisible' institution, may explain many policy outcomes. Many efforts to make policy responsive to the new demands of sustainable forest management have concentrated their recommendations on broad goals, without quite knowing how such changes could be brought about. Without basing such prescriptions on motivations for change within existing institutions, frustration at the lack of subsequent action is a common result.

Analysis of institutions in community forestry has blossomed in recent years (Ascher, 1995; Ostrom, 1991; Thomson, 1992; Thomson and Freudenberger, 1997) – with much to offer approaches which seek policy-level impact. Thomson and Freudenberger (1997) provide guidance on institutional analysis of community forestry, describing a series of steps in analysis and reform:

- Identifying the forest products that are involved in resource governance problems
- Analysing the characteristics of the products: whether a particular forest product is a private, common pool or public goods/ service. (Each of these types of goods and services creates a different kind of incentive affecting how people will behave toward the resource)
- Analysing the community's capacity for collective action
- Analysing the system of rules within the community, as well as outside rules that affect resource governance
- Identifying 'best bets' for improving resource management and the institutional adjustments that will be needed
- Planning and implementing institutional changes to suit the 'best bets'
- Managing institutional change and the consequences of change

The authors note, however, that these steps may be frustrated by various 'complicating issues', both internal and external to the community. *Internal issues* include: dominance by a few powerful individuals or interest groups; exclusion of women or minority interests; and competing factions based on economic interests. *External issues* include the limitations placed on decision-making and enforcement at the local level; and the bureaucratic imperatives of NGOs and government staff (Thomson and Freudenberger, 1997).

A4.9.3 Factors shaping institutional change

Policy processes inevitably lead to change. It is useful to know how institutions have dealt with change, and what their current capacities for change are.

It may be trite to say, but understanding and managing institutional change in forestry is both a science and an art. A functionalist analysis of institutional roles, functions, and efficiency is necessary for understanding the 'fit' of the institution to the job and for identifying the broad goals for change. An interpretative view of institutional histories, dynamics and actors is essential for understanding what kind of change is possible, who might lead it and how to get them involved. Both these functionalist and interpretative 'lenses' are necessary for understanding and guiding institutional change as a whole.

In a review of the issues connected to institutional change in public sector forestry, Bass *et al* (1998) present an analytical framework for describing institutional change (IC) based on five linked sets of issues, covering:

1. institutional context
2. pressures on institutions
3. state of the institution and its capacities
4. response – direction of change
5. institutional change management and methodologies

Splitting the categories up in this way allows for a cyclical approach, i.e. context>pressure>state>response>altered context. Different management actions and methodologies may be appropriate at different 'stages', and indicators of change may be developed for each of these stages. Some of the sets of the issues in Box A4.3 describe a spectrum, or degrees of magnitude on a single axis; others are merely empirical clusters of related issues.

The analytical framework in Box A4.3 was developed with a view to guiding those considering, or already engaged with, projects aimed at bringing about institutional change. It provides a way of describing institutional change processes and the contexts in which they operate. This it needs to incorporate information as described in Section A4.9.1-2. The framework was developed following the recognition that the theory available to those involved in institutional change in forestry is weak, that empricial lessons have not been fully drawn, and that the information base is poor. However, because development assistance is continuing to invest quite heavily in forest sector institutional change, further research and information-sharing is sorely needed (Bass *et al*, 1998).

Box A4.3 Pressure>state>response framework for analysing institutional change

A. 'Context': cultural/ political conditions surrounding institutions
- *Cultural factor influence*: degree to which cultural factors and especially the power structure determine what forest institutions do, and whether change is possible

- *Political influence*: degree to which politics dominates forest institutions and the scope for change, e.g. dominated by crisis politics, as opposed to incremental institutional reform in politically mature environments

- *Technical/ market influence*: degree to which forest institutions implement technocratically-developed, efficiency-driven policy that is responsive to markets and other needs

- *International agency/ policy influence*: some institutions, particularly in small and/ or poor states, can be considerably open to influence by international bodies such as aid agencies

B. Pressures: forces for institutional change

- *Motivations/ driving forces*: the key issue(s) for which there is pressure to change, e.g. imperatives stemming from: globalisation, finance/ efficiency, environment, social/ equity issues, and ethical issues/ (anti-)corruption

- *Actors* which are pushing for the above change(s), e.g: internal/ top of the hierarchy, internal/ lower in the hierarchy, other governmental, market actors, civil society actors, projects and one-off initiatives, and international bodies

- *Change agents and champions*: where is there capacity to lever change? For each, is the agent an individual, organisation, or institution?

- *Resistors to change*: where is there resistance to change? an individual, organisation, or institution? Is resistance active or passive?

- *Other factors enabling/ constraining change*:
  ○ legal scope for change e.g. resource ownership laws and legal mandates
  ○ concepts, capacities, skills, incentives and procedures that may help or hinder the ability to understand and undertake change
  ○ funding/ resource availability to (contemplate) making changes
  ○ perceptions of the costs – both of change, and of the status quo

- *Summary – degree of openness to change, from most to least open*:
  ○ unfrozen – widespread expectation of change
  ○ thawing– willingness to change amongst some influential actors
  ○ clashes between resistant and open partners
  ○ resistance all round

C. State: current institutional type, capacity and roles

- *Type*:
  ○ Organisations (central forest authority, decentralised forest authority and its organs, other governmental organisations involved in forestry, private sector forest bodies, civil society organisations)
  ○ Institutions (regulations – laws and rules, market institutions e.g. trading relationships and norms, civil society institutions, e.g. common property regimes and other traditions, societal norms, e.g. traditions, habits, hierarchies, and the forest sector as a whole)

- *Institutional capacities*: These do not concern the mandate alone, but also:
  ○ transparency
  ○ accountability
  ○ legitimacy and representativeness
  ○ learning processes, resilience, adaptability and longevity
  ○ commitment of leadership and others
  ○ enforceability of rules and effectiveness of incentives
  ○ relations with stakeholders and other institutions
  ○ skills and resources

- *Current priority roles of the above*:
  - ○ financial roles, e.g. earning timber revenue
  - ○ social roles, e.g. local (community) development
  - ○ environmental roles, e.g. biodiversity or water conservation
  - ○ development roles, e.g. supporting other sectors (agriculture or energy)
  - ○ political roles, e.g. controlling territory or certain people
  - ○ client orientation, e.g. big forestry companies or communities

D. Response: the scope/ trajectory of institutional change

This covers the degree/ scale of change, from (generally) easier to more ambitious:

- *improving efficiency* of one organisation in meeting existing objectives
- *changing objectives* of one organisation, including decentralisation
- *entering partnerships* between an organisation and other stakeholders
- *renegotiating specific institutional roles* within the sector
- *changing the institutional climate* – participation, devolution, legitimacy and accountability of different organisations, and the rules by which they operate

E. Institutional change management and methodologies

- *IC process with no formal project management.* Informal alliances, prejudices, market forces, and laissez-faire, normally giving rise to gradual change

- *IC process with formal (project) management.* This is normally the result of the perceived need to organise a response to the driving forces. IC management style may range:
  - ○ from top-down to bottom-up/ client-led
  - ○ from a process approach to output/ plan-led

- *IC methodologies*:
  - ○ coercive tactics: whistle-blowing , humiliation, disenfranchisement, imprisonment, certain donor conditionalities
  - ○ organisational analysis/ audits
  - ○ 'unfreezing' / awareness-raising/ visioning activities
  - ○ conflict resolution and consensus/ coalition-building
  - ○ coordination and participation mechanisms
  - ○ commercialisation/ privatisation
  - ○ organisational reform: structures, systems and procedures
  - ○ learning/ training: action learning; training in new functions; study tour and exchanges; pilot projects
  - ○ financial mechanisms

Source: Bass *et al.* (1998)

A4.10 Power analysis

Power is a touchy subject. However, as should be clear from our description of some approaches to stakeholder and institutional analysis above, there is a limit to how far progress can be made in either the analysis or the effective change of policy without broaching issues of power differences. Some stakeholders are usually losing out, or their contributions or negative influences are hidden; important issues are not being talked about; and problems might be solved if stakeholders were free to look at them in another way. Ways need to be found to get some of these issues 'out into the open' if they are going to be tackled.

The elements of power need to be unpacked. Power is not (like money) a single negotiable object. It is important to address the question of how stakeholders gain or lose power to influence the direction of the policy process. Filer with Sekhran (1998) identify four different types of power:

- *positional power*, which is the capacity to secure the sympathy and support of other stakeholders, on the assumption of some common interest;
- *bargaining power*, which is the capacity to extract resources or concessions from other stakeholders, by some combination of force and persuasion;
- *executive power*, which is the capacity to meet the needs and demands of other stakeholders, thus increasing one's authority over them; and
- *managerial power*, which is the capacity to control the productive activities of other stakeholders, and thus to determine the quantity and quality of their outputs.

Since each group of stakeholders is often internally divided, like a character played by several actors whose own attitudes and interests may be quite diverse, it is also possible to distinguish between *external* forms of power, which are exercised by one group of stakeholders over other groups of stakeholders, and *internal* forms of power, which are exercised by some members of a stakeholder group over other members of the same group. The power of each group can then be analysed and an indication of their overall 'weight' within the policy process given – the sum of all their influences over the direction of that process.

For policy analysis, a useful first step is to identify the relative degree of stakeholders' power, the source of that power, and the means by which power is exercised:

- *Degree of power*. Simple diagramming approaches can help here. Figure A4.10 is the product of a multi-stakeholder exercise in Pakistan, to map the proximity of stakeholders to the 'centre' of policy-making, i.e. an indication of the impact of their power. Figure A4.11, from Costa Rica, takes this approach a step further. It is an attempt by the Costa Rican country team to visualise the main actors' powers to influence policies affecting forest and people between pre-1950 and today. The diagrams indicate the relative influence of the different actors – their ability to create 'policy space' – the linkages between them, and the ways this pattern has changed over time.

- *Sources of power* may include:
  - ○ having useful personal contacts
  - ○ earning or otherwise gaining money
  - ○ possessing scientific knowledge
  - ○ holding an important job
  - ○ owning land
  - ○ controlling equipment or vehicles
  - ○ having authority to provide loans, allocate budgets or hire and fire employees
  - ○ securing international or political support

Documentary evidence and key informants are needed here.

Means to pursue interests may encompass:
  - ○ legal means, e.g. rights to resources or revenues
  - ○ illegal means, e.g. bribery or sabotage
  - ○ formal means, e.g. regulations or public meetings
  - ○ informal means, e.g. forming alliances or lobbying

Again, documentary evidence and key informants are the best source of this information. In Section A4.7 of the main text of this report an example is given from Papua New Guinea of narrative description of the exercise of stakeholder powers over time.

Figure A4.10. Stakeholder influences on policy, Pakistan

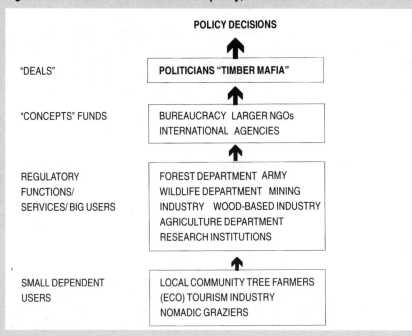

12 The size of the arrow indicates the degree of influence.

Figure A4.11 Power of different actor groups to influence forest policy, Costa Rica

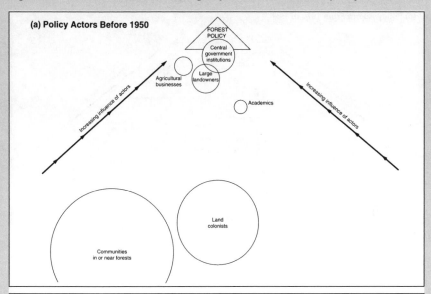

(a) Policy Actors Before 1950

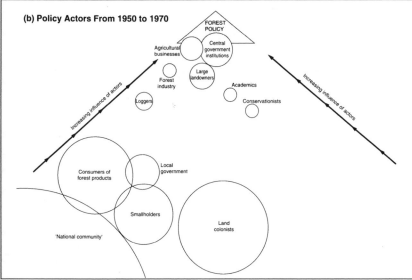

(b) Policy Actors From 1950 to 1970

Key

□ 'Forest Policy' - here refers to the combination of laws, decrees and government strategies that have strong impact on forests. Note that this 'grows' over the four periods.

□ Size of circle gives indication of number of people in the actor group.

□ Proximity or overlap of circles gives indication of an inter-relationship between actor groups.

□ Closeness to 'forest policy' gives indication of relative influence over policy formulating / implementing.

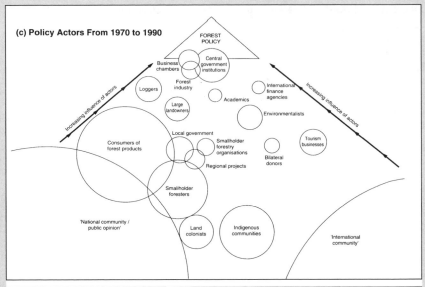

(c) Policy Actors From 1970 to 1990

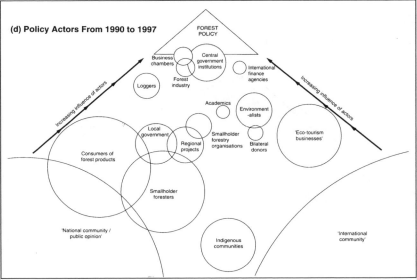

(d) Policy Actors From 1990 to 1997

Key

- □ 'Forest Policy' - here refers to the combination of laws, decrees and government strategies that have strong impact on forests. Note that this 'grows' over the four periods.
- □ Size of circle gives indication of number of people in the actor group.
- □ Proximity or overlap of circles gives indication of an inter-relationship between actor groups.
- □ Closeness to 'forest policy' gives indication of relative influence over policy formulating / implementing.

A4.11 National debate/ stakeholder validation workshops – developing conclusions and recommendations

Where improved national policy and process is the goal, the preliminary findings of policy work can be used as the raw material for national stakeholder workshops. Such workshops may need to bring key policy holders together with those, identified through the work as key stakeholders, who are currently marginalised by the formal policy system. If well facilitated, such workshops can result in sharpened findings and considerable levels of take-up by participants.

Refinement/ validation workshop exercises can also be highly productive when mixed international experience is brought together at the interim and preliminary stages of the findings. This is particularly important for international comparative work.

In Himachal Pradesh, for example, forest policy review workshops include:

- some which involve Forest Department staff only, so that findings and ideas can be internalised
- multi-stakeholder workshops, for debating findings and moving towards solutions; these will involve local-level stakeholders identified as key; one of these workshops will involve IIED as independent outside observer and facilitator, to help guide through sticky issues
- a multi-stakeholder validation workshop, at a stage where agreement has basically been produced, to confirm findings and the way forward, and to encourage subsequent commitments.

A5 Influencing policy – some tactics

A5.1 Revisit communication channels and uses of policy analysis information

Information may be used in a variety of ways. It can rarely be said to lead directly to the obvious, clear-cut event of 'policy change' It can help us to think about issues and define problems, rather than to seize on solutions. Thus information generated may be used as data, ideas or argument. The way in which it is communicated will have a major bearing on its actual use.

The critical concerns here are the ways in which information is generated, and how it is delivered to those who can use it to affect policy choices. Are we doing all we can to increase the probability that policy-makers and those affected by policy get taken up in the research process, and/ or that policy-makers use the information which researchers produce?

Communication is the 'heart beat' of an effective policy process. It can:

- prepare the ground: publicising the purpose/ scope of the policy analysis/ process, to elicit reactions through its consultative processes
- keep the story 'on the boil' – spreading results as they arise: the main issues identified, the main findings and recommendations on each issue, decisions made, etc.
- keep people informed about the 'next steps' at each stage
- rehearse the networking needed for shared action towards SFM

The major communication channels (newspapers, Internet, radio, lectures, training, etc) are populated by advocacy groups and the media as well as policy-makers. Consideration should be given to the nature and needs of information sources (information 'wholesalers') as distinct from intermediaries ('retailers') and the end-users of information. Often wholesalers have no idea how retailers package up and sell their products, let alone who is buying them. Consider also the time required to absorb the message; the timeliness – right place, right time; and how to 'hook' people.

A5.2 Target and package outputs

Content and packaging of outputs strongly influence the effectiveness of policy work. Audiences become hooked by the 'sizzle', not the 'sausage'. Some characteristics of 'successful' content may include:

* bringing out the hot news – the face of the issue most likely to attract attention
* trading off what are likely to be hot *future* issues – rather than just *'now'* preoccupations
* challenging conventional wisdom
* raising new issues, ideas or perspectives to show a clear 'angle'
* making information easily accessible, distinctive – provocative, surprising or cool
* selecting formats for different audiences – report, policy brief, video, training module
* selecting the style for different groups – amount of technical jargon, clarity, tone and design.

However, outputs may be high in content and well packaged, yet still fail to hit their target, even if policy-makers commissioned the work, if: results are not available when needed; there is too much information; findings are not adequately conclusive to enable the next step to be taken – and knowing what this next step is will always be key (more research, dialogue, option development, or direct implementation); the proposals are politically unfeasible; or priorities have not been sorted out through the process (a common failing).

Some of these characteristics may lend themselves to the development of indicators of the impact which outputs can have (see Section A6).

A5.3 Select and pursue tactics: some possibilities

In reviewing its own work to influence policy, IIED has noted how it has used a spectrum of tactics (Mayers and Bass 1998b). These are addressed in turn:

A5.3.1 Dump information near policy-makers

In some circumstances, analyses of particular policies, policy processes, outputs and impacts – or just empirical information which has 'policy implications'- can be influential even when not particularly actively sold to an audience. Such positive influence is likely if the work is seen to:

* Define a new debate.
* Reinvigorate a stale debate – a fresh or clearer look at a problem which had seemed intractable.
* Ask the right questions at the right time; and give a simple well-explained answer which can stand up to criticism.
* Have developed, or involved a methodology which stands up to peer review.
* Present the clearest possible picture of why current policy is a problem and what it could be like.

However, although nicely crafted reports are important – they show you have done your homework, and provide reference points – they often only have impact if they are 'pushed' further. 'Policy-makers' characteristically do not read very much. Indeed reports whose contents remain undigested may either provide the excuse to avoid taking action at policy level – to legitimise doing nothing – or leave policy-makers with the impression that the issue has somehow been sorted out by the initiative. This highlights the importance of 'going beyond the report'- into follow-up and advocacy.

A5.3.2 Draw policy-makers into the analysis itself

Policy-makers may not be like the 'end-users' of other forms of research. They may have no inkling that the work will be useful to them, or may actively resist it. It is often useful to draw formal policy-makers into the process, involving them at early and interim stages, as well as at the end, as discussed in Section A3.6:

- Choose research partners who are actively engaged in policy. Working with politically knowledgeable partners is vital for any policy analysis which seeks to be prescriptive.
- Try to make a policy impact right at the beginning – by raising the idea and engaging the right people

Involvement in policy research of policy 'end-users' of another type – those on the 'receiving end', affected by policy at local levels, is also vital. This may, in effect, be about bringing together the people who are suffering from bad policy. Overall, it is about beginning to create feedback loops in the policy process.

A5.3.3 Serve the existing policy machine

Direct engagement with the shaping of policy often occurs when invited in by policy-makers to flesh out the scale of a problem and prepare solutions, or to work out how policy can be improved. Policy-makers may commission research or even the writing of policy. Success here is likely to depend on whether there is real willingness and openness to consider change in the policy machine.

Yet there is a major concern with this type of work – getting locked in to servicing the machine, or only a part of it, and losing track of the vision of desirable change (both of who should constitute the 'policy machine' and what the policy should be). This kind of work may generate much material and consensual work but little vision. However, when done very comprehensively, the products of this research can become milestone documents – sources of reference material for subsequent debate – for years to follow.

A5.3.4 Stay 'connected', seize opportunities, 'gain power and influence'

Identifying or recognising leverage points in policy is an art in itself but, as we have noted, predicting when and where they occur is an even finer art. This is about the political tactics of networking, listening to politicians and the media to pick up nuances, making leading comments to elicit reactions, and keeping on the look-out for political opportunities. When the opportunity arises the game involves dropping the right phrase in the right ear at the right moment, at the right dinner, etc. For all this to be possible, an existing reputation is needed. Firm resolve is also needed – you need to be brave to change policy in this way; you need a certain kind of assertiveness.

As with servicing the policy machine, this approach may mean that desirable big policy changes are shelved in the short term, in favour of continuity of contacts, and making small gains when opportunities arise.

Box A5.1 How to have power and influence – the view from 'management science'

The worlds of business and organisational management studies have for years investigated, and provided guidance on, ways to develop the powers of influence. One award-winning book for business managers (Bragg, 1996) on how to be more influential, includes the following nuggets of wisdom:

Seven power levers open to managers:
1. resources
2. information
3. expertise
4. connections
5. coercion
6. position
7. personal power

Six principles of influence through which managers can activate power:
1. contrast
2. historical commitments and consistency
3. scarcity value
4. social proof
5. liking and ingratiation
6. emotion

Eight key tactics of influence:
1. pressure tactics
2. upward appeals
3. rational appeals
4. exchange tactics
5. coalition tactics
6. ingratiation tactics
7. inspirational appeals
8. consultation tactics

Four key steps to becoming an influential manager:

Step one – know yourself
Step two – identify your target
Step three – diagnose the system
Step four – decide on strategy and tactics

Source: Mary Bragg, 1996. Reinventing influence: how to get things done in a world without authority. Pitman, London

There is clearly a trade-off between getting/ staying politically connected – being a political animal – which can threaten an analyst/ institution's independence (because of the reactions it elicits from some quarters), and doing policy research. Ultimately, the capacity of an individual to do either probably depends on being able to do both. But in the meantime, an institution doing policy work may need to strive for a good balance of political animals and policy researchers.

There is a whole raft of literature on 'gaining power and influence', most written purely to further the reader's self-interest, rather than the public interest. But it can be instructive (Boxes A5.1 and A5.2).

Box A5.2 How to have power and influence – Machiavellian tactics

An examination of the lives of history's great strategists (Sun-tzu, Clausewitz), statesmen (Bismarck, Talleyrand), courtiers (Machiavelli, Castiglione, Gracián), seducers (Ninon de Lenclos, Casanova), and con artists ('Yellow Kid' Weil) suggests that decency, honesty and fairness can never help you win. You should instead be shrewd, ruthless, oily and amoral – and power will be yours (Greene and Jeffers, 1998).

First, conceal your intentions. If people do not know what you are up to, they cannot prepare a defence. Guide people down the wrong path. Be bold.

Next, get others to do your work for you, but always take the credit. Use other people's wisdom and knowledge to further your own cause. Not only will it save you time, it will give you a godlike aura of efficiency and speed.

Never trust friends. They are prone to attacks of envy. Instead, hire a former enemy who will be far more loyal because he has more to prove. If you have no enemies, make some. Then choose the most threatening ones and annihilate them totally. Keep your hands clean by using others as a screen to hide your involvement in shady deals.

Learn to exploit emotional weakness. Play on people's uncontrollable needs and insecurities and seduce them into becoming your loyal pawns. Appear to give your victims a choice while forcing them to choose between the lesser of two evils, both of which serve your own purpose. Put them on the horns of a dilemma: they will be gored whichever way they turn. But most importantly, never accept a free lunch. Learn to throw your money around and make sure it keeps circulating: generosity is the ultimate seal of power.

Source: R. Greene and J. Elffers, 1998.

A5.3.5 Convene better policy processes
Some initiatives aim to convene improved policy processes, to try and kick-start policy or to show what alternative policy processes could be like. They may attempt to involve previously unheard or under-represented groups. However, care needs to be taken not to 'jump in' to stakeholder policy processes before sufficient analysis has been done. Policy by brainstorm rarely works, and analysis can be used specifically to avoid this.

This approach is obviously more rewarding when following close on the heels of forest policy reviews which are full of expert analyses and depauperate of consultation. Since there are so many of these, having been spawned in recent years by TFAP, NEAP and other processes, it may be that this tactic will prove rather timely in the next couple of years.

In any case, in many contexts the policy analysis that is most needed draws on the same skills needed for better policy process, i.e. knowing who to talk to, what can and cannot be said, etc.

A5.3.6 Offer do-it-yourself policy review kits

This approach aims at helping to set the frameworks for policy research, and extending methodologies, but not actually doing the policy research itself. For example, setting out the sequence of questions which must be asked in order to arrive at a policy answer, and how they might be answered. Training in policy analysis and policy process may also be effective for others who are, or, more likely, may become, influential in government and civil society. This book, for example, has some of the ingredients to equip such a 'policy review kit'.

A5.3.7 Build constituencies

Some work appears to have little to do with policy, but aims to legitimise local views and support activities at local levels. However, some of this work explicitly recognises that the policy link – what is being contributed to, is a slow process of building a constituency through networks and support for community-level approaches. In other words, working outside the policy machine with constituencies whose agenda might one day include pushing the policy process.

With a long term view, work can contribute intellectual thinking – a body of understanding – which will bear fruit at policy level only in, say, fifteen or twenty years' time. For example, the several long-standing participatory forestry projects in Pakistan are fulfilling such a role. Indeed, there may be good grounds for arguing that the right foundation-laying and momentum-building activities are ultimately the strongest of all policy tactics. The key to success is being able to identify issues which are not completely off-the-wall today, but which will be especially meaningful in twenty years time; and working with credible people local to that issue to help build constituencies, and demonstrate what can be achieved in practice. There is little point in working outside the policy machine unless there is a chance that one day it will deal with these issues.

A5.3.8 Create vision

Big changes can be achieved through dramatic action – or alternatively through incremental progress. Both require vision. Tactics for generating vision differ from 'dumping information on policy-makers' by aiming for a more voluminous groundswell. It may require fostering an attitude of mind among politicians – introducing a 'rolling stone' and keeping it moving – or working with those who can frame opinion – the intermediaries such as NGO networks, rather than the decision-takers.

In some contexts, the business of 'capturing minds' may best be achieved by stimulating debate. Contentious messages can be important tools to provoke reaction. Broadcasting short and punchy messages – 'spinning a story' – at the right moment, is a key function of the policy advocate. Where there is much uncertainty and issues are complex – and much environment-development territory is like this – it is important that research provokes reaction, perhaps even if it is a bit 'wrong' so that others will join the fray and eventually get it 'right'. However, some positions become such strong narratives that they circumscribe what is possible in the future, even when consistently challenged or proven wrong, e.g. 'fuelwood crisis' and 'tragedy of the commons' narratives (see Sections A2.4-5). If stories are the building blocks of knowledge, they are crucial for better policy. In the 'information age' it has been argued that there is increasingly too much information ('informed bewilderment'), and not enough stories.

A5.3.9 Keep donors 'on tap', not 'on top'

In Sections 2.2, 4.2 and 5.2 we have discussed some of the roles and activities of donor organisations and external agencies in policy contexts. This is not the place for a thorough examination of the appropriate entry and exit points for external supporters of policy work. However a few general normative points from experience can be made. Few of the methods to analyse policy, and the tactics to influence it, are short-term endeavours; if donors seek to support policy work a medium- to long-term commitment – several years at least – is generally needed. But donors should be 'on tap' during the process, not 'on top' through inappropriate conditions on grants and loans.

Donor roles send signals about the purpose and 'ownership' of policy development. Donors should avoid involvement in defining strategy/ policy content.[13] Allying policy development processes too closely to immediate development assistance planning can dilute the likelihood that the policies developed will work. Similarly, if donors 'cherry-pick' only some activities arising from policy developments whilst ignoring others, they may skew internal capabilities adversely. To play an effective role in policy work donors might:

- be *facilitators*, more than issue experts
- pay for *secretariat* resources
- pay for time for *key informants and analysts* to explore key issues
- provide the *'lubricants'* – cross-institutional fora, quick money for transport and food, and process expertise
- assist in *methodology* development, training and use
- ask *'independent'* questions
- support involvement of *marginalised groups*
- support *risk-taking* and *experiments* that are otherwise impossible
- act as advocate for *international obligations*

If the above types of action are well-tailored to local circumstances, external agencies can help create political space in which internal advocates of policy change are able to manoeuvre.

---

13 The term 'facipulator' has been coined for the facilitator who turns manipulator.

A6 Track the impact

Rather occasionally, policy work has a direct and obvious impact on policy – and this can be easily identified. More often, impact is more 'sensed' than clearly seen. In general, it is necessary to at least be wary of the fact that there are many links, and the difficulty of pointing to any correlations, let alone causation.

As we have noted above, much policy work aims to contribute to a pool of knowledge and it is necessary to look ahead and acknowledge that work may only have impact in the long run. Thus, tracking the impact can be very difficult in a situation where bits of information are seeping in, uncatalogued, without citation, gradually forming a simple story and enlightening people, perhaps preparing for a change whose time has yet to come. Yet, if policy is to be improved, there is a vital need for information and monitoring systems which feed analysis of impact back into policy.

A6.1 Develop policy impact indicators

If we want to try and track whether policy work has impact, then for each of our influence tactics (Section A5) it may be possible to develop impact indicators. The types of impact that we might want to measure for each tactic include:

- *Outputs.* Content, quality and packaging of the information produced (research findings, stories, messages)
- *Process.* Delivery and use of the information in relation to policy making and implementing; contribution of affected actors to the process
- *Outcomes.* Appropriateness of chosen policies, and their effectiveness in moving practice towards or away from sustainable development goals.

Indicators of policy work *outputs* can be the most straightforward: numbers of research products distributed; qualities of the work identified in peer reviews, workshops, etc. (see Section A5.2 on the characteristics of effective outputs). In terms of *process*, it may be possible to identify what information was available before, during and after policy work, and thus to show that certain insights and arguments were not present in policy deliberations before, but were afterwards. However, policy processes are often notable for their lack of a paper trail. *Outcome* indicators can be of various sorts – at their simplest, 'real-life' indicators, monitorable by ordinary people, on key aspects of human and ecosystem well-being. At their most complex, they can include indicators on every single dimension of SFM (similar to certification) and end up reconfiguring national monitoring systems. The problem of correlation with single policies remains, however.

A6.2 Treat outcome indicators with caution

Outcomes are even more difficult to monitor! Where the policy analysts are also key policy protagonists – e.g. campesino forestry groups in Costa Rica doing their own research and pressing their case – identifying the outcomes should be possible. In other cases, however, there may be trouble if the analyst or institution stamps its signature on policy work. This project – *Policy that work*s – is perhaps a good example. It has had some impact because findings are owned by collaborators whose policy analysis capabilities have been used and, in some cases, built through the project, who are engaged with the holders of policy, and who are beginning to change policy. For policy change to work, 'ownership' of the change has to be in the right place.

It may not be wise for the policy analyst to become too closely associated with policy change. Ascribing the ideas or the groundwork to the analyst may jeopardise the relationship with other key policy actors and undermine their willingness to take the work on board. Thus, there may be very few benefits for authors and analysts in terms of citations of academic papers. Recognition is more likely to be based on word of mouth, repeat business, etc.

Much of the world as a whole – at least those in power – resist transparency of information on outcomes. Third parties can make progress in identifying indicators of policy outcome. IIED and others have been advocating systems that feed specific outcome indicators back into policy, e.g. through *pressure-state-response* models. But in general, the connection between policy work and policy change is likely to remain quite opaque – and often we may have to rely simply on the 'gut feeling' that policy work is worth doing!

Country team action frameworks

Action frameworks were developed, in each of the six country teams in the *Policy that works for forests and people* project, to summarise the plans for answering four main questions:

- **What is our message?** Key findings from the country study in distilled form, focused on the recommendations for improving policy processes

- **Who needs to hear it?** Key stakeholder groups who need to be informed of all, or of particular, findings

- **How do we get the message across?** Activities which need to be carried out over the next one year period, steered by the country teams and involving particular groupings of stakeholders.

- **How do we follow up?** Planned or notional activities which need to be carried out over a succeeding two-year period.

The six country team action frameworks follow.

PAKISTAN			
What is our message?	**Who needs to hear it?**	**How do we get the message across? PROJECT ACTIVITIES**	**How do we follow up? FUTURE ACTIVITIES**
[The bad news: command and control forest policy does not work (the timber mafia exploits the loopholes) and ignores farm forestry] [The good news: participatory rural development projects and cross- sectoral planning initiatives work; if prices are right, farmers plant trees] Create multi-stakeholder forest fora at national, provincial and lower levels Reorient forest policies towards: farm forestry and rural livelihoods; strengthening community organisation; and joint forest management, including private sector groups Develop simple, transparent information systems Reorganise forest authorities for decentralised capacity	Provincial Forest Departments	1. Facilitate workshops to debate and internalise the findings with policy development processes of provincial Forestry Departments in North West Frontier and Northern Areas	Establish a secretariat to monitor, record and disseminate findings on developing policy dialogues, and provide bridging agent between government, NGOs, private sector, Forestry Donors
	Office of the Inspector General of Forests Federal government politicians Forestry Development Corporations Wildlife department Agriculture department Army Mining industry Timber traders and wood-based industry International agencies NGOs Education institutions Research institutions	2. Host national workshop to launch report 3. Produce a short 'glossy' overview report in Urdu, and one-page flyers in both English and Urdu 4. Develop a video – illustrating multi-stakeholder fora and joint forest management 5. Conduct 'mini launches' at various forthcoming events/ meetings 6. Incorporate ideas on institutional changes in public-sector restructuring programmes	Coordinating Group and international obligations and opportunities Use findings to stimulate development of provincial forest policy in Punjab, Baluchistan and Sind Develop a policy analysis module for MSc course at Pakistan Forest Institute Work with Forestry Department and particular NGOs in Punjab and North West Frontier to build staff policy analysis capacities

PAPUA NEW GUINEA			
What is our message?	**Who needs to hear it?**	**How do we get the message across?** **PROJECT ACTIVITIES**	**How do we follow up?** **FUTURE ACTIVITIES**
For policy to have long-term 'bite', a new approach is needed which establishes a wide coalition of interests as a genuine 'policy community' Develop mechanisms for testing and publicising claims to productive innovation Experiment with new combinations of different scales of agricultural and forestry enterprise Generate a vision of the public interest and of private initiative through dialogue Install a brokering mechanism to connect needs with capacities	The 'six characters in search of an author': Loggers Donors Resource owners Public servants Politicians NGOs	1. Launch *Loggers, Donors and Resource Owners* with attendance of ministers. 2. Conduct 3-day seminar 'forest management: where do we go from here?' using *Policy that works* findings as the key material and incorporating debate on initiatives of UNDP, World Bank, national authority, NGOs and private sector 3. Disseminate findings through meetings with NGO network and newsletter, and through National Research Institute webpage 4. Prepare short briefing papers for private sector and donors 5. Collaborate with Institute of National Affairs to conduct briefings for the private sector and donors 6. Develop a format and materials for claim-testing fora based on comparative methodology developed by the *Policy that works* team, which enables stakeholders to openly debate and test claims made about forest development options 7. Conduct four claim-testing fora in areas which have been analysed in *Policy that works* or are subject to particular stakeholder claims 8. Complete analysis and publish results of survey of Rural Community Attitudes to Forestry and Conservation	Build on links made between national and local policy processes in claim-testing fora and establish the format and support mechanisms needed for stakeholder fora to become the accepted means for negotiating forest policy and development options Integrate *Policy That Works* outputs and approach into courses at the University of PNG and University of Science and Technology

INDIA			
What is our message?	**Who needs to hear it?**	**How do we get the message across? PROJECT ACTIVITIES**	**How do we follow up? FUTURE ACTIVITIES**
Joint forest management and farm forestry can only fully realise their potential (to provide equitable and sustainable livelihoods whilst providing environmental services) when actions are taken to tackle excessive powers over policy in three main areas: the inertia of 'fortress forestry' institutions; the over-influence of favoured forest industries; and the protectionist agenda which seeks to lock forests away from people's use. Specific actions are outlined to: Establish policy implementation and tracking capacity in the Ministry of Environment and Forests Improve economic and informational instruments for forest policy Bring momentum to institutional reforms Resolve conflicts between law and policy Create a framework for the union of conservation concerns and livelihood needs	Ministry of Environment and Forests State Forestry Departments and related organisations Large-scale forest industry Small-scale forest industry Social activists Wildlife conservationists Research and training institutions Donors and international organisations	1. Launch event for *Policy and Joint Forest Management* series of five analytical papers and *Policy that works* synthesis report 2. Internalisation workshop for selected national and state-level actors to debate findings and frame actionable responses 3. Briefing papers and findings summaries for different stakeholder groups 4. Articles for the print media 5. Presentations and distribution of *Policy that works* outputs by team members at a range of national, state and local-level fora	Video on the issues highlighted in the *Policy that works* study An adaptive research and outreach programme on: (a) Liberalisation of farm forestry, and associated requirements for: national policy change; deregulation of fibre prices; species research and technical support (b) Government-industry-community partnerships in forest reserve lands (c) Private sector wood fibre and NTFP production delivering social and environmental benefits

GHANA			
What is our message?	**Who needs to hear it?**	**How do we get the message across? PROJECT ACTIVITIES**	**How do we follow up? FUTURE ACTIVITIES**
Foster new links between agencies at national and district levels to create the conditions to look after/ grow tree outside reserves	Ministry of Lands and Forestry Forestry Department staff at all levels	1. Retreat for senior officers from Forestry Department and Ministry of Lands and Forestry to reflect on and internalise findings	Establish a Forestry and Environmental Justice Centre as a research, documentation, extension-brokering and claim-supporting organisation on forest, environment and people issues
Further develop mechanisms for multi-stakeholder negotiation	Executive branch of government Legislative branch of government	2. *Falling Into Place* launch event with attendance of ministers, forestry practitioners, influential members of civil society, TV, radio and press.	
Focus on reconciling equity and environmental quality objectives	Ministries and other government agencies in agriculture, mining, finance, local government and rural development	3. A 'ways forward' meeting of forest sector bodies and key stakeholders	Develop curriculum and source materials for a Forestry, Policy and Law course, and other courses at the University of Ghana
Seize opportunities in political and inter-sectoral processes to socialise forest policy	Forestry Commission Timber Industry Development Board	4. Individual briefings by team members of key private sector actors, legislators and public servants	Develop a *Policy that works* module in the curriculum of the Institute for Renewable Natural Resources
Foster partnerships between private sector, farmers and government	Effective mobilisers amongst NGOs and other civil society bodies	5. *Policy that works* team provide the panel for a popular TV programme 6. Articles by team members in the press	Integrate key findings into the curriculum of the Sunyani Forestry School
Invest greater trust in local institutions	Private sector timber producers and millers, and their associations	7. Lectures and talks by members of the team to civil society groups and symposia, and in their teaching roles at Sunyani Forestry School, Institute of renewable Resources and University of Ghana	
Resist efforts by some loggers and timber millers to obstruct implementation of the progressive new Timber Resources Management Act, 1998	Influential chiefs All actors in the National Development Planning process Trainees and students with futures in the above groups and organisations	8. Pamphlet in simple English, Twi and one other Ghanaian language on the main findings from *Falling Into Place* and the main provisions and innovations of the new Timber Resources Management Act	

ZIMBABWE			
What is our message?	**Who needs to hear it?**	**How do we get the message across? PROJECT ACTIVITIES**	**How do we follow up? FUTURE ACTIVITIES**
Promote policy accountability through opening up current and future policy consultations and enabling civil society initiatives Create favourable investment conditions for equitable private sector enterprises in communal and resettlement areas Concentrate central government support to forestry on forest extension Promote interim and long-term legislation for devolved, participatory natural resource management Fully assess the scope for a land and water tax Incorporate consideration of natural resources in resettlement schemes Support capacity of local institutions to deal with others and to manage resources Develop better information systems on natural resource assets, values and use Support adaptive research on negotiation processes, policy instruments and prediction of policy impacts Experiment further with models of forest co-management Allow some state land in reserves to be used for resettlement	Forest sector agencies	1. Retreat for officers from Forestry Commission with selected resource people to reflect on and internalise findings	Support new policy coordinator and team in Forestry Commission in work with NGOs and forest industry
	Central government agencies Land and agricultural agencies 'Opinion-formers': parliamentarians, researchers, managers of government agencies and NGOs 'Implementors': district and community-level government, NGO staff and local organisers Trainees and students with futures in the above groups and organisations	2. Prepare policy briefs of four types for: (a) parliamentary committees (b) managers of government agencies and NGOs (c) field staff of government agencies and NGOs, and for community representatives (d) donor field offices 3. Conduct launch of *Contesting inequality in access to forests* followed by 'what next' meeting 4. Edit and produce detailed analytical papers for key government departments, academic bodies, NGOs and private sector associations 5. Write articles for the press and journals 6. Conduct dissemination workshop for current 'policy-makers', 'opinion-formers' 7. Dissemination workshop for 'implementors'	Develop inter-organisational alliances for policy analysis and advocacy Curriculum and materials development, particularly by Institute for Environmental Studies and Centre for Applied Social Sciences at University of Zimbabwe, and forestry training bodies

COSTA RICA			
What is our message?	**Who needs to hear it?**	**How do we get the message across? PROJECT ACTIVITIES**	**How do we follow up? FUTURE ACTIVITIES**
Make policy processes affecting forests more inclusive Negotiate main forest goals at national level Work with real units of social organisation, not implanted structures	Ministers and key politicians in new government (February '98) Forestry business chambers	1. Use *Making space for better forestry* as basis for briefing papers and, 2. a video (for use in all activities below) 3. Individual briefings with new Minister and key political players 4. Minister presents *Making space* (English and Spanish versions) 5. Media launch	Convene a pilot multi-stakeholder policy process – to channel views, debate the *Making space* priorities, and move towards national goal negotiation Develop more focused radio and TV programmes Further develop capacity for policy analysis
Support capacity for local benefit forestry, and the fire-power of small-holder forestry alliances Improve analysis and information systems for policy	Forestry technicians: National System of Conservation Areas (SINAC) staff and forest regents Smallholder organisations: National Smallholder Forestry Assembly (JUNAFORCA) and others	6. Internalisation workshops (SINAC, JUNAFORCA) 7. Smallholders and indigenous leaders workshop	Support development of strategic information systems for policy
	Local associations	8. Local associations – regional workshops	
	Teachers	9. Teachers' union – speech-writing for congress and curriculum development	
	Church-goers	10. Church leaders' briefing	

Bibliography

Abbot, J. and Guijt, I. 1998. *Changing views on change: participatory approaches to monitoring the environment.* Discussion paper No.2, Sustainable Agriculture and Rural Livelihoods Programme, IIED, London.

Agarwal, A. and Narain, S. 1989. *Towards Green Villages.* Centre for Science and Environment, New Delhi.

Ahmed, J. and Mahmood, F. 1998. *Changing perspectives in forest policy. Policy that works for forests and people series no.1. Pakistan.* IUCN Pakistan and IIED, London.

Anderson, J. 1998. Accommodating multiple interests in local tree and forest management: some observations from a pluralistic perspective. *Proceedings of the Seminars on Co-Management in the Forest Context September 23-October 5, 1998.* International Agricultural Centre, Wageningen.

Anstey, S. 1999, forthcoming. Necessarily Vague: the political economy of community conservation in Mozambique. In: Hulme, D. and Murphree, M. (Eds) *African Wildlife and African Livelihoods: The Promise and Performance of Community Conservation.* James Currey, Oxford.

Aplet, G.H., Johnson, N., Olson, J.T. and Sample, V.A. 1993. *Defining sustainable forestry.* Island Press, Washington DC.

Apthorpe, R. 1997. Writing development policy and policy analysis plain and clear: on language, genre and power. In: Shore, C. and Wright, S. (Eds.) *op. cit.*

Ascher W. and Healy, R. 1990. *Natural resource policy making in developing countries: environment, economic growth, and income distribution.* Duke University Press, Durham, NC.

Ascher, W. 1995. *Communities and sustainable forestry in developing countries.* Institute for Contemporary Studies Press, San Francisco.

Associated Press. 1990. Loggers, environmentalists square off at ballot box. *Atlanta Journal,* December 31, cited in Cubbage et al, *op. cit.*

Baland, J.M. and Platteau, J.P. 1996. *Halting degradation of natural resources – is there a role for rural communities?* FAO, Clarendon Press, Oxford.

Banuri, T. 1996. Decentralisation and devolution. In: *Policy that works for forests and people:* proceedings of a planning workshop in Islamabad, November 30, 1995. IUCN Pakistan and IIED, London.

Baraclough, S. L. and Ghimire, K. 1995. *Forests and Livelihoods: The social dynamics of deforestation in developing countries.* UNRISD. Macmillan, London.

Barber, C.V., Johnson, N.C. and Hafild, E. 1994. *Breaking the Logjam: Obstacles to Forest Policy Reform in Indonesia and the United States.* World Resources Institute, Washington DC.

Bass, S. forthcoming, 1999. The importance of social values. In Evans, J. (Ed.) *The Forests Handbook.* Blackwell Science, Oxford.

Bass, S. 1998a. *International Regulation of Forest Issues: A Discussion Note for the IFF.* IIED, London.

Bass, S. 1998b. *Introducing Forest Certification.* Discussion Paper No. 1. European Forest Institute, Helsinki.

Bass, S. and Morrison, E. 1994. *Shifting cultivation in Thailand, Laos and Vietnam: regional overview and policy recommendations.* Forestry and Land Use Series No. 2. IIED, London.

Bass, S., Dalal-Clayton, B. and Pretty, J. 1995. *Participation in strategies for sustainable development.* IIED Environmental Planning Issues No. 7. IIED, London.

Bass S. and Thomson, K. 1997. *Forest security: challenges to be met by a global forest convention.* Forestry and Land Use Series No.10. IIED, London.

Bass, S. and Hearne, B. 1997. *Private sector forestry: a review of instruments for ensuring sustainability.* Forestry and Land Use Series No.11. IIED, London.

Bass, S., Mayers, J., Ahmed, J., Filer, C., Khare, A., Kotey, N.A., Nhira, C. and Watson, V. 1997. Policies Affecting Forests and People: Ten Elements That Work. *Commonwealth Forestry Review.* 76(3)

Bass, S., Balogun, P., Mayers, J., Dubois, O., Morrison, E. and Howard, W. 1998. *Institutional change in public sector forestry: a review of the issues.* IIED Forestry and Land Use Series No. 12. IIED, London.

Blaikie, P. 1985. *The political economy of soil erosion in developing countries.* Longman, London.

Bonnet, B. 1995. Instances décentralisées de décision, de régulation et de contrôle - Eléments de réflexion sur quelques expériences en cours dans le cadre des projets de gestion de terroir ou de développement local. In: *Rapport de la Journée d'Etude 1995 – Le Développement Local,* IRAM, Septembre 1995.

Bradley, P.N. and McNamara, K. 1993. *Living with Trees: policies for forestry management in Zimbabwe.* World Bank Technical Paper No. 210. World Bank, Washington DC.

Bragg, M. 1996. *Reinventing influence: how to get things done in a world without authority.* Pitman, London.

Brand D.G. and LeClaire A.M. 1994. The Model Forests Programme: International Cooperation to Define Sustainable Management. *Unasylva* 176 (45). FAO, Rome.

Bressers and Klok. 1988. Fundamentals for a theory of policy instruments. *International Journal of Social Economics,* 15(3/4): 22-41.

Bromley, D. W., Feeny, D., McKean, M.A., Peters, P., Gilles, J.L., Oakerson, R., Runge, C.F. and Thomson, J.T. (Eds). 1992. *Making the commons work: theory, practice and policy.* Institute for Contemporary Studies Press, San Francisco.

Brouwer, R. 1995. *Baldios* and common property resource management in Portugal. *Unasylva* 180, (46), 37-43.

Brown, C.L. and Valentine, J. 1994. The Process and Implications of Privatisation for Forestry Institutions: Focus on New Zealand. *Unasylva* 178 (45):11-19.

Bruce, J.W., Rudrappa, S. and Zongmin, L. 1995. Experimenting with approaches to common property forestry in China. *Unasylva* 180 (46), 44-49. FAO, Rome.

Buttoud, G. 1997. The influence of history in African forest policies: a comparison between Anglophone and Francophone countries. In: *Commonwealth Forestry Review* 76(1), 1997, pp. 46.

Callander, R. 1998. *How Scotland is Owned.* Canongate, Edinburgh.

Campbell, A. 1994. *Landcare: Communities Shaping the Land and the Future.* Allen and Unwin, Sydney.

Campbell, A. and Woodhill, J. 1997. The Policy Landscape and Prospects of Landcare. In: *IIED New Horizons: The Environmental, Economic and Social Impacts of Participatory Watershed Development.* Intermediate Technology Publications, London.

Carew-Reid J., Prescott-Allen R., Bass S. and Dalal-Clayton B. 1994. *Strategies for national sustainable development: a handbook for their planning and implementation.* Earthscan, London.

Carley, M. 1994. *Policy management systems and methods of analysis for sustainable agriculture and rural development.* IIED, London and FAO, Rome.

Carrere, R. and Lohmann, L. 1996. *Pulping the South: industrial tree plantations and the world paper economy.* Zed Books: London and New Jersey.

Carter, J. 1996. Recent approaches to participatory forest resource assessment. Rural Development Forestry Study Guide 2. Overseas Development Institute, London.

Castells, M. 1997. Writing in *The Guardian,* December 13, London.

Cernea, M.M. 1992. *A Sociological Framework: Policy, Environment, and the Social Actors for Tree Planting.* In: Sharma, 1992.

Chambers, R. 1993. *Challenging the Professions: Frontiers for rural development.* Intermediate Technology Publications, London.

Cirelli, M.T. 1993. Forestry legislation revision and the role of international assistance. *Unasylva* 175, Vol. 44, 10-15.

Clay, E.J. and Schaffer, B.B. (Eds). 1984. *Room for Manoeuvre: an exploration of public policy planning in agricultural and rural development.* Heinemann Educational Books, London.

Coakes, S. 1998. Valuing the social dimension: social assessment in the Regional Forestry Agreement process. *Australian Journal of Environmental Management.* 5:47-54.

Colchester, M. and Lohmann, L. (Eds). 1993. *The struggle for land and the fate of the forests.* The World Rainforest Movement, The Ecologist and Zed Books, London.

Colfer, C.J.P. 1995, 1998. *Who Counts Most in Sustainable Forest Management?* CIFOR Working Paper No.7 Centre for International Forestry Research, Bogor.

Contreras, A. 1999. *Underlying causes of deforestation and forest degradation.* Prepared for UNEP, by the Center for International Forestry Research (CIFOR), Bogor.

Cortner, H., Jensen, M. and Bright-Smith, D. 1995. Evaluating forest policies in the United States: Components of the Process and a Case Example. In: Solberg, B. and Pelli, P. (Eds). *Forest Policy Analysis – Methodological and Empirical Aspects.* EFI Proceedings No.2. European Forest Institute, Tampere.

Counsell, S. 1996. *The role of large corporations in the development of forest certification and product labelling schemes.* Unpublished Masters Thesis, Oxford Forestry Institute, Oxford.

Counsell, S. 1997. The Influences of the Private Sector in Forest-Related Policy. Draft for IIED's project *Policy that works for forests and people.* IIED, London.

Cubbage, F.W., O'Laughlin, J. and Bullock, C.S. 1993. *Forest Resource Policy.* Wiley, New York.

Cubbage, F.W. 1991. Public Regulation of Private Forestry: Proactive Policy Responses. *Journal of Forestry,* December 1991.

Dalal-Clayton, B., Bass, S., Robins, N. and Swiderska, K. 1998. *Rethinking Sustainable Development Strategies: Promoting Strategic Analysis, Debate and Action.* Environmental Planning Issues. No 19. IIED, London.

Dalal-Clayton ,D.B. and Dent, D. (1999, forthcoming). *Knowledge of the land: land resources information and its use in rural development.* Oxford University Press.

Dalal-Clayton, D.B., Dent, D. and Dubois, O. (Eds). (1999): *Local strategic planning and sustainable rural livelihoods: lessons learned and the way forward: an overview.* Report to UK Department for International Development. IIED, London.

Daniels, S. and Walker, G. 1997. Rethinking public participation in natural resource management: concepts from pluralism and five emerging approaches. *Paper prepared for the FAO Working Group on Pluralism and Sustainable Forestry and Rural Development,* FAO, Rome.

Dargavel, J. 1995. *Fashioning Australia's Forests.* Oxford University Press, Melbourne.

Dargavel, J., Dixon, K., and Semple, N. (Eds). 1988. *Changing tropical forests: historical perspectives on today's challenges in Asia, Australasia and Oceania.* Australian National University Press, Canberra.

Dargavel, J., Guijt, I., Kanowski, P., Race, D. and Proctor, W. 1998. *Australia: settlement, conflicts and agreements.* Prepared for *Policy that works for forests and people.* IIED, London (unpublished).

Dauvergne, P. 1995. *Shadows in the Forest: Japan and the politics of timber in southeast Asia*. PhD Thesis, Department of Politics, University of British Columbia, Canada.

Dauvergne, P. 1997. *Shadows in the Forest: Japan and the politics of timber in southeast Asia*. MIT Press, Cambridge, Massachusetts.

David, P. 1995. *Responsabilisation villageoise et transferts de compétence dans le cadre de la gestion des terroirs au PNGT*. Document de travail (2è version), Octobre 1995 (unpublished).

DFID, 1998. *National strategies for sustainable development: possible strategic environmental assessment tool*. Draft for consultation 31/12/98. Department for International Development, London (unpublished).

DFID, 1999. *Shaping Forest Management: how coalitions manage forests*. Department for International Development, London.

Do Dinh Sam, 1994. *Shifting cultivation in Vietnam: its social, economic and environmental values relative to alternative land use*. IIED Forestry and Land Use Series No. 3. IIED, London.

Dror, Y. 1986. *Policy-Making Under Adversity*. Transaction Books, New Brunswick.

Dubois, O. 1997. *Decentralisation and local management of forest resources in Sub-Saharan Africa: let it go or let it be?* Prepared for *Policy that works for forests and people*. IIED, London (unpublished).

Dubois, O. and Mayers, J. 1998. *Analysis of Sri Lanka's forest policy documents*. Draft. IIED, London (unpublished).

Dudley N., Jeanrenaud, J-P. and Sullivan, F. 1995. *Bad Harvest: the timber trade and the degradation of the world's forests*. WWF/Earthscan, London.

Duinker, P.N. 1998. Public participation's promising progress: advances in forest decision-making in Canada. *Commonwealth Forestry Review*. Vol.77(2) 107-112

Dye, T.R. 1976. *Policy Analysis*. University of Alabama Press, Tuscaloosa, Alabama.

Egbe, S. E. 1996. *Forest tenure and access to forestry resources in Cameroon: an overview*. Paper presented at the International Workshop on the Management of Land Tenure and Resource Access in West Africa, 18-22 November 1996, Gorée, Senegal (unpublished).

Ellefson, P.V. 1992. *Forest resources policy: process, participants, and programs*. McGrawHill, Inc., New York.

Ellefson, P. 1995b. *Forestry Research Undertaken by Private Organisations in Canada and the United States; a Review and Assessment*, in FAO, Rome.

Environmental Investigation Agency, 1996. *Corporate Power, Corruption and the Destruction of the World's Forests; the Case for a New Global Forest Agreement*, EIA, London.

Etzioni, A. 1971. Policy research. *The American Sociologist*, 6, 8-12

Evans, J. 1982. *Plantation Forestry in the Tropics*. Oxford University Press, Oxford.

Fairhead, J. and Leach, M. 1996. *Rethinking the Forest-Savanna Mosaic: Colonial science and its relics in West Africa*. In: Leach, M. and Mearns, R. (Eds) The Lie of the Land, *op.cit.* Falconer, J. 1990. *The major significance of minor forest products: the local use and value of forests in the West African humid forest zone*. Community Forestry Note 6. FAO, Rome.

FAO. 1987. *Guidelines for forest policy formulation*. FAO Forestry Paper 81. FAO, Rome.

FAO. 1993a. *Forestry policies in the Near East region: analysis and synthesis*. FAO Forestry Paper 111. FAO, Rome.

FAO. 1993b. *Forestry policies of selected countries in Asia and the Pacific*. FAO Forestry Paper 115. FAO, Rome.

FAO. 1997. *State of the Worlds Forests 1997*. FAO, Rome.

FAO. 1999a. *State of the World's Forests 1999*. FAO, Rome.

FAO. 1999b. *Global Forest Sector Outlook: The implications of future wood product market developments for sustainable forest management.* COFO-99/3. Presented at the Committee on Forestry, March 1999. FAO, Rome.

Farrington, J. and Bebbington, A. with Wellard, K. and Lewis, D.J. 1993. *Reluctant Partners? Non-governmental organizations, the state and sustainable agricultural development.* Routledge, London.

Fernandez Carro, O. and Wilson, R. 1992. Quality Management with Timber Crops, *TAPPI Journal,* February: 49-52, in Lohmann, 1996.

Filer, C. with Sekhran, N. 1998. *Loggers, donors and resource owners. Policy that works for forests and people series no. 2: Papua New Guinea.* National Research Institute, Port Moresby and IIED, London.

Filer, C. (Ed). 1997. *The Political Economy of Forest Management in Papua New Guinea.* NRI Monograph 32. National Research Institute, Port Moresby and IIED, London.

Financial Times, 1999. *Land and Freedom.* 6th February, 1999.

Forestry Commission, 1994. *Sustainable Forestry – the UK programme.* The Forestry Commission, Edinburgh.

Forestry Commission. 1998a. *UK Case Study: Six-country Initiative – Putting the IPF Proposals for Action into Practice.* The Forestry Commission, Edinburgh (unpublished).

Forestry Commission. 1998b. *The sustainable management of forests: a supplementary consultation paper to 'opportunities for change'.* The Forestry Commission, Edinburgh.

Foucault, M. 1977. *Discipline and Punish.* Penguin, Harmondsworth.

Fournier, Y. et Freudiger, P. 1995. Gestion de terroirs et développement local — Quels outils de financement pour les servir? In: *Rapport de la Journée d'Etude 1995 - Développement Local,* IRAM, Septembre 1995.

Fowler, A.F. 1998. Authentic NGO partnerships in the new policy agenda for international aid: dead end or light ahead?. *Development and Change.* Vol. 29: 137-159.

Foy, T.J., Pitcher, M.J. and Willis, C.B. 1998. Participatory development of forest policy: some practical lessons from recent South African experience. *Commonwealth Forestry Review.* Vol.77(2) 100-106.

Friends of the Earth. 1992. *Plunder in Ghana's Rainforest for Illegal Profit; an exposé of corruption, fraud and other malpractices in the international timber trade,* 2 Vols, Friends of the Earth, London.

Friends of the Earth International. 1997. *Cut and Run: Illegal Logging and Timber Trade in Four Tropical Countries,* Campaign Briefing, FoEI Amsterdam.

Gadgil, M. and Guha, R. 1995. *Ecology and Equity: the use and abuse of nature in contemporary India.* Routledge, London.

Gado, B.A. 1996. *Une instance locale de gestion et de régulation de la compétition foncière: rôle et limites des commissions foncières au Niger.* Communication présentée au Séminaire de Gorée sur l'Initiative Franco-Britannique relative au Foncier et la Gestion des Ressources Naturelles en Afrique de l'Ouest, 18-22 Novembre 1996 (unpublished).

Gane, M. 1987. *Preparing and presenting forest policies.* Proceedings of workshop on forestry, wildlife and national parks policy and legislation in the Eastern Caribbean. Castries, St. Lucia. 4-9 July 1987. FAO, Bridgetown.

Gentil, D. et Husson, B. 1995 . La décentralisation contre le développement local? In: *Rapport de la Journée d'Etude 1995 - Développement Local,* IRAM, Septembre 1995.

Gillis, M. 1992. *Forest Concession Management and Revenue Policies.* In N.P. Sharma (Ed). 1992.

Gluck, P. 1995. *Evolution of forest policy science in Austria.* In: Solberg, B. and Pelli, P. Forest policy analysis (*op. cit* – Cortner).

Gordon, I., Lewis, J. and Young, K. 1993. Perspectives on policy analysis. In: Hill, M. (Ed) 1993. *The Policy Process:* a Reader. Harvester Wheatsheaf, Hemel Hempstead.

Government of Indonesia and IIED. 1985. *A review of policies affecting the sustainable development of forest lands in Indonesia.* Government of Indonesia, Jakarta.

Government of Papua New Guinea. 1989. *Commission of Inquiry Into Aspects of the Timber Industry of Papua New Guinea.* Inquiry 1987-1989, Chaired by Justice Thomas Barnett. Port Moresby. 5,500pp.

Grayson, A.J. 1993. *Private Forestry Policy in Western Europe.* C.A.B. International, Wallingford, UK.

Grayson, A.J. and Maynard, W.B. (Eds). 1997. *The World's Forests – Rio + 5: International Initiatives Towards Sustainable Management.* Commonwealth Forestry Association, Oxford.

Greene, R. and Elffers, J. 1998. *The 48 laws of power.* Viking Penguin, New York and London.

Gregersen, H., Oram, P., Spears, J. (Eds). 1992. *Priorities for forestry and agroforestry policy research.* Report of an International Workshop, Washington, DC, 9-12 July 1991. International Food Policy Research Institute, Washington DC.

Gregerson, H., Arnold, J.E.M., Lundgren, A., Contreras, A., de Montalembert, M.R. and Gow, D. 1993. *Assessing forestry project impacts; issues and strategies.* FAO Forestry Paper 114. FAO, Rome.

Gregersen, H., Lundgren, A., Kengen, S.,and Byron, N. 1997. *Measuring and capturing forest values: issues for the decision-maker.* Paper for World Forestry Congress, Antalya, Turkey, October 1997. FAO, Rome.

Grieg-Gran, M., Westbrook, T., Mansley, M., Bass, S. and Robins, N. 1998. *Foreign Portfolio Investment and Sustainable Development: a study of the forest products sector in emerging markets.* Instruments for sustainable private sector forestry series. IIED, London.

Grimble, R., Chan, M. K., Aglionby, J. and Quan, J. 1995. *Trees and Trade-offs: A Stakeholder Approach to Natural Resource Management.* IIED Gatekeeper Series No.52, IIED, London.

Grindle, M. and Thomas, J. 1993. *Public Choices and Policy Change: The Political Economy of Reform in Developing Countries.* Johns Hopkins University Press, Baltimore, MD.

Groombridge, B. (Ed.) 1992. *Global biodiversity: status of the Earth's living resources.* World Conservation Monitoring Centre. Chapman and Hall, London.

Grove, R.H. 1995. *Green Imperialism: colonial expansion, tropical island Edens and the origins of environmentalism,* 1600-1860. Cambridge University Press, Cambridge.

Grundy, D. 1997. Developing forest policy in the UK. *Commonwealth Forestry Review,* 76 (1): 5-10.

Grut, M., Gray, J.A. and Egli, N. 1991. *Forest Pricing and Concession Policies: Managing the High Forests of West and Central Africa.* World Bank Technical Paper 143, Africa Technical Department Series, Washington DC.

Guèye, I., Kané, A. and Koné, O.N. 1994. Adapting forestry institutions to encourage people's participation in Senegal. *Unasylva* 178 (45). FAO, Rome.

Guha, R. 1983. Forestry in British and Post-British India: a historical analysis. *Economic and Political Weekly,* 29 October and 5-12 November.

Habermas, J. 1976. *Legitimation Crisis.* Heinemann, London.

Harris, B. 1991. *Child nutrition and poverty in South India: noon meals in Tamil Nadu.* Concept Publishing Company, New Delhi.

Harris, K.L. (Ed). *Making Forest Policy Work* 1996: Conference Proceedings of the Oxford Summer Course Programme 1996, Oxford Forestry Institute, Oxford.

Hecht, S. and Cockburn, A. 1989. *The Fate of the Forest: Developers, Destroyers and Defenders of the Amazon.* Verso, London.

Higman, S., Bass, S., Judd, N., Mayers, J. and Nussbaum, R. 1999. *The sustainable forestry handbook.* Earthscan, London.

Hill, M. 1997. *The Policy Process in the Modern State.* 3rd edition. Prentice Hall/Harvester Wheatsheaf, London.

Hoben, A. 1996. *The Cultural Construction of Environmental Policy: Paradigms and politics in Ethiopia*. In: Leach, M. and Mearns, R. (Eds) The Lie of the Land, *op. cit.*

Hobley, M. 1996b. *Participatory Forestry: The process of Change in India and Nepal*. ODI Rural Development Forestry Study Guide 3. Overseas Development Institute, London.

Hobley, M., Campbell, J.Y., and Bhotia, A. 1995. Community Forestry in India and Nepal, Learning from Each Other. *Himalayan Paryavaran* 78-97.

Hogwood, B.W. and Gunn, L.A. 1984. *Policy analysis for the real world*. Oxford University Press, Oxford.

Holmberg, J., Bass, S., and Timberlake, L. 1991. *Defending the future: a guide to sustainable development*. Earthscan, London.

Hummel, F.C. (Ed). 1984. *Forest Policy – a contribution to resource development*. Nijhoff-Junk, The Hague, Boston and Lancaster.

Humphreys, D. 1998. *Forest Politics: The Evolution of International Cooperation*. Earthscan, London.

Husch, B. 1987. *Guidelines for forest policy formulation*. FAO Forestry Paper No. 81. FAO, Rome.

Hyde, W.F. and Newman, D.H. with a contribution by Sedjo, R.A. 1991. *Forest Economics and Policy Analysis: An Overview*. World Bank Discussion Paper 134. World Bank, Washington, DC.

IIED. 1994. *Economic evaluation of tropical forest land use options: A review of methodology and applications*. Environmental Economics Programme, IIED, London.

IIED. 1995a. Literature review. *Policy that works for forests and people*. Draft IIED, London (unpublished).

IIED. 1995b. *Citizen action to lighten Britain's ecological footprints*. Report to the UK Dept of the Environment, February 1995. IIED, London.

IIED. 1996. *Towards a Sustainable Paper Cycle*. IIED, London.

IIED/WCMC 1996. *Forest Resource Accounting: Strategic Information for Sustainable Forest Management*. IIED, London/ World Conservation Monitoring Centre, Cambridge.

Inglis, A. and Guy, S. 1996. *Rural development forestry in Scotland: the struggle to bring international principles and best practices to the last bastion of British colonial policy*. Rural Development Forestry Network paper 20b, Overseas Development Institute, London.

Institute for Development Research. 1997. *Advocacy Sourcebook*. IDR, Boston, Mass.

IUCN Netherlands. 1994. *The Netherlands and the World Ecology*.

Jenkins, W.I. 1978. *Policy Analysis*. Martin Robertson, London.

Johnston, D.R., Grayson, A.J. and Bradley, R.T. 1967. *Forest Planning*. Faber & Faber, London.

Juniper, A. 1998. Effective campaigning. In: *Tropical Rain Forests: A Wider Perspective*. Edited by F.B. Goldsmith, Chapman and Hall, London.

Kaboré, C. 1995. L'apprentissage par les villageois de la gestion d'une collectivité locale au Burkina Faso. In: *Décentralisation de la Gestion Locale des Ressources Naturelles*. Numéro Spécial du Flamboyant No. 36 - Décembre 1995, pp. 9-12.

Kaimowitz, D. 1998. *Who determines conventional wisdom regarding policies towards forests*. Polex e-mail mailing list: D.Kaimowitz@cgnet.com

Kaimowitz, D. and Angelsen, A. 1998. *Economic Models of Tropical Deforestation: a Review*. CIFOR, Bogor.

Kaimowitz D. and Angelsen, A. 1999. *The World Bank and non-forest sector policies that affect forests*. CIFOR, Bogor.

Kanowski, P.J. 1998. Reflections on forestry and the forest products industries at the millennium. *Commonwealth Forestry Review*. Vol.77(2) 130-135.

Kanowski, P.J. and Potter, S.M. 1993. Making British forest policy work. *ForestryOxford* 66: 3, 233247; Oxford Forestry Institute, Oxford.

Karsenty, A. 1998. *Environmental taxation and economic instruments for forestry management in the Congo Basin*. IIED, London (unpublished).

Khare, A., Bathla, S., Palit, S., Sarin, M. and Saxena, N.C. 1999. *Joint forest management: policy, practice and prospects. Policy that works for forests and people series no.3. India.* IIED, London.

Korten, F.F. 1992. NGOs and the forestry sector: an overview. *Unasylva* 171 (43). FAO, Rome.

Kotey, E.N.A., Francois, J., Owusu J.G.K., Yeboah, R., Amanor, K. and Antwi, L. 1998. *Falling into Place. Policy that works for forests and people series no.4. Ghana.* IIED, London.

Kruiter, A. 1996. *Good governance for Africa: Whose governance?* Summary of the Conference organised by the University of Limburg and the European Centre for Development Policy Management, Maastricht, The Netherlands, 23-24 November 1995.

Landell-Mills N. and Ford, J. 1999. *Privatising sustainable forestry: a global review of trends and challenges.* IIED, London.

Lasswell, H.D. and Kaplan, A. 1950. *Power and Society: a framework for political inquiry.* Yale University Press.

Leach, M. and Mearns, R. (Eds). 1996. *The Lie of the Land: challenging received wisdom on the African environment.* International African Institute, James Currey, Oxford and Heinemann, Portsmouth.

Lee, K.N. 1993. *Compass and Gyroscope: Integrating Science and Politics for the Environment.* Island Press, Washington DC and Corvelo, California.

Lee, K. et al 1997 'Privatisation' in the United Nations System: Patterns of Influence in Three Intergovernmental Organisations, in *Global Society.*

Le Roy, E. 1995. Land-use plutôt que land-tenure: aux origines de la conception foncière du *common law.* In: *Décentralisation de la Gestion Locale des Ressources Naturelles.* Numéro Spécial du Flamboyant No. 36 – Décembre 1995, pp.7-8.

Lindblom, C.E. and Woodhouse, E.J. 1993. *The Policy Making Process.* Prentice Hall, Englewood Cliffs, New Jersey.

Liu Jinlong and Morrison, E. 1998. *China country paper.* Prepared for *Policy that works for forests and people.* IIED, London (unpublished).

Ljungman, L. 1994. The changing role of forestry institutions in former centrally planned economies of Eastern Europe. *Unasylva* 178 (45). FAO, Rome.

Lohmann, L. 1998. *Missing the Point of Development Talk: Reflections for Activists.* Briefing No.9, The Corner House, Sturminster Newton.

Long, N. and Long, A. (Eds). 1992. *Battlefields of Knowledge: the interlocking of theory and practice in social research and development.* Routledge, London.

Lukes, S. 1984. *Power: a Radical View.* MacMillan, London.

Mackay, D. 1995. *Scotland's Rural Land Use Agencies: the history and effectiveness in Scotland of the Forestry Commission, Nature Conservancy Council and Countryside Commission.* Scottish Cultural Press, Aberdeen.

Majchrzak, A. 1984. *Methods for Policy Research.* Applied Social Research Methods Series. Volume 3. Sage, London.

Malinowski, B. 1926. *Myth in Primitive Psychology.* Kegan Paul, London.

Malla, Y. 1998. Stakeholders' responses to changes in forest policies. In: *Proceedings of the Seminars on Co-management in the Forest Context.* September 23 to October 5, 1998, International Agricultural Centre, Wageningen.

Mannheim, K. 1936. *Ideology and Utopia.* Brace, Harcourt.

Mankin, W.E. 1998. *International Forest Policy Processes: an NGO perspective on their effectiveness. Policy that works for forests and people series no.9. Discussion Paper.* IIED, London.

Marcuse, P. 1998. *Sustainability is not enough.* Columbia University, New York (unpublished).

Markopoulos, M. 1998. The Impacts of Certification on Community Forest Enterprises; a Case Study of the Lomerío Community Forest Management Project, Bolivia. IIED Forestry and Land Use Series No. 13. IIED, London.

Marr, A. and Hutton, W. 1998. So who can lead us out of this mess? *The Observer*. 6 September, London.

Marshall, G. 1990. *The Barnett Report: A Summary of the Report of the Commission of Inquiry into Aspects of the Timber Industry in Papua New Guinea*. Asia Pacific Action Group/ Rainforest Information Centre, Lismore, Australia.

Martin, P. and Woodhill, J. 1995. Landcare in the Balance. Government Roles and Policy Issues in Sustaining Rural Environments. *Australian Journal of Environmental Management* 2: 173-83.

Marx, K. and Engels, F. 1998. *The Communist Manifesto. A Modern Edition* (first published 1848). Verso, London.

Mather, A.S. 1990. *Global Forest Resources*. Belhaven, London.

Maughan, J. 1994. *Taming troubled waters*. Ford Foundation, Washington.

Mayers, J. 1998. *Policy that works for forests and people:* findings from a six-country study. In: *Proceedings of the Seminars on Co-management in the Forest Context.* September 23 - October 5, 1998, International Agricultural Centre, Wageningen.

Mayers, J. 1999. 'Our biggest challenge is the people thing': forests, land and policy in Scotland. Prepared for *Policy that works for forests and people*, IIED, London.

Mayers, J. forthcoming,1999. *Forestry Company-Community Partnerships: a review of international experience – some best bets and dilemmas*. Department of Water Affairs and Forestry, Pretoria, and IIED, London.

Mayers, J. and Peutalo, B. 1995. *NGOs in the Forest: Participation of NGOs in National Forestry Action Programmes, New Experience in Papua New Guinea*. Forestry and Land Use Series No.8. IIED, London.

Mayers, J. and Kotey, E.N.A. 1996. *Local Institutions and Adaptive Forest Resource Management in Ghana*. Forestry and Land Use Series No.7. IIED, London.

Mayers, J, Howard, C., Kotey, N.A., Prah, E. and Richards, M. 1996. *Incentives for Sustainable Forest Management: a study in the tropical high forest of Ghana*. Forestry and Land Use Series No.6. IIED, London.

Mayers, J. and Bass, S. 1998a. The role of policy and institutions. In: *Tropical Rain Forests: A Wider Perspective*. Edited by F.B. Goldsmith, Chapman and Hall, London.

Mayers J. and Bass, S. 1998b. *Policy for real*. IIED, London (unpublished).

McKean, M. and Ostrom, E. 1995. Common property regimes in the forest: just a relic from the past? *Unasylva* 180 (46). FAO, Rome.

Menzies, N.K. and Peluso, N.L. 1991. Rights of access to upland forest resources in southwest China. *Journal of World Forest Management*, Volume 6: 1-20.

Menzies, N. 1993. Putting people back into forestry: some reflections on social and community forestry. *Forestry and Society newsletter.* Volume 1 (1). June 1993.

Merlo, M. and Paveri, M. 1998. *Formation and implementation of forest policies: a focus on the policy tools mix.* Paper prepared for the World Forestry Congress, FAO, Rome.

Mortimore, M. 1996. Land tenure and resource access in Anglophone West Africa. In Foncier rural, ressources renouvelables et développement – analyse comparative des différentes approches, Novembre 1996.

Munasingh, M. and Cruz, W. 1994. *Economy-Wide Policies and the Environment*. Environmental paper 10, Environment Department, World Bank, Washington DC.

Nhira, C., Baker, S., Gondo, P., Mangono, J.J. and Marunda, C. 1998. *Contesting inequality in access to forests. Policy that works for forests and people series no.5. Zimbabwe.* Centre for Applied Social Sciences and Forestry Commission, Harare and IIED, London.

North, R. D. 1995. *Life on a modern planet.* A manifesto for progress. Manchester University Press.

Oldfield, S., Lusty, C. and MacKinven, A. 1998. *The World List of Threatened Trees.* World Conservation Press, World Conservation Monitoring Centre, Cambridge.

Olowu, D. 1990. *The failure of current decentralisation programmes in Africa.* In: Wunsh, J.S. and Olowu, D. (Eds). Failure of the centralised state: Institutions and self-governace in Africa. Westview, Boulder Co.

Osborn, D. 1999. *Towards Earth Summit III in 2002.* Millennium Papers No.1. UNED-UK, London.

Ostrom, E. 1991. *Governing the commons: the evolution of institutions for collective action.* Cambridge University Press, Cambridge.

Ostrom, E. and Wertime, M.B. 1994. *IFRI Research Strategy.* International Forestry Resources and Institutions (IFRI) Research Program and Database. Workshop in Political Theory and Policy Analysis. Indiana University, Bloomington.

Palo M. and J. Uusivuori. 1999. *World Forests, Society and Environment Vol. 1.* Kluwer, AH Dordrecht.

Panayotou, T. 1992. *Getting Incentives Right: Economic Instruments for Environmental Management in Developing Countries.* Harvard Institute for International Development.

Panayotou T. 1998. *Instruments of change: motivating and financing sustainable development.* Earthscan, London.

Parren, M.P.E. 1994. *French and British colonial forest policies: Past and present implications for Côte d'Ivoire and Ghana.* Working paper No. 188, African Studies Center, Boston University, Boston.

Pearce, D.W. 1994. *Blueprint 3.* Earthscan, London.

Peluso, N.L. 1992. *Rich forests, poor people: forest access control and resistance in Java.* University of California Press.

Pénélon, A. 1996. *Règles d'utilisation du terroir villageois et règles d'appropriation des ressources dans deux villages de l'Est-Cameroun – Cas de la forêt communautaire: un outil de gestion nouveau certes, mais accessible?* Communication présentée au cours du Colloque panafricain sur 'La Gestion Communautaire des Ressources Naturelles et le Développement Durable', Harare, Zimbabwe, 24-27 Juin 1996.

Perlin, F. 1989. *A forest journey: The role of wood in the development of civilisation.* New York.

Peters, C., Gentry, A. and Mendelsohn, R. 1989. Valuation of an Amazonian Rainforest. *Nature*, 339, 655-656.

Poffenberger, M. (Ed). 1990. *Keepers of the forest: land management alternatives in southeast Asia.* Kumarian Press, Inc., West Hartford, Connecticut.

Poore, D. 1984. *Can government policies be changed – the role of the forestry policy review?* IIED, London.

Poore, D., Burgess, P., Palmer, J., Rietbergen, S. and Synnott, T. 1989. *No timber without trees: sustainability in the tropical forest.* Earthscan, London.

Poulantzas, N. 1973. *Political Power and Social Classes.* New Left Books. London.

Pretty, J.N. 1995. *Regenerating Agriculture: policies and practice for sustainability and self-reliance.* Earthscan, London.

Pretty, J.N., Guijt, I., Thompson, J. and Scoones, I. 1995. *Participatory Learning and Action: A Trainers Guide.* IIED Participatory Methodology Series. IIED, London.

Pryor, S. 1998. *Development of certification in the UK: reconciling civil society and government needs.* Paper delivered to workshop on Globalisation and Localisation: Exploring Tensions and Synergies in Forestry Policy, 19-25 November, CIFOR, Bogor.

Rees, J. 1990. *Natural Resources: Allocation, Economics and Policy.* 2nd Edition. Routledge, London.

Repetto, R. 1990. *Macroeconomic policies and deforestation.* Paper prepared for UNU/WIDER Project The Environment and Emerging Development Issues. WRI, Washington.

Repetto, R. and Gillis, M. (Eds). 1988. *Public Policies and the Misuse of Forest Resources.* Cambridge University Press, Cambridge.

Rerkasem, K. and Rerkasem, B. 1994. *Shifting cultivation in Thailand: its current situation and dynamics in the context of highland development.* Forestry and Land Use Series No. 4. IIED, London.

Ribot, J. C. 1995a. *Local forest control in Burkina Faso, Mali, Niger, Senegal and the Gambia: a review and critique of new participatory policies.* Regional synthesis report for the World Bank, January 1995.

Rich, B. 1994. *Mortgaging the Earth: The World Bank, Environmental Impoverishment and the Crisis of Development.* Earthscan, London.

Roberts, B. 1999. Forest policy formulation and implementation in Sweden. Prepared for *Policy that works for forests and people.* IIED, London (unpublished).

Robin, L. 1998. *Defending the Little Desert: the rise of ecological consciousness in Australia.* Melbourne University Press.

Roe, E. 1994. *Narrative Policy Analysis: Theory and Practice.* Duke University Press, Durham and London.

Röling, N. and Jiggins, J. 1998. *The Soft Side of Land: Instalment.* An incomplete exploration of the implications of seeing ecological sustainability as emerging from human learning and interaction. Proceedings of the Seminars on Co-Management in the Forest Context, September 23 - October 5, 1998. International Agricultural Centre, Wageningen.

Romm, J. 1985. *Developing Models for Effective Analysis of Forest Policy.* Paper for Conference on Renewable Resource Problems in Asia. Sapporo, June 24-28, 1985.

Sargent C. and Bass S. (Eds). 1992. *Plantation Politics: Forest Plantations in Development*, Earthscan, London.

Schaffer, B.B. 1984. *Towards Responsibility: Public Policy in Concept and Practice.* In: Clay, E.J. and Schaffer, *op.cit.*

Schama, S. 1995. *Landscape and Memory.* HarperCollins, London.

Schanz, H. 1997. On the role of the state in German forestry: implications for forest policy and forest policy science. In: Tikkanen, I., Gluck, P. and Solberg, B. (Eds)

Scoones, I. and Matose, F. 1993. Local Woodland Management: Constraints and Opportunities for Sustainable Resource Use. In: Bradley, P.N. and McNamara, K. (Eds) *Living with Trees: Policies for Forestry Management in Zimbabwe.* World Bank Technical Paper No. 210: 157-198, Washington DC.

Scoones, I. 1996. Range Management Science and Policy: Politics, Polemics & Pasture in Southern Africa. In: Leach, M. and Mearns, R., *op.cit.*

Scoones, I. 1998. *Sustainable rural livelihoods: a framework for analysis.* Working paper No. 72. Institute of Development Studies, Brighton.

Scott, J.C. 1985. *Weapons of the Weak: Everyday Forms of Peasant Resistance.* Yale, New Haven.

Scottish Office, 1999. *Recommendations for Action.* HMSO, London.

Sharma, N.P. (Ed). 1992. *Managing the World's Forests: Looking for a Balance Between Conservation and Development.* The World Bank, Washington DC.

Shepherd, G. (Ed). 1992. *Forest Policies, Forest Politics.* ODI Agricultural Occasional Paper 13. Overseas Development Institute, London.

Shi Kunshan, Li Zhiyong, Lin Fengming and Sheng Rui. 1998. *The development of China's forestry: review and prospects.* Country report for Asia-Pacific Forestry Outlook Study, FAO, Rome.

Shiva, V. 1987. *Forestry Crisis and Forestry Myths: a critical review of 'Tropical Forests a Call for Action'.* World Rainforest Movement, Penang, Malaysia.

Shore, C. and Wright, S. 1997. *Policy – a new field of anthropology.* In: Shore, C. and Wright, S. (Eds.) Anthroplogy of policy: critical perspectives on governance and power. Routledge, London.

Sizer, N. and Rice, R. 1995. *Backs to the wall in Suriname: Forest policy in a country in crisis.* World Resources Institute, Washington DC.

SRDFP, 1994. *Communities and Woodlands: proceedings of the Scottish Rural Development Forestry Programme National Seminar, Laggan.* Scottish Rural Development Forestry Programme. Edinburgh (unpublished).

Stuart, M. and Moura-Costa, P. 1998. *Greenhouse Gas Mitigation: a review of international policies and initiatives. Policy that works for forests and people series no.8: Discussion paper.* IIED, London.

Sulieman, M.S. 1997. *Managing the Forest Resources Battlefield: Appreciating Actors' Roles and Realities.* Paper prepared for the FAO Working Group on Pluralism and Sustainable Forestry and Rural Development, FAO, Rome.

Swift, J. 1996. *Desertification: narratives, winners and losers.* In: Leach, M. and Mearns, R. (Eds) The Lie of the Land *op.cit.*

Tapp, N. 1996. Social aspects of China fir plantations in China. *Commonwealth Forestry Review* 75 (4).

Tarasofsky R. 1995. *The International Forests Regime: Legal and Policy Issues.* IUCN and WWF, Gland, Switzerland.

Thomson, J.T. 1992. *A framework for analysing institutional incentives in community forestry.* Community Forestry Note 10. FAO, Rome.

Thomson, J.T. and Coulibaly, C. 1994. *Decentralisation in the Sahel: regional synthesis.* SAH/D (94) 427. Paper prepared for the Regional Conference on Land Tenure and Decentralisation in the Sahel, June, Praia, Cape Verde. OECD/Club du Sahel and CILSS, Paris and Ouagadougou.

Thomson, J.T. and Freudenberger, K. S. 1997. *Crafting institutional arrangements for community forestry.* Community Forestry Field Manual 7. FAO, Rome.

Thomson, K. 1996. Global initiatives in forest policy: context setting. In Harris K. L. 1996. *Making forest policy work.* Oxford Forestry Institute.

Tikkanen, I. and Solberg, B. 1995. Evolution of forest policy science in Finland and Norway. In: Solberg, B. and Pelli, P. (Eds). *Forest Policy Analysis – Methodological and Empirical Aspects.* EFI Proceedings No.2. European Forest Institute, Tampere.

Tompkins, S. 1989. *Forestry in Crisis: the battle for the hills.* Christopher Helm, London.

Tribe, J. (forthcoming). The law of the jungles: Regional Forestry Agreements. *Journal of Planning, Environment and Law.*

Unasylva. 1998. Issue on Accommodating Multiple Interests in Forestry. 194. Vol. 49. 1998/3. FAO, Rome.

UNCED. 1992. *Agenda 21.* United Nations Conference on Environment and Development. UN General Assembly, New York.

UNCED. 1992. *Non-Legally Binding Authoritative Statement of Principles for a Global Consensus on the Management, Conservation and Sustainable Development of All Types of Forests.* United Nations Conference on Environment and Development. UN General Assembly, New York.

Upton, C. and Bass, S. 1995. *The Forest Certification Handbook.* Earthscan, London.

Utting, P. 1993. *Trees, people and power: social dimensions of deforestation and forest protection in Central America.* Earthscan, London.

Vanclay, F. 1992. *The social context of farmers' adoption of environmentally sound farming practices.* In: Lawrence. G., Vanclay, F. and Furze, B. (Eds) Agriculture, environment and society: Contemporary issues for Australia. Macmillan, South Melbourne.

Watson, V., Cervantes, S., Castro, C., Mora, L., Solis, M., Porras, I. and Cornejo, B. 1998. *Making space for better forestry. Policy that works for forests and people series no.6. Costa Rica.* Centro Cientifico Tropical, San José and IIED, London.

WCED. 1987. *Our Common Future.* World Commission on Environment and Development/ Oxford University Press, Oxford.

WCFSD. 1999. *Our Forests: Our Future.* World Commission on Forests and Sustainable Development. Cambridge University Press, Cambridge.

Weiss, C. 1986. Research and policy-making: a limited partnership. In: F.Heller (Ed.) *The Use and Abuse of Social Science*, Sage, London.

Westoby, J.C. 1989. *An Introduction to World Forestry.* Blackwell, Oxford.

Wibe, S. and Jones, T. (Eds). 1992. *Forests: Market and Intervention Failures.* Earthscan, London.

Wiersum, F. 1997. *Normative pluriformity in forest management: professional and community perspectives.* Paper prepared for a workshop on Pluralism and Sustainable Forestry and Rural Development, FAO, Rome.

Wightman, A. 1996. *Who Owns Scotland?* Canongate, Edinburgh.

Wildavsky, A. 1987. *Speaking Truth to Power: The Art and Craft of Policy Analysis.* Transaction Publishers, New Brunswick and Oxford.

Wilson, E.O. 1988. The current state of biological diversity. pp. 1 – 18. In: Wilson E.O. and Peter, F.M. (Eds) *Biodiversity.* National Academy Press. Washington DC.

World Bank. 1991. *The Forest Sector: A World Bank Policy Paper.* World Bank. Washington DC.

World Bank. 1992. *World Development Report 1992. Development and the environment.* World Bank, Washington, DC.

Worrell, A.C. 1970. *Principles of Forest Policy.* McGraw Hill, New York.

WRI. 1998. *The Last Frontier Forests: ecosystems & economies on the edge.* World Resources Institute, Washington DC.

Yacouba, M. 1997. *Décentralisation et développement local: quelques axes de réflexion.* In The Sahel – Natural Resource Management Projects, Energy provision, Decentralisation, Empowerment and Capacity Building, SEREIN Occasional Paper No 5., pp. 187-202.

Index

Page numbers in *italic* refer to Boxes, Tables and Figures. Page numbers followed by 'n' denote footnotes.

Other books in the *Policy That Works* series (edited by James Mayers) are available from:

Publications, International Institute for Environment and Development
3 Endsleigh Street, London WC1H 0DD, UK
Tel:+44 171 388 2117 Fax:+44 171 388 2826 e-mail:bookshop@iied.org

no. 1 *Changing Perspectives on Forest Policy*:Pakistan country study. Javed Ahmed and Fawad Mahmood

no. 2 *Loggers, Donors and Resource Owners*:Papua New Guinea country study. Colin Filer with Nikhil Sekhran

no. 3 *Joint Forest Management:policy, practice and prospects*:India country study. Arvind Khare, Seema Bathla, S Palit, Madhu Sarin and NC Saxena

no. 4 *Falling into Place*:Ghana country study. Nii Ashie Kotey, Johnny Francois, JGK Owusu, Raphael Yeboah, Kojo S Amanor and Lawrence Antwi

no. 5 *Contesting Inequality in Access to Forests*:Zimbabwe country study. Calvin Nhira, Sibongile Baker, Peter Gondo, JJ Mangono and Crispen Marunda

no. 6 *Making Space for Better Forestry*:Costa Rica country study. Vicente Watson, Sonia Cervantes, Cesar Castro, Leonardo Mora, Magda Solis, Ina T. Porras and Beatriz Cornejo

[no. 7 *Series Overview*. James Mayers and Stephen Bass – out of print, replaced by this book]

no. 8 *Climate Change Mitigation by Forestry:a review of international initiatives*. Marc D. Stuart and Pedro Moura-Costa

no. 9 *Entering the Fray. International forest policy processes:an NGO perspective on their effectiveness*. William E. Mankin

no.10 Participation in the Caribbean:a review of Grenada's forest policy process. Stephen Bass

no.11 Forestry Tactics:lessons from Malawi's National Forestry Programme. James Mayers, John Ngalande, Pippa Bird and Bright Sibale

A CD Rom *Policy That Works for Forests and People*, containing all the above studies, is also available

Other Forestry titles from Earthscan

The Sustainable Forestry Handbook:Second Edition. Sophie Higman, James Mayers, Stephen Bass, Neil Judd and Ruth Nussbaum

The Forest Certification Handbook:Second Edition. Ruth Nussbaum and Markku Simula

Local Forest Management:The Impacts of Devolution Policies. David Edmunds and Eva Wollenberg

5944 343